Chapter 4
$A \times B$	the Cartesian product of A and B, p. 72
$R(x)$	the R-relative set of x, p. 81
$R(A)$	the R-relative set of A, p. 81
$\mathbf{M}_R$	the matrix of R, p. 84
R^∞	the connectivity relation of R, the transitivity closure of R, p. 89

Chapter 5
Δ	the relation of equality, p. 98
$\equiv \pmod{n}$	congruence mod n, p. 108
A/R	a partition of a set A determined by the equivalence relation R on A, p. 110
$S \circ R$	the composition of R and S, p. 118

Chapter 6
1_A	the identity function on A, p. 136
f^{-1}	the inverse function of the function f, p. 140
$\begin{pmatrix} a_1 & a_2 \ldots & a_n \\ p(a_1) & p(a_2) \ldots & p(a_n) \end{pmatrix}$	permutation of the set $A = \{a_1, a_2, \ldots, a_n\}$, p. 140

Chapter 7
$\leq$	a partial order relation, p. 169
$a \vee b$, LUB(a, b)	the least upper bound of a and b, p. 187
$a \wedge b$, GLB(a, b)	the greatest lower bound of a and b, p. 187
a'	the complement of a, p. 192

Introductory
Discrete Structures
with Applications

Introductory Discrete Structures with Applications

Bernard Kolman

Robert C. Busby

Drexel University

Prentice-Hall, Inc., Englewood Cliffs, New Jersey 07632

Library of Congress Cataloging-in-Publication Data

Kolman, Bernard, (date)
  Introductory discrete structures with applications.

  Bibliography: p.
  Includes index.
  1. Mathematics—1961- . 2. Electronic data
processing—Mathematics. I. Busby, Robert C. II. Title
QA39.2.K65 1987   510   86–25139
ISBN 0–13–500794–1

Editorial/production supervision: Maria McColligan
Interior design: Anne T. Bonanno
Cover design: Christine Gehring-Wolf. Adapted by Anne T. Bonanno.
Cover photo: *Ondocto-Re*, by Vasarely, courtesy of Vasarely Center, New York.
Manufacturing buyer: John B. Hall

Printed in the United States of America
10 9 8 7 6 5 4 3 2 1

ISBN 0-13-500794-1 01

Prentice-Hall International (UK) Limited, *London*
Prentice-Hall of Australia Pty. Limited, *Sydney*
Prentice-Hall Canada Inc., *Toronto*
Prentice-Hall Hispanoamericana, S. A., *Mexico*
Prentice-Hall of India Private Limited, *New Delhi*
Prentice-Hall of Japan, Inc., *Tokyo*
Prentice-Hall of Southeast Asia Pte. Ltd., *Singapore*
Editora Prentice-Hall do Brasil, Ltda., *Rio de Janeiro*

To Judith
B.K.

To Patricia
R.C.B.

Contents

Part II Applications

Preface

Within the past few years discrete mathematics has come to be recognized as essential to the education of computer scientists, mathematicians, and students in a wide variety of related areas. Thus, a need has emerged for a text that can be used as an introduction to discrete mathematics for a diverse audience of beginning students. Such a book should cover the necessary theoretical material in as straightforward and concrete a fashion as is possible. It should limit its coverage to material that is appropriate for a one-semester first course in the subject. The book should offer a broad spectrum of applications of the material, thus providing the flexibility needed by students with widely differing backgrounds and interests.

This book has been written to meet the above objectives.

Organization

The book is divided into two parts. Part 1 (Chapters 1 to 7) is devoted to the basic theoretical material. We have limited both areas covered and depth of coverage to what is appropriate for an introductory course. We have stressed the essential unity of the various topics, thus minimizing the number of definitions and concepts. Wherever possible, we have emphasized the computational and geometrical aspects of the material, keeping the level of abstraction to a minimum. Thus, we have omitted the proofs of certain difficult or less important results, choosing instead to illustrate them with examples. Proofs that are included are presented at a level that is suitable for the beginning student.

Part II (Chapters 8 to 12) presents a menu of applications of the first seven chapters. The chapters in Part II covering applications to mathematics, computer science, and probability theory are largely independent of one another. Depending on the desired level and emphasis of the course, the instructor may choose to cover various chapters from Part II. This can be done after Part I, or each selected application can be covered after the required material from Part I has been developed. The chart shown on page xiv shows the way in which the applications chapters logically depend on Part I.

Content

Part I

Chapter 1 is a brief introduction to sets and sequences. *Chapter 2* contains material on combinatorics. It includes counting methods from set theory, standard material on permutations and combinations, and a section devoted to an elementary presentation of the pigeonhole principle. *Chapter 3* covers some basic algebraic ideas, including a section on induction and recursion, and material on ordinary and Boolean matrix algebra. *Chapter 4* gives an introduction to relations, and to the algebraic and geometric ways of representing them. *Chapter 5* discusses some of the important properties that relations may possess. It devotes a separate section to a careful and elementary treatment of equivalence relations and their associated partitions. *Chapter 6* presents the notion of a function as a certain type of relation. This allows us to use pedagogical ideas developed for relations to reduce to computation various function-related concepts. This approach eliminates certain logical difficulties that beginning students have with the traditional presentation of functions. *Chapter 7* deals with partially ordered sets, including lattices and finite Boolean algebras. The latter are introduced in a simplified way that avoids the axiomatization inherent in other approaches.

Part II

Chapter 8 provides an application of Chapter 7 to the design and construction of logical circuits. *Chapter 9*, on trees and searching, illustrates how certain relations can be used in computer science. *Chapter 10* shows how the properties and algebra of relations can be used to give a precise yet elementary development of that part of linguistics that is widely used in computer science. *Chapter 11* develops the theory of finite state machines as an application of relations and functions. *Chapter 12* expands the material on set theory and counting techniques to provide an introduction to finite probability theory.

There is an appendix which provides an introduction to logic and techniques of proof. It contains exercises and can be included in the course, if desired.

Exercises

The book has a wide variety of exercises at all levels, which have been grouped into two classes. The first class, "Exercises," contains computational exercises. The second class "Theoretical Exercises," contains exercises that can be used to vary the level of the course. Answers to the odd-numbered exercises appear at the end of the book. Answers to all exercises and solutions to all theoretical exercises appear in the Instructor's Manual which is available (to instructors only) gratis from the publisher. The Instructor's Manual also contains a large test bank with solutions.

End of Chapter Material

Every chapter contains a summary of Key Ideas for Review, references for further reading, review exercises (odd answers in the back of the book), and a chapter test (all answers in the back of the book).

Acknowledgments

We are pleased to express our thanks to the following reviewers, whose suggestions, comments, and criticisms greatly improved the manuscript: Bette Warren, Eastern Michigan University; Thomas C. Upson, Rochester Institute of Technology; Jerry Bloomberg, Essex Community College; and Nina Edelman, Spring Garden College.

We wish to express our thanks to Andrew Rippert for proofreading galleys and for checking the solutions to many of the exercises, to Amelia Maurizio and Susan Gershuni for help with typing. Finally, we thank Maria McColligan, our production editor; David Ostrow, mathematics editor; and the entire staff of Prentice-Hall, Inc. for their support, encouragement, creative imagination, and unfailing cooperation during the conception, design, production, and marketing phases of this book. We wish to express our sincere gratitude to all those mentioned above.

B.K.
R.C.B.

Logical Dependence of Chapters

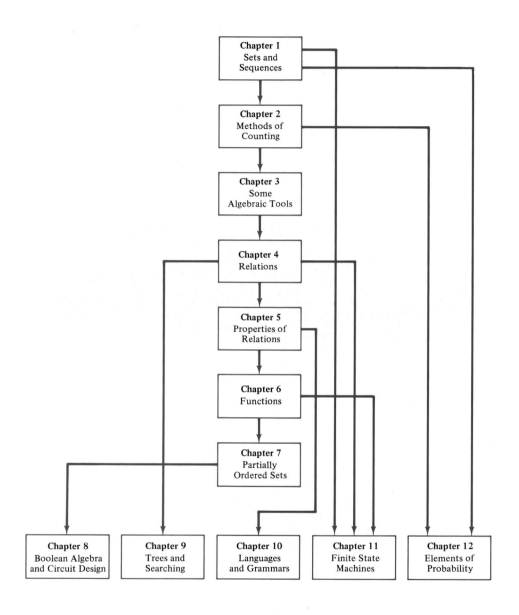

Introductory Discrete Structures with Applications

Sets and Sequences

Prerequisites:

There are no formal prerequisites, but the reader is encouraged to consult the Appendix on logic as needed.

In this chapter we introduce the basic language of discrete mathematics. We start with sets, subsets, and their operations, notions with which many of you are already familiar.

1.1 Sets and Subsets

Sets

A **set** is any well-defined collection of objects called the **elements** or **members** of the set. For example, the set of all wooden chairs, the set of all one-legged black birds, or the set of real numbers between zero and one. Almost all mathematical objects are first of all sets, regardless of any additional properties they may possess. Thus set theory is, in a sense, the foundation on which virtually all of mathematics is constructed. In spite of this, set theory (at least the informal brand we need) is quite easy to learn and use.

One way of describing a set that has a finite number of elements is by listing the elements of the set between braces. Thus the set of all positive integers that are less than 4 can be written as

$$\{1, 2, 3\} \tag{1}$$

The order in which the elements of a set are listed is not important. Thus $\{1, 3, 2\}, \{3, 2, 1\}, \{3, 1, 2\}, \{2, 1, 3\}$, and $\{2, 3, 1\}$ are all representations of the set given in (1). Moreover, repeated elements in the *listing* of the elements of a set can be ignored. Thus $\{1, 3, 2, 3, 1\}$ is another representation of the set given in (1).

We use uppercase letters such as A, B, C, to denote sets, and lowercase letters such as a, b, c, x, y, z, t, . . . , to denote members (or elements) of sets.

We indicate the fact that x is an element of the set A by writing

$$x \in A$$

and we indicate the fact that x is not an element of A by writing

$$x \notin A$$

Example 1   Let

$$A = \{1, 3, 5, 7\}$$

Then $1 \in A$, $3 \in A$ but $2 \notin A$.

Sometimes it is inconvenient or impossible to describe a set by listing all its elements. Another useful way to define a set is by specifying a property that the elements of the set have in common. We use the notation $P(x)$ to denote a proposition (sentence or statement) P concerning the variable object x. The set defined by $P(x)$, written

$$\{x \mid P(x)\}$$

is just the collection of all objects x for which P is sensible and true. For example,

$$\{x \mid x \text{ is a positive integer less than 4}\}$$

is the set

$$\{1, 2, 3\}$$

described in (1) by listing its elements.

Example 2   The set consisting of all the letters in the word "byte" can be denoted by

$$\{b, y, t, e\}$$

or by

$$\{x \mid x \text{ is a letter in the word ``byte''}\}$$

Example 3   We introduce here several sets (and their notations) that will be used throughout this book.

    (a) $Z^+ = \{x \mid x \text{ is a positive integer}\}$

Thus Z^+ consists of the numbers used for counting: 1, 2, 3, ...

    (b) $N = \{x \mid x \text{ is a positive integer or zero}\}$

Thus N consists of the positive integers and zero: 0, 1, 2, ...

    (c) $Z = \{x \mid x \text{ is an integer}\}$

Thus Z consists of all the integers: . . . , -3, -2, -1, 0, 1, 2, 3, ...

    (d) $\mathbb{R} = \{x \mid x \text{ is a real number}\}$

    (e) The set that has no elements in it is denoted by the symbol $\varnothing$.

Example 4

$$\varnothing = \{x \mid x \text{ is a real number and } x^2 = -1\}$$

since the square of a real number x is always nonnegative.

    Sets are completely known when their members are all known. Thus we say that two sets A and B are **equal** if they have the same elements and we write

$$A = B$$

Example 5   If

$$A = \{1, 2, 3\}$$

and

$$B = \{x \mid x \text{ is a positive integer and } x^2 < 12\}$$

then

$$A = B$$

Example 6   If

$$A = \{\text{ALGOL, FORTRAN, BASIC}\}$$

and

$$B = \{\text{BASIC, FORTRAN, ALGOL}\}$$

then

$$A = B$$

Subsets

If every element of A is also an element of B, that is, if whenever $x \in A$, then $x \in B$, we say that A is a **subset** of B, or that A is **contained** in B, and we write

$$A \subseteq B$$

If A is not a subset of B, we write

$$A \nsubseteq B$$

(see Fig. 1).

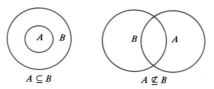

$$A \subseteq B \qquad\qquad A \nsubseteq B \qquad\qquad \textbf{Figure 1}$$

Diagrams, such as those in Fig. 1, which are used to show relationships between sets are call **Venn diagrams** after the British logician John Venn. Venn diagrams will be used extensively in the next section.

Example 7 We have

$$Z^+ \subseteq Z$$

Moreover, if Q denotes the set of all rational numbers, then

$$Z \subseteq Q$$

Example 8 Let

$$A = \{1, 2, 3, 4, 5, 6\}, \qquad B = \{2, 4, 5\}, \qquad C = \{1, 2, 3, 4, 5\}$$

Then

$$B \subseteq A, \qquad B \subseteq C, \qquad C \subseteq A$$

However,

$$A \nsubseteq B, \qquad A \nsubseteq C, \qquad C \nsubseteq B$$

Example 9   If A is any set, then $A \subseteq A$. That is, every set is a subset of itself.

Example 10   Let A be a set and let

$$S = \{\{A\}\}$$

Then, $\{A\}$ is an element of S, that is, $\{A\} \in S$. On the other hand, it is not true that $\{A\} \subseteq S$ or that $A \subseteq S$.

Since an implication is true if the hypothesis is false (see the Appendix at the end of the book), it follows that $\varnothing \subseteq A$.

It is easy to show (Exercise T1) that $A = B$ if and only if $A \subseteq B$ and $B \subseteq A$.

The set of everything, it turns out, cannot be considered defined without destroying the logical structure of mathematics. To avoid this and other problems, which need not concern us here, we will assume that for each discussion there is a "universal set" U (which may vary with the discussion) containing all objects for which our discussion is meaningful. Any other set S mentioned in the discussion will automatically be assumed to be a subset of U. Thus, if we are discussing real numbers and we mention sets A and B, then A and B must (we assume) be sets of real numbers, not matrices, electronic circuits, or rhesus monkeys. In most problems, the universal set will be apparent from the setting of the problem. In Venn diagrams, the universal set U will be denoted by a rectangle, while sets within U will be denoted by circles as shown in Fig. 2.

A set A is called **finite** if it has n distinct elements, where $n \in N$. In this case, n is called the **cardinality** of A and is denoted by $|A|$. Thus the sets of Examples 1,

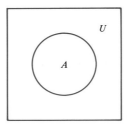

Figure 2

2, 4, 5, and 6 are finite. A set that is not finite is called **infinite.** The sets introduced in Examples 3 (except for $\varnothing$) and 7 are infinite.

If A is a set, then the set of all subsets of A is called the **power set** of A and is denoted by $P(A)$.

Example 11   Let $A = \{1, 2, 3\}$. Then $P(A)$ consists of the following subsets of A: $\varnothing$, $\{1\}$, $\{2\}$, $\{3\}$, $\{1, 2\}$, $\{1, 3\}$, $\{2, 3\}$, $\{1, 2, 3\} = A$.

It can be shown (Exercise 18 of Section 3.1) that if A has k elements, then $P(A)$ has 2^k elements.

EXERCISE SET 1.1

1. Let $A = \{1, 2, 4, a, b, e\}$. Answer each of the following as true or false.
    (a) $2 \in A$ (b) $3 \in A$ (c) $c \notin A$
    (d) $\varnothing \in A$ (e) $\varnothing \notin A$ (f) $A \in A$

2. Let $A = \{x \mid x \text{ is a real number and } x < 6\}$. Answer each of the following as true or false.
    (a) $3 \in A$ (b) $6 \in A$ (c) $5 \notin A$
    (d) $8 \notin A$ (e) $-8 \in A$ (f) $3.4 \notin A$

3. In each part, write the set of letters in each word by listing the elements in the set.
    (a) AARDVARK   (b) BOOK
    (c) MISSISSIPPI

4. In each part, write the set by listing the elements.
    (a) The set of all positive odd integers that are less than 10.
    (b) $\{x \mid x \in Z \text{ and } x^2 < 12\}$

5. In each part, write the set in the form $\{x \mid P(x)\}$, where $P(x)$ is a property that describes the elements of the set.
    (a) $\{2, 4, 6, 8, 10\}$
    (b) $\{a, e, o, i, u\}$
    (c) $\{1, 4, 9, 16, 25, 36\}$
    (d) $\{-2, -1, 0, 1, 2\}$

6. Let $A = \{1, 2, 3, 4, 5\}$. Which of the following sets equal A?
    (a) $\{4, 1, 2, 3, 5\}$ (b) $\{2, 3, 4\}$
    (c) $\{1, 2, 3, 4, 5, 6\}$

    (d) $\{x \mid x \text{ is an integer and } x^2 \leq 25\}$
    (e) $\{x \mid x \text{ is a positive integer and } x \leq 5\}$
    (f) $\{x \mid x \text{ is a positive rational number and } x \leq 5\}$

7. Which of the following sets are empty?
    (a) $\{x \mid x \text{ is a real number and } x^2 - 1 = 0\}$
    (b) $\{x \mid x \text{ is a real number and } x^2 + 1 = 0\}$
    (c) $\{x \mid x \text{ is a real number and } x^2 = -9\}$
    (d) $\{x \mid x \text{ is a real number and } x = 2x + 1\}$
    (e) $\{x \mid x \text{ is a real number and } x = x + 1\}$

8. List all subsets of the set $\{a, b\}$.

9. List all subsets of the set $\{$ALGOL, BASIC, FORTRAN$\}$.

10. List all subsets of the set $\varnothing$.

11. Let $A = \{1, 2, 5, 8, 11\}$. Answer each of the following as true or false.
    (a) $\{5, 1\} \subseteq A$
    (b) $\{8, 1\} \in A$
    (c) $\{1, 6\} \not\subseteq A$
    (d) $\{1, 8, 2, 11, 5\} \not\subseteq A$
    (e) $\varnothing \subseteq A$
    (f) $\{2\} \subseteq A$
    (g) $A \subseteq \{11, 2, 5, 1, 8, 4\}$
    (h) $\{3\} \notin A$

12. Let $A = \{x \mid x \text{ is an integer and } x^2 < 16\}$. Answer each of the following as true or false.
    (a) $\{0, 1, 2, 3\} \subseteq A$
    (b) $\{-3, -2, -1\} \subseteq A$

(c) $\varnothing \subseteq A$

(d) $\{x \mid x$ is an integer and $|x| < 4\} \subseteq A$

(e) $A \subseteq \{-3, -2, -1, 0, 1, 2, 3\}$

13. Let

$A = \{1\},$ $B = \{1, a, 2, b, c\},$ $C = \{b, c\},$

$D = \{a, b\},$ $E = \{1, a, 2, b, c, d\}.$

In each part, replace the symbol $\square$ by the symbol $\subseteq$ or $\not\subseteq$ to give a correct statement.

(a) $A \square B$ (b) $\varnothing \square A$ (c) $B \square C$

(d) $C \square E$ (e) $D \square C$ (f) $B \square E$

14. In each part, find the set with the smallest number of elements that contains the given sets as subsets.

(a) $\{a, b, c\}, \{a, d, e, f\}, \{b, c, e, g\}$

(b) $\{1, 2\}, \{1, 3\}, \varnothing$

(c) $\{1, a\}, \{2, b\}$

15. Use Fig. 3 to answer each of the following as true or false.

(a) $A \subseteq B$ (b) $B \subseteq A$

(c) $C \subseteq B$ (d) $x \in B$

(e) $x \in A$ (f) $y \in B$

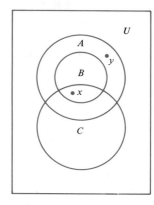

Figure 3

16. If $A = \{1, 2\}$, find $P(A)$.

THEORETICAL EXERCISE

T1. Show that $A = B$ if and only if $A \subseteq B$ and $B \subseteq A$.

1.2 Sequences

Some of the most important sets arise in connection with sequences. A **sequence** is simply a list of objects in order: a "first term" "second term" "third term" and so on. The list may stop after n steps, $n \in N$, or it may go on forever. In the first case we say that the sequence is **finite**, and in the second case we say that it is **infinite.** The terms may all be different, or some may be repeated.

Example 1 The sequence

$$1, 0, 0, 1, 0, 1, 0, 0, 1, 1, 1$$

is a finite sequence with repeated terms. The digit zero, for example, occurs as the second, third, fifth, seventh, and eighth terms of the sequence.

Example 2 The list

$$1, 4, 9, 16, 25, \ldots$$

is the infinite sequence of the squares of all positive integers. The three dots in the expression above mean "and so on," that is, continue the pattern established by the first few terms.

If the successive terms of a sequence can be determined by a formula, then we can use a more compact notation. The formula for the nth term of the sequence, in terms of the variable n, is put in parentheses. Then the permitted values of n are listed outside the parentheses.

Example 3   The sequence of Example 2 can be written equivalently as $(n^2)_{1 \leq n \leq \infty}$ and similarly, the notation $(1/n)_{2 \leq n \leq \infty}$ represents the infinite sequence

$$\tfrac{1}{2}, \tfrac{1}{3}, \tfrac{1}{4}, \ldots$$

Example 4   The finite sequence

$$1, 2, 4, \ldots, 256$$

can be denoted by $(2^n)_{0 \leq n \leq 8}$, while the expression $(a_n)_{1 \leq n \leq 200}$ represents the finite sequence

$$a_1, a_2, a_3, \ldots, a_{200}$$

Example 5   An ordinary English word such as "sturdy" can be viewed as the finite sequence

$$s, t, u, r, d, y$$

composed of letters from the ordinary alphabet. In such examples, it is customary to omit the commas and write the word in the usual way if no confusion results. Similarly, even a meaningless word such as "abacabcd" may be regarded as a finite sequence of length 8. Sequences of letters or other symbols, written without the commas, are also referred to as **strings.**

Example 6   An infinite string such as abababab . . . may be regarded as the infinite sequence

$$a, b, a, b, a, b, \ldots$$

Example 7   The sentence "now is the time for the test" can be viewed as the finite sequence of English words now, is, the, time, for, the, test

Here the terms of the sequence are themselves words of varying length, so we would not be able to simply omit the commas. The custom, as we indicated initially, is to use spaces instead of commas.

The **set corresponding to a sequence** is simply the set whose elements are the distinct terms in the sequence. Note that an essential feature of a sequence is the order in which the terms are listed in the sequence. However, the order in which the elements of a set are listed in the set is of no significance at all.

Example 8 (a) The set corresponding to the sequence in Example 2 is

$$\{1, 4, 9, 16, 25, \ldots\}.$$

(b) The set corresponding to the sequence in Example 6 is simply $\{a, b\}$.

The idea of a sequence is also important in computer science, where a sequence is sometimes called a **linear array** or **list.** We will make a slight but useful distinction between a sequence and an array, and use a slightly different notation. If we have a sequence $S = s_1, s_2, \ldots$, we think of all of the terms of S as completely determined. The term s_4, for example, is some fixed term of S, located in position four. Moreover, if we change any of the terms s_i, we think of the result as a new sequence, and we probably will rename it. Thus if we begin with the finite sequence

$$S = 0, 1, 2, 3, 2, 1, 1$$

and we change the 3 to 4, getting

$$0, 1, 2, 4, 2, 1, 1$$

we would think of this as a different sequence, say S'.

An array, on the other hand, may be viewed as a "sequence of positions," which we represent below as boxes.

The positions form a finite or infinite list, depending on the desired size of the array. Elements from some set may be assigned to the positions of the array S. The element assigned to position n will be denoted by $S(n)$, and the sequence $S(1), S(2), S(3), \ldots$ will be called the **sequence of values** of the array S. The point is that S is considered to be a well-defined object, even if some of the positions have not been assigned values, or if some values are changed during the discussion.

A set is called **countable** if it is the set corresponding to some sequence. Informally, this means that the members of the set can be arranged in a list, with a first, second, third, . . . , term, and the set can therefore be "counted." We shall show in Section 1.6 that all finite sets are countable. However, not all infinite sets are countable. A set that is not countable is called **uncountable.**

The most accessible example of an uncountable set is the set of all real numbers between zero and one, that is, the set of all real numbers in the open interval $(0, 1)$. Depending on your definition of real numbers, it is either an axiom or a theorem that a real number in $(0, 1)$ is just an infinite decimal. $a_1 a_2 a_3 \ldots$, where a_i is an integer such that $0 \leq a_i \leq 9$. We shall now show that this set is uncountable. We will prove this result by contradiction; that is, we will show that the countability of this set implies an impossible situation (see the Appendix).

Assume that the set of all such decimals is countable. Then we can form the following list (sequence), containing all such decimals:

$$d_1 = .a_1 a_2 a_3 \ldots$$
$$d_2 = .b_1 b_2 b_3 \ldots$$
$$d_3 = .c_1 c_2 c_3 \ldots$$
$$\vdots$$

Then every infinite decimal must appear somewhere on this list. We shall establish a contradiction by constructing an infinite decimal that is not on this list. Now construct a decimal x as follows: $x = .x_1 x_2 x_3 \ldots$, where x_1 is 1 if $a_1 = 2$, otherwise $x_1 = 2$; x_2 is 1 if $b_2 = 2$, otherwise $x_2 = 2$; $x_3 = 1$ if $c_3 = 2$, otherwise $x_3 = 2$. This process can clearly be continued indefinitely. The resulting number x is certainly an infinite decimal of 1's and 2's and is certainly in $(0, 1)$, but by its construction it differs from each number in the list at some digit. Thus x is not on the list, a contradiction to our assumption. Hence no matter how the list is constructed, there is some real number in $(0, 1)$ that is not in the list.

It can be shown that the set of all rational numbers is countable.

Given a set A, we can construct the set A^* consisting of all finite sequences of elements of A. Often, the set A is not a set of numbers but some set of symbols. In this case A is called an **alphabet,** and the finite sequences in A^* are called **words** from A, or sometimes **strings** from A. For this case in particular, the sequences in A^* are *not* written with commas between the elements. We assume that A^* contains the **empty sequence** or **empty string,** containing no symbols, and we denote this string by Λ. This string will be useful in Chapters 10 and 11.

Example 9   Let $A = \{a, b, c, \ldots, z\}$ be the usual English alphabet. Then A^* consists of all ordinary words, such as ape, sequence, antidisestablishmentarianism, and so on, as well as "words" such as yxaloble, zigadongdong, coccaaaa, and pqrst. All finite sequences from A are in A^*, whether they have meaning or not.

If $w_1 = s_1 s_2 \cdots s_n$ and $w_2 = t_1 t_2 \cdots t_k$ are elements of A^*, for some set A, we define the **catenation** of w_1 with w_2 as the sequence $s_1 s_2 s_3 \cdots s_n t_1 t_2 \cdots t_k$. The catenation of w_1 with w_2 is written $w_1 \cdot w_2$, and is another element of A^*. Note that if w belongs to A^*, then $w \cdot \Lambda = w$ and $\Lambda \cdot w = w$. This property is convenient and is one of the main reasons for introducing the empty string Λ.

Example 10   Let $A = \{$John, Sam, Jane, swims, runs, well, quickly, slowly$\}$. Then A^* consists of finite sequences formed from these words. Thus A^* contains legitimate sentences, such as "Jane swims quickly" and "Sam runs well," as well as nonsense sentences such as "Well swims Jane slowly John." Here we separate the terms in each sequence by blanks. This is often done when the elements of A are "words."

EXERCISE SET 1.2

In Exercises 1–6, give the set corresponding to the sequence.

1. 2, 1, 2, 1, 2, 1, 2, 1
2. 0, 2, 4, 6, 8, 10, . . .
3. 1, 3, 5, 7, 9, 11, 13, . . .
4. $a\,a\,b\,b\,c\,c\,d\,d\,e\,e \ldots z\,z$
5. $a\,b\,b\,c\,c\,c\,d\,d\,d\,d$
6. 1, 2, 2, 3, 3, 3, 4, 4, 4, 4, 5, 5, 5, 5, 5, 6, . . .

In Exercises 7–10, write out the first four, beginning with $n = 1$, terms of the sequence whose general term is given.

7. 5^n
8. $2^n - 1$
9. $3n^2 + 2n - 6$
10. $n/(n + 1)$
11. Give three different sequences that have $\{x, y, z\}$ as a corresponding set.
12. Give three different sequences that have $\{1, 2, 3, \ldots\}$ as a corresponding set.

In Exercises 13–18, write a formula for the nth term of the sequence.

13. 1, 3, 5, 7, . . .
14. 0, 3, 8, 15, 24, 35, . . .
15. 1, -1, 1, -1, 1, -1, . . .
16. 0, 2, 0, 2, 0, 2, . . .
17. 1, 4, 7, 10, 13, 16, . . .
18. 1, $\frac{1}{2}$, $\frac{1}{4}$, $\frac{1}{8}$, $\frac{1}{16}$, . . .

1.3 Operations on Sets

In this section we discuss several operations which will combine given sets to yield new sets. These operations, which are analogous to the familiar operations on the real numbers, will play a key role in the many applications and ideas that follow.

If A and B are sets, we define their **union** as the set consisting of all elements that belong to A *or* B and denote it by $A \cup B$. Thus

$$A \cup B = \{x \mid x \in A \text{ or } x \in B\}$$

Observe that $x \in A \cup B$ if $x \in A$ or $x \in B$ or x belongs to both A and B.

Example 1 Let

$$A = \{a, b, c, e, f\}, \qquad B = \{b, d, r, s\}$$

Find $A \cup B$.

Solution. Since $A \cup B$ consists of all the elements that belong to either A or B (or both), then

$$A \cup B = \{a, b, c, d, e, f, r, s\}$$

We can illustrate the union of two sets by a Venn diagram as follows. If A and B are the sets given in Fig. 1(a), then $A \cup B$ is the set of points in the shaded region as indicated in Fig. 1(b).

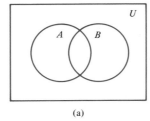 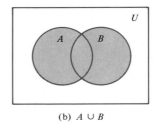

(a)   (b) $A \cup B$ **Figure 1**

If A and B are sets, we define their **intersection** as the set consisting of all elements that belong to *both A and B* and denote it by $A \cap B$. Thus

$$A \cap B = \{x \mid x \in A \text{ and } x \in B\}$$

Example 2 Let

$$A = \{a, b, c, e, f\}, \qquad B = \{b, e, f, r, s\}, \qquad C = \{a, t, u, v\}$$

Find $A \cap B$, $A \cap C$, and $B \cap C$.

Solution. The elements b, e, and f are the only ones that belong to both A and B, so

$$A \cap B = \{b, e, f\}$$

Similarly,

$$A \cap C = \{a\}$$

There are no elements that belong to both B and C, so

$$B \cap C = \varnothing$$

Two sets that have no common elements, such as B and C in Example 2, are called **disjoint** sets.

We can illustrate the intersection of two sets by a Venn diagram as follows. If A and B are the sets given in Fig. 2(a), then $A \cap B$ is the set of points in the shaded region as indicated in Fig. 2(b). Figure 3 illustrates a Venn diagram of two disjoint sets.

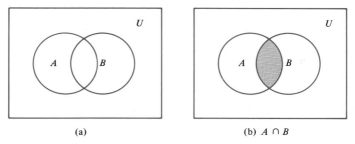

(a)   (b) $A \cap B$

Figure 2

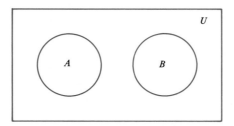

Figure 3

The operations of union and intersection can be defined for three or more sets in the obvious manner. Thus

$$A \cup B \cup C = \{x \mid x \in A \text{ or } x \in B \text{ or } x \in C\}$$

and

$$A \cap B \cap C = \{x \mid x \in A \text{ and } x \in B \text{ and } x \in C\}$$

The shaded region in Fig. 4(b) is the union of the sets A, B, and C shown in Fig. 4(a), and the shaded region in Fig. 4(c) is the intersection of the sets A, B, and C shown in Fig. 4(a).

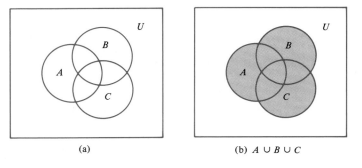

(a)   (b) $A \cup B \cup C$

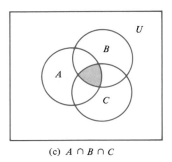

(c) $A \cap B \cap C$

Figure 4

In general, if $A_1, A_2, \ldots, A_n$ are subsets of U, then $A_1 \cup A_2 \cup \cdots \cup A_n$ will be denoted by $\bigcup_{k=1}^{n} A_k$, and $A_1 \cap A_2 \cap \cdots \cap A_n$ will be denoted by $\bigcap_{k=1}^{n} A_k$.

Example 3 Let

$$A = \{1, 2, 3, 4, 5, 7\}, \qquad B = \{1, 3, 8, 9\}, \qquad C = \{1, 3, 6, 8\}$$

Then $A \cap B \cap C$ is the set of elements that belong to A, B, and C. Thus

$$A \cap B \cap C = \{1, 3\}$$

If A and B are two sets, we define the **complement of B with respect to A** as the set of all elements that belong to A but not to B and denote it by $A - B$. Thus

$$A - B = \{x \mid x \in A \text{ and } x \notin B\}$$

Example 4   Let

$$A = \{a, b, c\} \qquad \text{and} \qquad B = \{b, c, d, e\}$$

Then

$$A - B = \{a\} \qquad \text{and} \qquad B - A = \{d, e\}$$

If A and B are the sets in Fig. 5(a), then $A - B$ and $B - A$ are the sets of points in the shaded regions in Fig. 5(b) and (c), respectively.

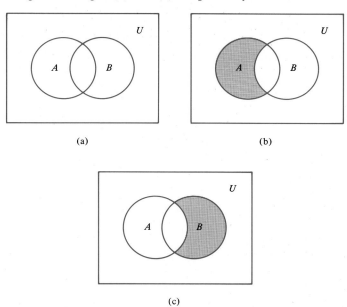

(a)   (b)

(c)

Figure 5

If U is a universal set containing A, then $U - A$ is called the **complement** of A and is denoted by $\overline{A}$. Thus

$$\overline{A} = \{x \,|\, x \notin A\}$$

Example 5   Let $U = Z$ and

$$A = \{x \,|\, x \text{ is an integer and } x \geq 4\}$$

Then

$$\overline{A} = \{x \mid x \text{ is an integer and } x < 4\}$$

If A is the set in Fig. 6, its complement is the shaded region in that figure.

If A and B are two sets, we define their **symmetric difference** as the set consisting of all elements that belong to A or to B, but not to both A and B, and we denote it by $A \oplus B$. Thus

$$A \oplus B = \{x \mid (x \in A \text{ and } x \notin B) \text{ or } (x \in B \text{ and } x \notin A)\}$$

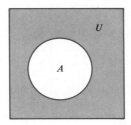

Figure 6

Example 6 Let

$$A = \{a, b, c, d\} \qquad \text{and} \qquad B = \{a, c, e, f, g\}$$

Then

$$A \oplus B = \{b, d, e, f, g\}$$

If A and B are as indicated in Fig. 7(a), their symmetric difference is the shaded region shown in Fig. 7(b). It is easy to see that

$$A \oplus B = (A - B) \cup (B - A)$$

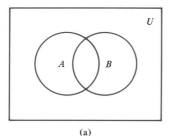 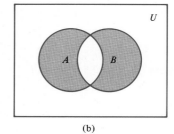

(a)   (b)

Figure 7

Algebraic Properties of Set Operations

The operations on sets that we have just defined satisfy many algebraic properties, some of which resemble the algebraic properties satisfied by the real number system. All the principal properties listed here can be proved using the definitions given and the rules of logic. We shall prove only some of the properties and leave proofs of the remaining ones as exercises for the reader. Venn diagrams are often useful to suggest or to justify the method of proof.

Theorem 1   The operations on sets defined above satisfy the following properties:

Commutative Properties
  1. $A \cup B = B \cup A$
  2. $A \cap B = B \cap A$

Associative Properties
  3. $A \cup (B \cup C) = (A \cup B) \cup C$
  4. $A \cap (B \cap C) = (A \cap B) \cap C$

Distributive Properties
  5. $A \cap (B \cup C) = (A \cap B) \cup (A \cap C)$
  6. $A \cup (B \cap C) = (A \cup B) \cap (A \cup C)$

Idempotent Properties
  7. $A \cup A = A$
  8. $A \cap A = A$

Properties of the Complement
  9. $\overline{(\overline{A})} = A$
  10. $A \cup \overline{A} = U$
  11. $A \cap \overline{A} = \varnothing$
  12. $\overline{\varnothing} = U$
  13. $\overline{U} = \varnothing$
  14. $\overline{A \cup B} = \overline{A} \cap \overline{B}$ ⎫
  15. $\overline{A \cap B} = \overline{A} \cup \overline{B}$ ⎬ DeMorgan's laws

Properties of the Universal Set
  16. $A \cup U = U$
  17. $A \cap U = A$

Properties of the Null Set
  18. $A \cup \varnothing = A$
  19. $A \cap \varnothing = \varnothing$

Proof. We prove property 14 and leave the proofs of the remaining properties as an exercise for the reader.

Suppose that $x \in \overline{A \cup B}$. Then $x \notin A \cup B$, so $x \notin A$ and $x \notin B$. This means that $x \in \overline{A} \cap \overline{B}$, so

$$\overline{A \cup B} \subseteq \overline{A} \cap \overline{B}$$

Conversely, suppose that $x \in \overline{A} \cap \overline{B}$. Then $x \notin A$ and $x \notin B$, so $x \notin A \cup B$, or $x \in \overline{A \cup B}$. Hence

$$\overline{A} \cap \overline{B} \subseteq \overline{A \cup B}$$

It then follows that $\overline{A \cup B} = \overline{A} \cap \overline{B}$.

Characteristic Functions

A very useful concept for sets is the characteristic function. We treat the general concept of function in Section 6.1, but for now we can proceed intuitively, and think of a function on a set as a rule that assigns some "value" to each element of the set. If A is a subset of a universal set U, the **characteristic function** f_A of A is defined as follows:

$$f_A(x) = \begin{cases} 1 & \text{if } x \in A \\ 0 & \text{if } x \notin A \end{cases}$$

We may add and multiply characteristic functions, since their values are numbers, and these operations sometimes help us prove theorems about properties of subsets.

Theorem 2   Characteristic functions of subsets satisfy the following properties.

(a) $f_{A \cap B} = f_A f_B$; that is, $f_{A \cap B}(x) = f_A(x) f_B(x)$ for all x.
(b) $f_{A \cup B} = f_A + f_B - f_A f_B$; that is, $f_{A \cup B}(x) = f_A(x) + f_B(x) - f_A(x) f_B(x)$ for all x.
(c) $f_{A \oplus B} = f_A + f_B - 2 f_A f_B$; that is, $f_{A \oplus B}(x) = f_A(x) + f_B(x) - 2 f_A(x) f_B(x)$ for all x.

Proof. (a) $f_A(x) f_B(x)$ equals 1 if and only if both $f_A(x)$ and $f_B(x)$ are equal to 1, and this happens if and only if x is in A and in B, that is in $A \cap B$. Since $f_A f_B$ is 1 on $A \cap B$ and 0 otherwise, it must equal $f_{A \cap B}$.

(b) If $x \in A$, then $f_A(x) = 1$, so $f_A(x) + f_B(x) - f_A(x) f_B(x) = 1 + f_B(x) - f_B(x) = 1$.

Similarly, when $x \in B$, $f_A(x) + f_B(x) - f_A(x) f_B(x) = 1$. If x is not in A or B, then $f_A(x)$ and $f_B(x)$ are 0, so $f_A(x) + f_B(x) - f_A(x) f_B(x) = 0$. Thus $f_A + f_B - f_A f_B$ is 1 on $A \cup B$ and 0 otherwise, so it equals $f_{A \cup B}$.

We leave parts (c) and (d) as exercises.

Representing Sets by Sequences of 0's and 1's

For various purposes, including representing a set in a computer, the elements of the set must be arranged in a sequence. The particular sequence selected is of no importance. When we list the set $A = \{a, b, c, \ldots, r\}$, we normally assume no particular ordering of the elements in A. In the remainder of this section, we identify the set A with the sequence $a, b, c, \ldots, r$. That is, we assume that a is the first term, b is the second one, c is the third one, and so on.

When a universal set U is finite, say $U = \{x_1, x_2, \ldots, x_n\}$, and A is a subset of U, then the characteristic function f_A assigns 1 to an element x_i that belongs to A and 0 to an element x_i that does not belong to A. Thus f_A can be represented by a sequence of 0's and 1's of length n.

Example 7   Let $U = \{1, 2, 3, 4, 5, 6\}$, $A = \{1, 2\}$, $B = \{2, 4, 6\}$, and $C = \{4, 5, 6\}$. Then $f_A(x)$ has value 1 for $x = 1, 2$, and is otherwise 0. Hence f_A corresponds to the sequence 1, 1, 0, 0, 0, 0. In a similar way, the finite sequence 0, 1, 0, 1, 0, 1 represents f_B and 0, 0, 0, 1, 1, 1 represents f_C.

Any set with n elements can be arranged in a sequence of length n (as we show in Section 3.1) so each of its subsets corresponds to a sequence of zeros and ones of length n, representing the characteristic function of that subset. This fact allows us to represent a universal set in a computer as an array A of length n. Assignment of a zero or one to each location $A[k]$ of the array specifies a unique subset of U.

Example 8   Let $U = \{a, b, e, g, h, r, s, w\}$. The array A of length 8, shown in Fig. 8, represents U, since $A[k] = 1$ for $1 \leq k \leq 8$.

1	1	1	1	1	1	1	1

Figure 8

If $A = \{a, e, r, w\}$, then

$$A[k] = \begin{cases} 1 & \text{if } k = 1, 3, 6, 8 \\ 0 & \text{if } k = 2, 4, 5, 7 \end{cases}$$

Hence the array shown in Fig. 9 represents the subset A.

1	0	1	0	0	1	0	1

Figure 9

EXERCISE SET 1.3

In Exercises 1–4, let

$$U = \{a, b, c, d, e, f, g, h, k\}, \qquad A = \{a, b, c, g\},$$

$$B = \{d, e, f, g\}, \qquad C = \{a, c, f\}, \qquad D = \{f, h, k\}$$

1. Compute
   (a) $A \cup B$ (b) $B \cup C$
   (c) $A \cap C$ (d) $B \cap D$
   (e) $A - B$ (f) $\bar{A}$
   (g) $A \oplus B$ (h) $A \oplus C$

2. Compute
   (a) $A \cup D$ (b) $B \cup D$
   (c) $C \cap D$ (d) $A \cap D$
   (e) $B - C$ (f) $C - B$
   (g) $\bar{B}$ (h) $C \oplus D$

3. Compute
   (a) $A \cup B \cup C$ (b) $A \cap B \cap C$
   (c) $A \cap (B \cup C)$ (d) $(A \cup B) \cap C$
   (e) $\overline{A \cup B}$ (f) $\overline{A \cap B}$

4. Compute
   (a) $A \cup \varnothing$ (b) $A \cap U$
   (c) $B \cup B$ (d) $C \cap \varnothing$
   (e) $\overline{C \cup D}$ (f) $\overline{C \cap D}$

In Exercises 5–8, let

$$U = \{1, 2, 3, 4, 5, 6, 7, 8, 9\}$$

$$A = \{1, 2, 4, 6, 8\}, \qquad B = \{2, 4, 5, 9\}$$

$$C = \{x \mid x \text{ is a positive integer and } x^2 \le 16\},$$

$$D = \{7, 8\}$$

5. Compute
   (a) $A \cup B$ (b) $A \cup C$
   (c) $A \cap C$ (d) $C \cap D$
   (e) $A - B$ (f) $B - A$
   (g) $\bar{A}$ (h) $A \oplus B$

6. Compute
   (a) $A \cup D$ (b) $B \cup C$
   (c) $A \cap D$ (d) $B \cap C$

(e) $C - D$ (f) $\bar{C}$
(g) $C \oplus D$ (h) $B \oplus C$

7. Compute
   (a) $A \cup B \cup C$ (b) $B \cup C \cup D$
   (c) $A \cap B \cap C$ (d) $B \cap C \cap D$
   (e) $A \cup A$ (f) $A \cap \bar{A}$

8. Compute
   (a) $\overline{A \cup B}$ (b) $\overline{A \cap B}$
   (c) $A \cap (B \cup C)$ (d) $(A \cup B) \cap D$
   (e) $A \cup \bar{A}$ (f) $A \cap (\bar{C} \cup D)$

In Exercises 9 and 10, let

$$U = \{a, b, c, d, e, f, g, h\}$$

$$A = \{a, c, f, g\}, \qquad B = \{a, e\}, \qquad C = \{b, h\}$$

9. Compute
   (a) $\bar{A}$ (b) $\bar{B}$
   (c) $\overline{A \cup B}$ (d) $\overline{A \cap B}$
   (e) $\bar{U}$ (f) $A - B$

10. Compute
    (a) $\bar{A} \cap \bar{B}$ (b) $\bar{B} \cup \bar{C}$
    (c) $\overline{A \cup A}$ (d) $\overline{C \cap C}$
    (e) $A \oplus B$ (f) $B \oplus C$

11. Let

    $U = $ set of all real numbers
    $A = \{x \mid x \text{ is a solution of } x^2 - 1 = 0\}$
    $B = \{-1, 4\}$

    Compute
    (a) $\bar{A}$ (b) $\bar{B}$
    (c) $\overline{A \cup B}$ (d) $\overline{A \cap B}$

In Exercises 12 and 13, refer to Fig. 10.

12. Answer the following as true or false.
    (a) $y \in A \cap B$ (b) $x \in B \cup C$
    (c) $w \in B \cap C$ (d) $u \notin C$

13. Answer the following as true or false.
    (a) $x \in A \cap B \cap C$ (b) $y \in A \cup B \cup C$
    (c) $z \in A \cap C$ (d) $v \in B \cap C$

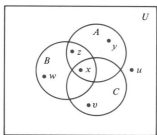

Figure 10

14. Let

$$U = \{a, b, c, d, e, f, g, h, i, j, k\}$$

$$A = \{a, c, e, h, j, k\}, \qquad B = \{a, b, c, d, h, i\}$$

$$C = \{a, k\}, \qquad D = \{a, c, d, e, j, h, k\}$$

In each of the following, represent the indicated set by an array of 0's and 1's.

(a) A

(b) $A \cup B$

(c) $A \cap C$

(d) $(A \cup C) \cap B$

(e) $(A \cap B) \cap \bar{D}$

15. Let

$$U = \{\text{ALGOL, FORTRAN, BASIC, COBOL,}$$
$$\text{ADA, PL1, PASCAL, LISP}\},$$

$$B = \{\text{ALGOL, BASIC, ADA}\},$$

$$C = \{\text{ALGOL, ADA, PASCAL, LISP}\},$$

$$D = \{\text{FORTRAN, BASIC, ADA, PL1,}$$
$$\text{PASCAL}\},$$

$$E = \{\text{ALGOL, COBOL, ADA, PASCAL, LISP}\}$$

In each of the following, represent the given set by an array of zeros and ones.

(a) $B \cup C$ (b) $C \cap D$

(c) $B \cup (D \cap E)$ (d) $\bar{B} \cup E$

(e) $\bar{C} \cap (B \cup E)$

16. Let

$$U = \{b, d, e, f, h, k, m, n\}$$

$$B = \{b\}, \quad C = \{d, f, m, n\}, \quad D = \{d, k, n\}$$

(a) What is $f_B(b)$?

(b) What is $f_C(e)$?

(c) Find the sequences of length 8 that correspond to f_B, f_C, and f_D.

(d) Represent $B \cup C$, $C \cup D$, and $C \cap D$ by arrays of zeros and ones.

THEORETICAL EXERCISES

T1. Prove that $A \subseteq A \cup B$.

T2. Prove that $A \cap B \subseteq A$.

T3. Prove that if $C \subseteq A$ or $C \subseteq B$, then $C \subseteq A \cup B$.

T4. Prove that if $A \subseteq C$ and $B \subseteq C$, then $A \cup B \subseteq C$.

T5. Prove that $A - A = \varnothing$

T6. Prove that $A - B = A \cap \bar{B}$.

T7. Prove that $A - (A - B) \subseteq B$.

T8. Prove that $A \oplus B = B \oplus A$.

T9. If $A \cup B = A \cup C$, must $B = C$? Explain.

T10. If $A \cap B = A \cap C$, must $B = C$? Explain.

T11. Prove that if $A \subseteq B$ and $C \subseteq D$, then $A \cup C \subseteq B \cup D$ and $A \cap C \subseteq B \cap D$.

T12. Prove Theorem 2(c).

T13. Using characteristic functions prove that

$$(A \oplus B) \oplus C = A \oplus (B \oplus C)$$

T14. When is $A - B = B - A$? Explain.

T15. Let A and B be subsets of a universal set U. Show that $A \subseteq B$ if and only if $\bar{B} \subseteq \bar{A}$.

KEY IDEAS FOR REVIEW

☐ Set: a well-defined collection of objects.

☐ $\varnothing$ (empty or null set): the set with no elements.

☐ Equal sets: sets with the same elements.

☐ $A \subseteq B$ (A is a subset of B): Every element of A is also an element of B.
☐ $|A|$ (cardinality of A): the number of elements in A.
☐ Infinite set: see page 6.
☐ $P(A)$ (power set of A): the set of all subsets of A.
☐ Sequence: list of objects in a definite order.
☐ Set corresponding to a sequence: see page 9.
☐ Countable set: a set that corresponds to a sequence.
☐ The interval $(0, 1)$ is an uncountable set.
☐ $A \cup B$ (union of A and B): $\{x \mid x \in A \text{ or } x \in B\}$.
☐ $A \cap B$ (intersection of A and B): $\{x \mid x \in A \text{ and } x \in B\}$.
☐ Disjoint sets: two sets with no elements in common.
☐ $A - B$ (complement of B with respect to A): $\{x \mid x \in A \text{ and } x \notin B\}$.
☐ $\bar{A}$ (complement of A): $\{x \mid x \notin A\}$.
☐ $A \oplus B$ (symmetric difference of A and B): $\{x \mid (x \in A \text{ and } x \notin B) \text{ or } (x \in B \text{ and } x \notin A)\}$.
☐ Algebraic properties of set operations: see page 17.
☐ Characteristic function of a set A:

$$f_A(x) = \begin{cases} 1 & \text{if } x \in A \\ 0 & \text{if } x \notin A \end{cases}$$

REVIEW EXERCISES

1. In each part, write the given set by listing the elements.
   (a) $\{x \mid x \in Z^+ \text{ and } x^2 < 10\}$.
   (b) The set of positive prime integers less than 50.
   (c) $\{x \mid x > 0 \text{ and } x^2 - x - 2 = 0\}$.
   (d) $\{x \mid x \in Z, x^3 < 220, x > 0, \text{ and } x \neq 5\}$.
   (e) The set of fractions $x = n/m$ with $|x| < \frac{1}{2}$, where n and m are positive integers and $m < 10$.
   (f) The set of distinct letters in the word antidisestablishmentarianism.

2. Which of the following sets are equal to $\{a, b, c\}$?

   $A = \{a, b, c\}$, $\qquad B = \{b, a, c\}$,

   $C = \{c, a, b, d\}$, $\qquad D = \{a, b, c, a, b\}$

3. Let

   $A = \{1, 2, 3, 4\}$, $\qquad B = \{2, 4, 8, 9\}$,

   $C = \{3, 7, 8\}$, $\qquad D = \{2, 3, 4, 7, 8\}$

   In each part replace the symbol ☐ by $\subseteq$, $\supseteq$, $\in$, or $\notin$ to give a correct statement.
   (a) $A \,\square\, B$ $\qquad$ (b) $\varnothing \,\square\, A$ $\qquad$ (c) $C \,\square\, \varnothing$
   (d) $2 \,\square\, A$ $\qquad$ (e) $3 \,\square\, B$ $\qquad$ (f) $4 \,\square\, \varnothing$
   (g) $C \,\square\, D$ $\qquad$ (h) $D \,\square\, A$

4. Give the set corresponding to the sequence $-2, 1, 0, 1, 2, -2, 1, 0, 1, 2, -2, \ldots$.

5. Write a formula for the nth term of the sequence $1, 3, 8, 15, 31$.

6. Let

   $U = \{s, t, u, v, w, x, y, z\}$, $\qquad A = \{t, u, x, y\}$

   $B = \{u, v, w, x, z\}$ $\qquad C = \{s, v, y\}$

   Compute
   (a) $A \cup B$ $\qquad\qquad$ (b) $A - B$
   (c) $(A \cap B) \cup C$ $\qquad$ (d) $B \oplus C$
   (e) $\bar{A} \cap (\bar{B} \cup \bar{C})$

7. Let

$$U = \{1, 2, 3, 4, 5, 6, 7\}, \qquad A = \{1, 3, 5, 7\},$$
$$B = \{1, 2, 3, 4\}, \qquad C = \{2, 4, 6\},$$
$$D = \{3, 4, 5, 7\}$$

Compute
   (a) $A \cap C$ (b) $(A \cap B) \cup (C \cap D)$
   (c) $A \cap B \cap D$ (d) $C \cup (B - A)$

8. Let $A = \{a, b, x, y\}$, $B = \{x, y, u, s, t\}$. Verify by direct computation that
   (a) $A \cap B \subseteq A$
   (b) $A \subseteq A \cup B$

9. Express the shaded region shown in terms of unions and intersections of the sets A, B, and C (several answers are possible).

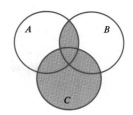

10. Let

$$U = \{2, 4, 6, 8, 10, 12, 14, 16\}$$
$$A = \{2, 6, 8, 14\}, \qquad B = \{6, 8, 10, 12\},$$
$$C = \{10, 12, 14, 16\}$$

In each of the following, represent the given set by an array of 0's and 1's.
   (a) $(A \cap B) \cup C$ (b) $A \cap (\overline{C} \cup B)$
   (c) $A \oplus C$ (d) $A - (B \cup C)$

CHAPTER TEST

1. Let $A = \{1, 2, 3, 4\}$ and $B = \{1, x, y, 4\}$. For what values of x and y are A and B equal sets?

2. Which of the following sets are empty?
   (a) $\{x \mid x \text{ is an even integer and } x \text{ is prime}\}$.
   (b) $\{x \mid x \text{ is a real number and } x^2 - 2 = 0\}$.
   (c) $\{x \mid x \text{ is an integer and } x^2 - 2 = 0\}$.
   (d) $\{x \mid x \text{ is a real number and } x^2 + 6 = 1\}$.
   (e) $\{x \mid x \text{ is an integer and } 2x^2 + 3x + 1 = 0\}$.

3. Let

$$U = \{1, 2, 3, 4, 5, 6, 7, 8, 9\}$$
$$A = \{1, 2, 5, 6, 9\} \qquad B = \{4, 5, 6\}$$
$$C = \{1, 3, 5, 7\} \quad D = \{6, 9\}$$

Compute
   (a) $A - B$
   (b) $A \cap (B \cup D)$
   (c) $(C \cup B) \cap (D \cap A)$
   (d) $D - (B \cap (A \cup D))$

4. Give the set corresponding to the sequence 0, 1, 7, 0, 0, 1, 1, 7, 7, 0, 0, 0, 1, 1, 1, 7, 7, 7, 0, . . .

5. Write a formula for the nth term of the sequence 1, 4, 7, 10, 13, 16, . . .

6. Answer each of the following as true (T) or false (F).
   (a) $\varnothing \in \varnothing$
   (b) $\varnothing \subseteq \varnothing$
   (c) $\{x, y, z\} \cap \{z, y, a\} = \{z, y\}$
   (d) If $A = \{a, b, c, d\}$, $B = \{b, c, d\}$, then $B \oplus A = \{a, d\}$

   In parts (e), (f), and (g) let D be the shaded area shown below
   (e) $D = A - (B - C)$
   (f) $D = A - (C - B)$
   (g) $D = A - (A \cap C)$

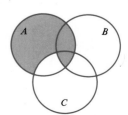

2

Methods of Counting

Prerequisites: Chapter 1.

In this chapter we deal with ways of counting the number of elements in a set and in a sequence. We also discuss permutations and combinations, and the pigeonhole principle, notions that have a wide variety of applications.

2.1 The Addition Principle

Suppose that A and B are finite subsets of a universal set U. It is frequently useful to have a formula for $|A \cup B|$, the cardinality of the union. If A and B are disjoint, that is, if $A \cap B = \varnothing$, then each element of $A \cup B$ appears in either A or B but not both; therefore, $|A \cup B| = |A| + |B|$. If A and B overlap, as shown in Fig. 1, then $A \cap B$ belong to both sets, and the sum $|A| + |B|$ counts the elements of the elements in $A \cap B$ twice. To correct for this duplication, we subtract $|A \cap B|$. Thus, we have the following theorem, sometimes called the addition principle.

Theorem 1 If A and B are finite sets, then $|A \cup B| = |A| + |B| - |A \cap B|$.

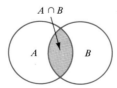

Figure 1

Example 1   Let

$$A = \{a, b, c, d, e\} \quad \text{and} \quad B = \{c, e, f, h, k, m\}$$

Verify Theorem 1.

Solution. We have

$$A \cup B = \{a, b, c, d, e, f, h, k, m\} \quad \text{and} \quad A \cap B = \{c, e\}$$

Also,

$$|A| = 5, |B| = 6, \quad |A \cup B| = 9, \quad |A \cap B| = 2$$

Then

$$|A \cup B| = 9 = |A| + |B| - |A \cap B| = 5 + 6 - 2$$

If A and B are disjoint sets, $A \cap B = \varnothing$, $|A \cap B| = 0$, so the formula in Theorem 1 now becomes

$$|A \cup B| = |A| + |B|$$

This special case can be stated in a way that is quite useful in a variety of counting situations.

The Addition Principle for Combinatorics

If a task T_1 can be performed in exactly n ways and independently a task T_2 can be performed in exactly m ways, then the number of ways of performing task T_1 *or* task T_2 is $n + m$.

Example 2   Suppose that there are seven bicycle trails and five horse trails in a certain state park. Then the total number of ways to take a ride is $7 + 5 = 12$.

The situation for three sets is more complicated, as we show in Fig. 2. We state the following three-set addition principle without comment.

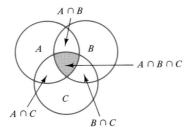

Figure 2

Theorem 2 Let A, B, and C be finite sets. Then

$$|A \cup B \cup C|$$
$$= |A| + |B| + |C| - |A \cap B| - |B \cap C| - |A \cap C| + |A \cap B \cap C|$$

Example 3 Let

$$A = \{a, b, c, d, e\}, \qquad B = \{a, b, e, g, h\}, \qquad C = \{b, d, e, g, h, k, m, n\}$$

Verify Theorem 2.

Solution. We have

$$A \cup B \cup C = \{a, b, c, d, e, g, h, k, m, n\}$$
$$A \cap B = \{a, b, e\}, \qquad A \cap C = \{b, d, e\}, \qquad B \cap C = \{b, e, g, h\},$$
$$A \cap B \cap C = \{b, e\}$$

so

$$|A| = 5, \qquad |B| = 5, \qquad |C| = 8$$
$$|A \cup B \cup C| = 10, \qquad |A \cap B| = 3, \qquad |A \cap C| = 3, \qquad |B \cap C| = 4,$$
$$|A \cap B \cap C| = 2$$

Then

$$|A \cup B \cup C| = |A| + |B| + |C| - |A \cap B| - |B \cap C| - |A \cap C|$$
$$+ |A \cap B \cap C|$$
$$10 = 5 + 5 + 8 - 3 - 3 - 4 + 2$$

Example 4 A computer company must hire 25 programmers to handle systems programming tasks and 40 programmers for applications programming. Of those hired, 10 will be expected to perform tasks of each type. How many programmers must be hired?

Solution. Let A be the set of systems programmers hired, and let B be the set of applications programmers hired. We must have $|A| = 25$ and $|B| = 40$, and $|A \cap B| = 10$. Thus the number that must be hired is $|A \cup B| = |A| + |B| - |A \cap B| = 25 + 40 - 10 = 55$.

Example 5 A survey is taken on methods of commuter travel. Each respondent is asked to check BUS, TRAIN, or AUTOMOBILE, as a major method of traveling to work. More than one answer is permitted. The results reported were as follows:

(a) 30 people checked BUS.

(b) 35 people checked TRAIN.

(c)   100 people checked AUTOMOBILE.

(d)   15 people checked BUS and TRAIN.

(e)   15 people checked BUS and AUTOMOBILE.

(f)   20 people checked TRAIN and AUTOMOBILE.

(g)   5 people checked all three methods.

How many respondents completed their surveys?

Solution. Let A, B, and C be the sets of people who checked BUS, TRAIN, and AUTOMOBILE, respectively (See Fig. 3). We know that $|A| = 30$, $|B| = 35$, $|C| = 100$, $|A \cap B| = 15$, $|A \cap C| = 15$, $|B \cap C| = 20$, and $|A \cap B \cap C| = 5$. The total number of people responding is then $|A \cup B \cup C| = (30 + 35 + 100) - (15 + 15 + 20) + 5 = 120$.

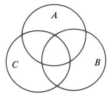

Figure 3

EXERCISE SET 2.1

1. Verify Theorem 1 for the following sets.
    (a) $A = \{1, 2, 3, 4\}$, $B = \{2, 3, 5, 6, 8\}$
    (b) $A = \{a, b, c, d, e, f\}$, $B = \{a, c, f, g, h, i, r\}$
    (c) $A = \{x \mid x$ is a positive integer $< 8\}$,
         $B = \{x \mid x$ is an integer such that $2 \le x \le 5\}$

2. Verify Theorem 1 for the following disjoint sets.
    (a) $A = \{1, 2, 3, 4\}$, $B = \{5, 6, 7, 8, 9\}$
    (b) $A = \{a, b, c, d, e\}$, $B = \{f, g, r, s, t, u\}$
    (c) $A = \{x \mid x$ is a positive integer and $x^2 \le 16\}$,
         $B = \{x \mid x$ is a negative integer and $x^2 \le 25\}$

3. Suppose that a student is asked to choose a positive integer less than 50, which must be either a multiple of 11 or a multiple of 5. How many numbers can be selected?

4. Verify Theorem 2 for the following sets.
    (a) $A = \{a, b, c, d, e\}$, $B = \{d, e, f, g, h, i, k\}$,
         $C = \{a, c, d, e, k, r, s, t\}$
    (b) $A = \{1, 2, 3, 4, 5, 6\}$, $B = \{2, 4, 7, 8, 9\}$,
         $C = \{1, 2, 4, 7, 10, 12\}$
    (c) $A = \{x \mid x$ is a positive integer $< 8\}$,

$B = \{x \mid x$ is an integer such that $2 \le x \le 4\}$,
$C = \{x \mid x$ is an integer such that $x^2 < 16\}$

5. Verify Theorem 2 for the following sets.
    (a) $A = \{1, 2, 3, 4, 5\}$, $B = \{6, 7, 8, 9\}$,
         $C = \{1, 3, 4, 6, 8, 10, 12\}$
    (b) $A = \{a, b, c, d, e, f\}$, $B = \{a, b, d, y, w\}$,
         $C = \{x, y, z, w\}$
    (c) $A = \{x \mid x$ is an integer such that $|x| < 4\}$,
         $B = \{x \mid x$ is a positive integer and $x \le 5\}$,
         $C = \{-1, 2, 5, 9\}$

6. Suppose that $|A| = 5$, $|B| = 6$, and $|A \cup B| = 8$. Find $|A \cap B|$.

7. Suppose that $|A| = 2$, $|A \cup B| = 6$, and $|A \cap B| = 2$. Find $|B|$.

8. Suppose that $|B| = 4$, $|A \cup B| = 10$, and $A \cap B = \varnothing$. Find $|A|$.

9. If $|A| = |A \cup B|$, what is the relationship between A and B?

10. If $|A| = |A \cap B|$, what is the relationship between A and B?

11. If $|A| = 5$, $|A \cup B| = 9$, and $|A \cap B| = 3$, what is $|B|$?

12. Suppose that $|A| = 10$, $|B| = 7$, $|C| = 10$, $|A \cap B| = 5$, $|A \cap C| = 7$, $|B \cap C| = 4$, and $|A \cup B \cup C| = 14$. Find $|A \cap B \cap C|$.

13. Suppose that $A \cap B = \varnothing$, $B \cap C = \varnothing$, $|A| = 3$, $|C| = 5$, $|A \cap C| = 2$, and $|A \cup B \cup C| = 10$. Find $|B|$.

14. In a survey of 260 college students, the following data were obtained:

    64 had taken a mathematics course.

    94 had taken a computer science course.

    58 had taken a business course.

    28 had taken both a mathematics and a business course.

    26 had taken both a mathematics and a computer science course.

    22 had taken both a computer science and a business course.

    14 had taken all three courses.

    (a) How many students were surveyed who had taken none of the three courses?

    (b) Of the students surveyed, how many had taken only a computer science course?

15. A survey of 500 television watchers produced the following information:

    285 watch football games.

    195 watch hockey games.

    115 watch basketball games.

    45 watch football and basketball games.

    70 watch football and hockey games.

    50 watch hockey and basketball games.

    50 do not watch any of the three games.

    (a) How many people in the survey watch all three games?

    (b) How many people watch exactly one of the three games?

16. In a psychology experiment, the subjects under study were classified according to body type and sex as follows:

	Endomorph	Ectomorph	Mesomorph
Male	72	54	36
Female	62	64	38

    (a) How many male subjects are there?

    (b) How many subjects are ectomorphs?

    (c) How many subjects are either female or endomorphs?

    (d) How many subjects are not male mesomorphs?

    (e) How many subjects are either male, ectomorph, or mesomorph?

2.2 Counting Sequences and Subsets (Permutations and Combinations)

We begin with a simple but general result that we will use frequently in this section and elsewhere.

Theorem 1 Suppose that two tasks T_1 and T_2 are to be performed in sequence. If T_1 can be performed in n_1 ways, and for each of these ways T_2 can be performed in n_2 ways, then the sequence $T_1 T_2$ can be performed in $n_1 n_2$ ways.

Proof. Each choice of a method for performing T_1 will result in a different way of performing the task sequence. There are n_1 such methods, and for each of these we may choose n_2 methods of performing T_2. Thus, in all, there will be $n_1 n_2$ ways of performing the sequence $T_1 T_2$. See Fig. 1 for the case where $n_1 = 3$ and $n_2 = 4$.

Theorem 1 is sometimes called the **multiplication principle of combinatorics.** It is an easy matter to extend the result as follows.

Possible ways of performing task 1

Possible ways of performing task 2

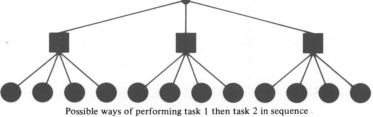

Possible ways of performing task 1 then task 2 in sequence

Figure 1

Theorem 2   Suppose that tasks T_1, T_2, . . . , T_k are to be performed in sequence. If T_1 can be performed in n_1 ways, and for each of these ways T_2 can be performed in n_2 ways, and for each of these $n_1 n_2$ ways of performing $T_1 T_2$ in sequence, T_3 can be performed in n_3 ways, and so on, then the sequence $T_1 T_2 \cdot \cdot \cdot T_k$ can be performed in exactly $n_1 n_2 \cdot \cdot \cdot n_k$ ways.

We omit the proof.

Example 1   A label identifier, for a computer program, consists of one letter followed by three digits. If repetitions are allowed, how many distinct label identifiers are possible?

Solution. There are 26 possibilities for the beginning letter and there are 10 possibilities for each of the three digits. Thus, by the extended multiplication principle, there are

$$26 \times 10 \times 10 \times 10 = 26,000 \text{ possible label identifiers}$$

Example 2   Let A be a set with n elements. How many subsets does A have?

Solution. We know from Section 1.3 that each subset of A is determined by its characteristic function, and if A has n elements, this function may be described as an array of 0's and 1's, having length n. The first element of the array can be filled in two ways (with a 0 or a 1), and this is true for all succeeding elements as well. Thus, by the extended multiplication principle, there are $2 \cdot 2 \cdot \cdot \cdot 2 = 2^n$ ways of filling in the array, and therefore 2^n subsets of A.

We now turn our attention to the following two counting problems. Let A be any set with n elements, and suppose that $1 \le r \le n$.

Problem 1 How many different sequences, each of length r, can be formed using elements from A, if
(a) elements in the sequence may be repeated, or
(b) all elements in the sequence must be distinct?

Problem 2 How many different subsets of A are there, each having r elements?

First we consider Problem 1, and we note that any sequence of length k can be formed by filling in r boxes.

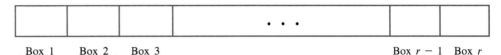

Box 1   Box 2   Box 3   Box $r-1$ Box r

in order from left to right with elements of A [or in case (a) of Problem 1, with copies of elements of A].

Let T_1 be the task "fill box 1," let T_2 be the task "fill box 2," and so on. Then the combined task $T_1 T_2 \cdots T_r$ represents the formation of the sequence.

Case (a). T_1 can be accomplished in n ways, since we may copy any element of A for the first position of the sequence. The same is true for each of the tasks T_2, T_3, $\cdots$, T_r. Then, by the extended multiplication principle, the number of sequences that can be formed is

$$\underbrace{n \cdot n \cdot \ldots \cdot n = n^r}_{r \text{ factors}}$$

We have therefore proved the following result.

Theorem 3 Let A be a set with $|A| = n$, and $1 \le r \le n$. Then the number of sequences of length r (allowing repetitions) that can be formed from elements of A is n^r.

Example 3 How many three letter "words" can be formed from letters in the set $\{a, b, y, z\}$ (repeated letters allowed)?

Solution. Here $n = 4$ and $r = 3$, so the numbers of such words is $4^3 = 64$ by Theorem 3.

Now we consider case (b) of Problem 1. Here also T_1 can be performed in n ways, since any element of A can be chosen and placed in the first position. Whichever element is chosen, only $(n - 1)$ elements remain, so that T_2 can be performed in

$n - 1$ ways, and so on, until finally, T_r can be performed in $(n - r + 1)$ ways. Thus, by the extended multiplication principle, a sequence of r distinct elements from A can be formed in $n(n - 1)(n - 2) \cdots (n - r + 1)$ ways.

     A sequence of r distinct elements of A is often called a permutation of A taken r at a time. This terminology is standard, and therefore we adopt it, but it is confusing. A better terminology might be a "permutation of r elements chosen from A." Many sequences of interest are permutations of some set of n objects taken r at a time. The discussion above shows that the number of such sequences depends only on n and r, not on A. This number is often written P_r^n and is called the **number of permutations of n objects taken r at a time.** We have just proved the following result.

Theorem 4   If $1 \le r \le n$, then P_r^n, the number of permutations of n objects taken r at a time, is $n(n - 1)(n - 2) \cdots (n - r + 1)$.

Example 4   Let $A = \{1, 2, 3, 4\}$. Then the sequences 124, 421, 341, and 243 are some permutations of A taken three at a time. The sequences 12, 43, 31, 24, and 21 are examples of different permutations of A taken two at a time. By Theorem 4, the total number of permutations of A taken three at a time is $P_3^4 = 4 \cdot 3 \cdot 2 = 24$. The total number of permutations of A taken two at a time is $P_2^4 = 4 \cdot 3 = 12$.

     When $r = n$, we are counting the distinct arrangements of the elements of a set A, with $|A| = n$, into sequences of length n. Such a sequence is simply called a **permutation** of A. In Chapter 3 we will use the term "permutation" in a slightly more complicated way, to increase its utility. The number of permutations of A is thus $P_n^n = n(n - 1)(n - 2) \cdots 2 \cdot 1$, if $n \ge 1$. This number is also written $n!$, and is called n **factorial.**

Example 5   Let $A = \{a, b, c\}$. Then the possible permutations of A are the sequences abc, acb, bac, bca, cab, and cba.

     If we agree to define $0!$ as 1, then for every $n \ge 0$ we see that the number of permutations of n objects is $n!$. If $n \ge 1$ and $1 \le r \le n$, we can give a more compact form for P_r^n as follows:

$$P_r^n = n \cdot (n - 1) \cdots (n - r + 1)$$

$$= \frac{n \cdot (n - 1) \cdots (n - r + 1) \cdot (n - r) \cdot (n - r - 1) \cdots 2 \cdot 1}{(n - r) \cdot (n - r - 1) \cdots 2 \cdot 1}$$

$$= \frac{n!}{(n - r)!}$$

Example 6   Let A consist of all 52 cards in an ordinary deck of playing cards. Suppose that this deck is shuffled and a hand of five cards is dealt. A list of cards in this hand, in the order in which they were dealt, is a permutation of A, taken five at a time. Examples would include AH, 3D, 5C, 2H, JS; 2H, 3H, 5H, QH, KD; JH, JD, JS, 4H, 4C; and 3D, 2H, AH, JS, 5C. Note that the first and last hands are the same, but they represent different permutations since they were dealt in a different order. The number of permutations of A taken five at a time is

$$P_5^{52} = \frac{52!}{47!} = 52 \cdot 51 \cdot 50 \cdot 49 \cdot 48 = 311{,}875{,}200$$

This is the number of five-card hands that can be dealt if we count the order in which they are dealt.

Example 7   If A is the set of Example 5, then $n = 3$ and the number of permutations of A is $3! = 6$. Thus all permutations of A are listed in Example 5, as stated there.

Example 8   How many "words" of three distinct letters can be formed from the alphabet $\{a, b, y, z\}$?

Solution. The number is

$$P_3^4 = \frac{4!}{(4 - 3)!} = \frac{4!}{1!} = 24$$

We next turn to the solution of the second problem. Again, we assume that $1 \le r \le n$. The traditional combinatorial name for an r-element subset of an n-element set A is a **combination of A, taken r at a time.**

Example 9   Let $A = \{1, 2, 3, 4\}$. The following are all the distinct combinations of A, taken three at a time: $A_1 = \{1, 2, 3\}$, $A_2 = \{1, 2, 4\}$, $A_3 = \{1, 3, 4\}$, $A_4 = \{2, 3, 4\}$. Note that these are subsets not sequences. Thus $A_1 = \{2, 1, 3\} = \{2, 3, 1\} = \{1, 3, 2\} = \{3, 2, 1\} = \{3, 1, 2\}$. In other words, when it comes to combinations, unlike permutations, the order of the elements is irrelevant.

Example 10   Let A be the set of Example 6. Then a combination of A taken five at a time is just a hand of five cards regardless of how these cards were dealt.

We now want to count the number of r-element subsets of an n-element set A. This is most easily accomplished by using the information on permutations that we have developed so far. Observe that each permutation of the elements of A, taken r at a time, can be produced by performing the following two tasks in sequence.

Task 1. Choose a subset B of A containing r elements of A.

Task 2. Choose a particular permutation of B.

We are trying to compute the number of ways of choosing B. Call this number c. Then task 1 can be performed in c ways, and task 2 can be performed in $r!$ ways. Thus the total number of ways of performing both tasks, which is P_r^n, is seen, by the multiplication principle, to equal $cr!$. Hence

$$c \cdot r! = P_r^n = \frac{n!}{(n-r)!}$$

Therefore,

$$c = \frac{n!}{r!\,(n-r)!}$$

We have therefore proved the following result.

Theorem 5   Let A be a set with $|A| = n$, and let $1 \le r \le n$. Then the number of combinations of the elements of A, taken r at a time, that is, the number of r-element subsets of A, is

$$\frac{n!}{r!(n-r)!}$$

Note again that the number of combinations of A, taken r at a time, does not depend on A, but only on n. This number is written C_r^n or sometimes $\binom{n}{r}$, and is called the **number of combinations of n objects taken r at a time**. We have

$$\binom{n}{r} = C_r^n = \frac{n!}{r!(n-r)!}$$

Example 11   Compute the number of distinct five card hands which can be dealt from a deck of 52 cards.

Solution. This number is

$$C_5^{52} = \frac{52 \cdot 51 \cdot 50 \cdot 49 \cdot 48 \cdot (47!)}{5!\,(47!)} = \frac{52!}{5!\,47!} = \frac{52 \cdot 51 \cdot 50 \cdot 49 \cdot 48}{1 \cdot 2 \cdot 3 \cdot 4 \cdot 5} = 2,598,960$$

Compare this number with the number computed in Example 6.

In general, when order matters, we count the number of sequences or permutations; when order does not matter, we count the number of subsets or combinations.

Some problems require that the counting of permutations and combinations be combined or supplemented by direct use of the multiplication principle.

Example 12   Suppose that a valid computer password consists of four characters, the first of which is a letter chosen from the set $A = \{A, B, C, D, E, F\}$ and the remaining three characters are letters chosen from the English alphabet or digits chosen from the set $T = \{0, 1, 2, 3, 4, 5, 6, 7, 8, 9\}$. How many different passwords are there?

Solution. A password can be constructed by performing the following tasks T_1 and T_2 in succession:

Task T_1. Choose a starting letter from the set A.

Task T_2. Choose a sequence of letters and digits (repeats are allowed).

Task T_1 can be performed in $C_1^6 = 6$ ways. Since there are 26 letters and 10 digits that can be chosen for each of the remaining three characters, and since repeats are allowed, task T_2 can be performed in $36^3 = 46,656$ ways (by Theorem 3). By the multiplication principle there are $(6)(46,656) = 279,936$ different passwords.

Example 13   How many different seven-person committees can be formed each containing three female members from an available set of 20 females and four male members from an available set of 30 males?

Solution. In this case a committee can be formed by performing the following two tasks in succession:

Task 1. Choose three females from a set of 20 females.

Task 2. Choose four males from a set of 30 males.

Here order does not matter in the individual choices, we are merely counting the number of possible subsets. Thus, task T_1 can be performed in

$$\binom{20}{3} = \frac{20!}{3!\,17!} = \frac{20 \cdot 19 \cdot 18}{1 \cdot 2 \cdot 3} = 1140 \text{ ways}$$

and task T_2 can be performed in

$$\binom{30}{4} = \frac{30!}{4!\,26!} = \frac{30 \cdot 29 \cdot 28 \cdot 27}{1 \cdot 2 \cdot 3 \cdot 4} = 27,405 \text{ ways}$$

By the multiplication principle there are

$$(1140)(27,405) = 31,241,700$$

different committees.

EXERCISE SET 2.2

1. A bank password consists of two letters of the English alphabet followed by two digits from 0 to 9. How many different passwords are there?

2. In a psychological experiment a subject must arrange a cube, square rectangle, triangle, and circle in a row. How many different arrangements are possible?

3. A coin is tossed four times and the result of each toss is recorded. How many different sequences of heads and tails are possible?

4. A menu is to consist of a soup dish, a main course, a dessert, and a beverage. Suppose that the chef can select from four soup dishes, five main courses, three desserts, and two beverages. How many different menus can be prepared?

5. A die is tossed four times and the numbers shown are arranged in a sequence. How many different sequences are there?

6. Compute
   (a) P_4^4 (b) P_5^6 (c) P_2^7
   (d) P_{n-1}^n (e) P_{n-2}^n (f) P_{n-1}^{n+1}

7. How many permutations are there of each of the following sets?
   (a) $\{r, s, t, u\}$ (b) $\{1, 2, 3, 4, 5\}$
   (c) $\{a, b, 1, 2, 3, c\}$

8. For each set A, find the number of permutations of A taken r at a time.
   (a) $A = \{1, 2, 3, 4, 5, 6, 7\}$, $r = 3$
   (b) $A = \{a, b, c, d, e, f\}$, $r = 2$
   (c) $A = \{x \mid x$ is an integer and $x^2 < 16\}$, $r = 4$

9. In how many ways can six men and six women be seated in a row if
   (a) any person may sit next to any other person?
   (b) men and women must occupy alternate seats?

10. How many different arrangements of the letters in the word "bought" can be formed if the vowels must be kept next to each other?

11. Find the number of different permutations of the letters in the word "group".

12. Find the number of distinguishable permutations of the letters in the word "Boolean." (*Hint:* For example a permutation that only interchanges the two o's is not distinguishable from the permutation that leaves the word unchanged.)

13. In how many ways can six people be seated in a circle?

14. A bookshelf is to be used to display six new books. Suppose that there are eight computer science books and five French books for display. If we are to show four computer science books and two French books, how many different displays are there if we require that the books in each subject must be kept together?

15. Compute
   (a) C_7^7 (b) C_4^7 (c) C_5^6
   (d) C_{n-1}^n (e) C_{n-2}^n (f) C_{n-1}^{n+1}

16. In how many ways can a committee of six people be selected from a group of 10 people if one person is to be designated as chair?

17. In how many ways can a committee of three faculty members and two students be selected from seven faculty members and eight students?

18. In how many ways can a six-card hand be dealt from a deck of 52 cards?

19. At a certain college, the housing office has decided to appoint for each floor one male and one female residential advisor. How many different pairs of advisors can be selected for a seven-story building from 12 male candidates and 15 female candidates?

20. A microcomputer manufacturer who is designing an advertising campaign is considering six magazines, three newspapers, two television stations, and four radio stations. In how many ways can six advertisements be run if

   (a) all six are to be in magazines?

   (b) two are to be in magazines, two are to be in newspapers, one is to be on television, and one is to be on radio?

21. How many different eight-card hands with five red cards and three black cards can be dealt from a deck of 52 cards?

22. (a) Find the number of subsets of a set containing four elements.

   (b) Find the number of subsets of a set containing n elements.

23. An urn contains 15 balls, eight of which are red and seven are black. In how many ways can five balls be chosen so that

   (a) all five are red?

   (b) all five are black?

   (c) two are red and three are black?

   (d) three are red and two are black?

THEORETICAL EXERCISES

T1. Show that $C_r^{n+1} = C_{r-1}^n + C_r^n$.

T2. Show that $C_r^n = C_{n-r}^n$.

2.3 The Pigeonhole Principle

A method of solving certain counting problems involves an investigation of how objects may be placed into compartments. It is customary to state the main result in a picturesque and informal way. The objects are thought of as "pigeons" and the compartments as "pigeonholes". We then have the following result.

Theorem 1 (The Pigeonhole Principle)
If n pigeons are assigned to m pigeonholes, and $m < n$, then at least one pigeonhole contains two or more pigeons.

We shall use the pigeonhole principle to solve a diverse number of interesting counting problems.

Example 1 If eight people are chosen in any way whatsoever, at least two of them will have been born on the same day of the week.

Here each person (pigeon) is assigned the day of the week (pigeonhole) on which he or she was born. Since there are eight people and only seven days of the week, the pigeonhole principle tells us that at least two people must be assigned to the same day.

Example 2 Show that if any 11 numbers are chosen from the set $\{1, 2, \ldots, 20\}$, then one of them will be a multiple of the other.

Solution. Every positive integer n can be written as $n = 2^k m$, where m is odd and $k \geq 0$. This can be seen by simply factoring out all powers of 2 (if any) from n. In this case let us call m the odd part of n. If 11 numbers are chosen from the set $\{1, 2, \ldots, 20\}$, then two of them must have the same odd part. This follows from the pigeonhole principle since there are 11 numbers (pigeons), but only 10 odd numbers from 1 to 20 (pigeonholes) that can be the odd parts of these numbers.

Let n_1 and n_2 be the two chosen numbers that have the same odd parts. We must have $n_1 = 2^k m$ and $n_2 = 2^{k'} m$, for some k and k'. If $k \geq k'$, then n_1 is a multiple of n_2; otherwise, n_2 is a multiple of n_1.

Example 3 Show that if any five numbers from 1 to 8 are chosen, then two of them will add to 9.

Solution. Construct four different sets, each containing two numbers that add to 9, as follows; $A_1 = \{1, 8\}$, $A_2 = \{2, 7\}$, $A_3 = \{3, 6\}$, and $A_4 = \{4, 5\}$. Each of the five numbers chosen will be assigned to the set that contains it. Since there are only four sets, the pigeonhole principle tells us that two of the chosen numbers will be assigned to the same set. These two numbers will add to 9.

Example 4 Consider the area shown in Fig. 1. It is bounded by a regular hexagon whose sides have length 1. Show that if any seven points are chosen within this area, then two of them must be no farther apart than 1 unit.

Solution. Suppose that the area is divided into six equilateral triangles, as shown in Fig. 2. If seven points are chosen, we can assign each one to a triangle that contains it. If the point belongs to several triangles, assign it arbitrarily to one of them. Then the seven points are assigned to six triangles, so by the pigeonhole principle, at least two points must belong to the same triangle. These two cannot be more than one unit apart.

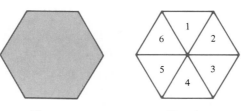

Figure 1   Figure 2

Example 5 Shirts numbered consecutively from 1 to 20 are worn by 20 members of the bowling league. If any three of these members are chosen to be a team, the sum of their shirt numbers will be used as a code number for this team. Show that if any eight of the

20 are preselected, then from these eight one may form at least two different teams having the same code number.

Solution. From the preselected eight bowlers, we can form a total of

$$\binom{8}{3} = \frac{8!}{3!\,5!} = 56$$

different teams. The largest possible team code number is $18 + 19 + 20 = 57$, and the smallest possible is $1 + 2 + 3 = 6$. Thus only the 52 code numbers between 6 and 57 inclusive are available for 56 possible teams. By the pigeonhole principle, at least two of the possible teams will have the same code number.

The Extended Pigeonhole Principle

Note that if there are m pigeonholes and more than $2m$ pigeons, three or more pigeons will have to be assigned to at least one pigeonhole. We now prove a generalization of this result.

First a word about notation. If n and m are integers, then $[n/m]$ will stand for the largest integer less than or equal to the rational number n/m. Thus $[\frac{3}{2}] = 1$, $[\frac{9}{4}] = 2$, and $[\frac{6}{2}] = 3$. We now have the following result.

Theorem 2   **(The Extended Pigeonhole Principle)**
If n pigeons are assigned to m pigeonholes, then one of the pigeonholes must contain at least $[(n - 1)/m] + 1$ pigeons.

Proof. If each pigeonhole contains no more than $[(n - 1)/m]$ pigeons, then there are at most $m \cdot [(n - 1)/m] \le m \cdot (n - 1)/m = n - 1$ pigeons in all. This contradicts our assumptions, so one of the pigeonholes must contain at least $[(n - 1)/m] + 1$ pigeons.

Example 6   We give an extension of Example 1. Show that if any 30 people are selected, then one may choose a subset of five so that all five were born on the same day of the week.

Solution. Assign each person to the day of the week on which he or she was born. Then $n = 30$ pigeons are being assigned to $m = 7$ pigeonholes. By the extended pigeonhole principle, at least $[(n - 1)/m] + 1 = [29/7] + 1 = 5$ of the people must have been born on the same day of the week.

Example 7   Show that if 30 dictionaries in a library contain a total of 61,327 pages, then one of the dictionaries must have at least 2045 pages.

Solution. Let the pages be the pigeons and the dictionaries be the pigeonholes.

Assign each page to the dictionary in which it appears. Then by the extended pigeon-hole principle, one dictionary must contain at least $[61,326/30] + 1 = 2045$ pages.

Applications

At this point, applications Chapter 12 can be covered if desired.

EXERCISE SET 2.3

1. If 13 people are assembled in a room, show that at least two of them must have their birthday in the same month.

2. Show that if any seven numbers from 1 to 12 are chosen, then two of them will add to 13.

3. Let T be an equilateral triangle whose sides have length 1. Show that if any five points are chosen lying on or inside T, then two of them will be no more than $\frac{1}{2}$ unit apart.

4. Show that if any eight positive integers are chosen, two of them will have the same remainder when divided by 7.

5. Show that if seven colors are used to paint 50 bicycles, at least eight bicycles must have the same color.

6. Ten people volunteer for a three-person committee. Every possible committee of three that can be formed from these ten names is written on a slip of paper, one slip for each possible committee, and the slips are put into 10 hats. Show that at least one hat contains 12 or more slips of paper.

7. Six friends discover that they have a total of $21.61 with them during a trip to the movies. Show that one or more of them must have at least $3.61.

8. A store has an introductory sale on 10 types of candy bars. A customer may choose one bar of any three different types and will be charged no more than $1 for the purchase. Show that although different choices may cost different amounts, there must be at least two different ways to choose so that the cost will be the same for both choices.

9. Show that there are at least six different ways to choose three numbers from 1 to 10 so that all choices have the same sum.

KEY IDEAS FOR REVIEW

☐ Theorem (The Addition Principle): If A and B are finite sets, then

$$|A \cup B| = |A| + |B| - |A \cap B|$$

☐ Theorem (The Three-Set Addition Principle): If A, B, and C are finite sets, then

$$|A \cup B \cup C| = |A| + |B| + |C| - |A \cap B| - |B \cap C|$$
$$- |A \cap C| + |A \cap B \cap C|$$

☐ Theorem (The Multiplication Principle): Suppose that two tasks T_1 and T_2 are to be performed in sequence. If T_1 can be performed in n_1 ways, and, for each of these ways T_2 can be performed in n_2 ways, then the sequence $T_1 T_2$ can be performed in $n_1 n_2$ ways.

☐ Permutation: an arrangement of the n elements of a set A into a sequence of length n.

☐ Permutation of an n-element set A, taken r at a time ($1 \leq r \leq n$): a sequence of length r formed from distince elements of A.

☐ Combination of an n-element set A, taken r at a time ($1 \leq r \leq n$): a subset of A having exactly r elements.

☐ Theorem: If A is set with $|A| = n$ and $1 \leq r \leq n$, then the number of permutations of A taken r at a time is

$$P_r^n = n(n-1)(n-2) \cdots (n-r+1) = \frac{n!}{(n-r)!}$$

☐ Theorem: If A is a set with $|A| = n$ and $1 \leq r \leq n$, then the number of combinations of the elements of A, taken r at a time, is

$$C_r^n = \frac{n!}{r!\,(n-r)!}$$

☐ The Pigeonhole Principle: see page 36.
☐ The Extended Pigeonhole Principle: see page 38.

REVIEW EXERCISES

1. Verify Theorem 2 for the following sets.
   (a) $A = \{A, B, D, G, H\}$, $B = \{C, D, E, F, K\}$,
   $C = \{F, K, X, Y, Z\}$
   (b) $A = \{1, 2, 3, 5, 7, 11\}$, $B = \{3, 11, 13, 17\}$,
   $C = \{2, 11, 17, 19, 23\}$

2. If $|A| = 10$, $|B| = 3$, and $|A \cup B| = 6$, find $|A \cap B|$.

3. Suppose that

   $$|A| = 19, \quad |B| = 18, \quad |A \cap B| = 12$$
   $$|A \cap C| = 6, \quad |B \cap C| = 9,$$
   $$|A \cap B \cap C| = 5, \quad |A \cup B \cup C| = 28$$

   Compute $|C|$.

4. A certain personal computer is available with three possible operating systems, four programming languages, and three different printers. If a configuration consists of a computer, an operating system, a programming language, and a printer, how many different configurations are there?

5. How many permutations are there for each of the following sets?
   (a) {Mary, had, a, little, lamb}
   (b) {B, C, A, D}
   (c) {4, 5, 2, 3, 6, 7}

6. Compute
   (a) P_3^5 (b) P_5^{12} (c) P_4^8
   (d) C_{200}^{200} (e) C_{77}^{79} (f) C_4^5

7. How many six-card hands with four black cards and six red cards can be dealt from a standard 52-card deck?

8. An urn contains six distinguishable red balls and four distinguishable black balls. Suppose that five balls are selected.
   (a) In how many ways can the five balls be selected?
   (b) In how many ways can the five balls be selected if three must be red and two must be black?

(c) In how many ways can the five balls be selected if they must all be red?

9. A group consists of five men and six women. How many ways are there to fill a row of eight chairs with members of this group if the two middle chairs must always be occupied by women?

10. Show that if 102 students take a test, then at least two students must receive the same grade. (Possible grades run from 0 to 100.)

11. A florist grows 30 different types of flowers. Show that if she grows a total of 350 flowers, then it is possible to supply at least a dozen of one type of flower.

CHAPTER TEST

1. Let A, B, and C be finite sets with

$$|A| = 8, \qquad |B| = 11, \qquad |C| = 9$$
$$|A \cap B \cap C| = 3, \qquad |A \cap C| = 5,$$
$$|B \cap C| = 6, \qquad |A \cup B \cup C| = 16$$

Find $|A \cap B|$.

2. Twenty balls numbered 1 to 20 are placed in a hat. Three balls are removed consecutively and the resulting sequence of numbers is recorded. How many sequences can be formed in this way?

3. How many different five-card hands with two red cards and three black cards can be dealt from a deck of 52 cards?

4. A bank assigns a code word of length 5 to each account. The code word must consist of three digits (0 to 9) followed by two letters. The digits may be repeated but the letters must be distinct. How many code words can be formed by the bank?

5. A parking lot holds 100 cars. Six attendants are on duty during the day to park cars. Show that on any given day at least one attendant must park 17 or more cars.

6. Answer each of the following as true (T) or false (F).
   (a) If $|A \cup B| = 11$, $|A| = 3$, and $|B| = 8$, then $|A \cap B| = \varnothing$.
   (b) The number of four-element subsets of the set $\{a, b, c, d, e, f, g\}$ is 36.
   (c) For any $n \geq 1$, $\binom{n}{1} = n$.
   (d) If each of 105 students is assigned a two-digit code number (the digits need not be distinct), then it is possible that every student will have a different code number.

3

Some Algebraic Tools

Prerequisites: Chapter 1.

In this chapter we present several algebraic tools that will be used throughout the book. We start with a discussion of induction and recursion, central ideas in computer science, and illustrate some of their applications. Next we review some of the basic divisibility properties of the integers. Finally, we introduce matrices and their operations.

3.1 Induction and Recursion

Mathematical Induction

Let n_0 be a fixed integer (positive, negative, or zero). Suppose that for each integer $n \geq n_0$ we have a corresponding statement $P(n)$, and we wish to show that $P(n)$ is true for all $n \geq n_0$. The following result shows how this can be done. Suppose that

(a) $P(n_0)$ is true
(b) If $P(k)$ is true for some $k \geq n_0$, then $P(k + 1)$ must also be true.

Then $P(n)$ is true for all $n \geq n_0$.

This result is called the **principle of mathematical induction.** Thus to prove the truth of the statements $P(n)$ for all $n \geq n_0$, using the principle of mathematical induction, we must begin by proving directly that the first proposition $P(n_0)$ is true. This is called the **basis step** of the induction, and it is generally easy.

Then we must prove that if, purely hypothetically, statement $P(k)$ is assumed to be true for some $k \geq n_0$, then the next statement $P(k + 1)$ would have to be true as well. This is called the **induction step,** and some work will usually be required to show that it is true.

Example 1   Show, by mathematical induction, that for all $n \geq 1$,

$$1 + 2 + \cdots + n = \frac{n(n + 1)}{2}$$

Solution. Let $P(n)$ be the statement

$$1 + 2 + \cdots + n = \frac{n(n + 1)}{2}$$

In this example, $n_0 = 1$.

Basis Step. We must first show that $P(1)$ is true. We see that $P(1)$ is

$$1 = \frac{1(1 + 1)}{2}$$

which is clearly true.

Induction Step. Assume now that we know $P(k)$ to be true for some $k \geq 1$. That is, we assume that for some fixed $k \geq 1$,

$$1 + 2 + \cdots + k = \frac{k(k + 1)}{2} \tag{1}$$

We now wish to prove the validity of the statement $P(k + 1)$

$$1 + 2 + \cdots + (k + 1) = \frac{(k + 1)((k + 1) + 1)}{2}$$
$$= \frac{(k + 1)(k + 2)}{2}$$

Since (1) is assumed true, we may add $(k + 1)$ to both sides and preserve equality. Thus we have

$$1 + 2 + \cdots + k + (k + 1) = [1 + 2 + \cdots + k] + (k + 1)$$

$$= \frac{k(k + 1)}{2} + (k + 1) \qquad \text{[the expression in brackets is replaced by}$$

$$= \frac{k(k + 1) + 2(k + 1)}{2} \qquad \frac{k(k + 1)}{2} \text{ using}$$

$$= \frac{(k + 1)(k + 2)}{2} \qquad \begin{array}{l}\text{Equation (1)]}\\ \text{(factoring)}\end{array}$$

which proves the validity of $P(k + 1)$. By the principle of mathematical induction, it follows that $P(n)$ is true for all $n \geq 1$.

Example 2 Let A_1, A_2, A_3, $\cdots$, A_n be any n subsets of a set U. We show by mathematical induction that

$$\overline{\left(\bigcup_{j=1}^{n} A_j \right)} = \bigcap_{j=1}^{n} \overline{A}_j$$

[extended De Morgan's law (see Section 1.3)].

Let $P(n)$ be the statement that the given equality holds for any n subsets of U. We prove by mathematical induction that $P(n)$ is always true, for $n \geq 1$.

Basis Step. $P(1)$ is equivalent to the statement $\overline{A}_1 = \overline{A}_1$, which is obviously true.

Induction Step. Suppose that $P(k)$ is true for some $k \geq 1$ and let $A_1, \ldots, A_k, A_{k+1}$ be any $k + 1$ subsets of U. Then let $B = A_1 \cup A_2 \cup \cdots \cup A_k$. We have

$$\overline{\left(\bigcup_{j=1}^{k+1} A_j \right)} = \overline{B \cup A_{k+1}}$$

$$= \overline{B} \cap \overline{A}_{k+1} \qquad \text{(by Section 1.3)}$$

$$= \left(\bigcap_{j=1}^{k} \overline{A}_j \right) \cap \overline{A}_{k+1} \qquad \text{(by induction assumption)}$$

$$= \bigcap_{j=1}^{k+1} \overline{A}_j$$

Thus $P(k + 1)$ is true, so by the principle of mathematical induction, $P(n)$ is true for all $n \geq 1$.

Example 3   We show by mathematical induction that any finite, nonempty set is countable (that is, can be arranged in a list).

Let $P(n)$ be the statement that if A is any set with $|A| = n \geq 1$ (see Section 1.1), then A is countable.

Basis Step. Here $n_0 = 1$, so we let A be any set with one element, say $A = \{x\}$. In this case x forms a sequence all by itself whose set is A, so $P(1)$ is true.

Induction Step. Suppose that $P(k)$ is true, for some $k \geq 1$, that is, if A is any set with $|A| = k$, then A is countable. Now choose any set A with $|A| = k + 1$, and choose any element x in A. Since $A - \{x\}$ is a set with k elements, the induction hypothesis [that is the truth of $P(k)$] tells us that there is a sequence $x_1, x_2, \cdots, x_k$ having $A - \{x\}$ as a corresponding set. The sequence $x_1, x_2, \cdots, x_k, x$ then has A as a corresponding set, so A is countable. Since A can be any set with $k + 1$ elements, $P(k + 1)$ is true. Thus by the principle of mathematical induction, $P(n)$ is true for all $n \geq 1$.

Example 4   Show by mathematical induction that for all $n \geq 5$, $n^2 < 2^n$.

Solution. Let $P(n)$ be the statement $n^2 < 2^n$. We prove by mathematical induction that $P(n)$ is always true, for $n \geq 5$.

Basis Step. We must first show that $P(5)$ is true. We see that $P(5)$ is

$$5^2 < 2^5$$

which is true.

Induction Step. Suppose that $P(k)$ is true for some $k \geq 5$. Then

$$k^2 < 2^k \qquad (1)$$

Since $k \geq 5$, it is easily seen that $2k + 1 < 2k + 2k = 4k < k^2$. Thus we have

$$
\begin{aligned}
(k + 1)^2 &= k^2 + 2k + 1 \\
&< k^2 + k^2 = 2k^2 \quad \text{(by inequality above)} \\
&< 2 \cdot 2^k \qquad\qquad \text{[by inequality (1)]} \\
&= 2^{k+1}
\end{aligned}
$$

Thus $P(k + 1)$, which is the statement $(k + 1)^2 < 2^{k+1}$, is true. By the principle of mathematical induction, $P(n)$ is true for all $n \geq 5$.

Warning. In proving results by induction, you should not start by assuming that $P(k + 1)$ is true and attempting to manipulate this result until you arrive at a true statement. This common mistake is always an incorrect use of the principle of mathematical induction.

Recursion

It often happens that we want to define a finite or countable set of objects—for example, the elements in a sequence, the members of a set, the operations that an algorithm performs on its inputs, or the values of a function defined for certain integers—but that we cannot easily give a definition that applies all at once to the entire set. Instead, we define the objects in a sequence of steps or rules, where each step or rule defines some of the objects in terms of others that were previously defined. In this case, we say that the definition—and also the sequence, set, algorithm, or function that it defines—is **recursive.**

Example 5   Consider the following definition of the factorial function:

$$1! = 1$$
$$n! = n(n - 1)! \qquad \text{for } n > 1$$

Thus, for example

$$2! = 2 \cdot 1! = 2 \cdot 1$$
$$3! = 3 \cdot 2! = 3 \cdot 2 \cdot 1$$
$$\vdots$$

Notice that in the second part of the definition, factorial is used as part of the definition of $n!$. Thus this definition is recursive. Note that it is not circular because by the time one wants to define $n!$ (if we look at the definition in increasing order of n). We have already defined $(n - 1)!$, so, we can use its value.

We now consider another slightly different example of recursion.

Example 6   Suppose that we define a sequence $(F_n)_{1 \le n < \infty}$, of integers as follows:

$$F_1 = 1$$

$$F_2 = 1$$

$$F_n = F_{n-1} + F_{n-2} \qquad \text{if } n > 2$$

Since, in the third line, the definition refers to earlier versions of itself, it is recursive. It is easy to compute the first few terms in this sequence $1, 1, 2, 3, 5, 8, \ldots$, called the **Fibonacci sequence.**

The main connection between recursion and induction is that objects that are defined recursively are often defined by means of a natural sequence, so induction is frequently the best (possibly the only) way to prove results about recursively defined objects.

Example 7   Consider our recursive definition of $n!$ given in Example 5. Suppose we wish to prove that for all $n \geq 1$,

$$n! \geq 2^{n-1} \tag{2}$$

We proceed by mathematical induction. Let $P(n)$ by the statement given by equation (2). Here $n_0 = 1$.

Basis Step. $P(1)$ is the statement

$$1! \geq 2^0 = 1$$

Since 1! equals 1, this statement is true.

Induction Step. Suppose that $P(k)$ is true for some $k \geq 1$. That is, for this value of k.

$$k! \geq 2^{k-1}$$

Then by the recursive definition,

$$
\begin{aligned}
(k + 1)! &= (k + 1) \times k! \\
&\geq (k + 1) \times 2^{k-1} \\
&\geq 2 \times 2^{k-1} \qquad (\text{since } k \geq 1, \text{ we have } k + 1 \geq 2) \\
&= 2^k
\end{aligned}
$$

and $P(k + 1)$ is true. By the principle of mathematical induction, it follows that $P(n)$ is true for all $n \geq 1$.

EXERCISE SET 3.1

In Exercises 1-17, prove that the statement is true by using mathematical induction.

1. $2 + 4 + 6 + \cdots + 2n = n(n + 1)$
2. $1^2 + 3^2 + 5^2 + \cdots + (2n - 1)^2$
   $= \dfrac{n(2n + 1)(2n - 1)}{3}$
3. $2 + 5 + 8 + \cdots + (3n - 1) = \dfrac{n(3n + 1)}{2}$
4. $4 + 8 + 12 + \cdots + 4n = 2n(n + 1)$
5. $1 + 2 + 2^2 + 2^3 + \cdots + 2^n = 2^{n+1} - 1$
6. $5 + 10 + 15 + \cdots + 5n = \dfrac{5n(n + 1)}{2}$
7. $1^2 + 2^2 + 3^2 + \cdots + n^2 = \dfrac{n(n + 1)(2n + 1)}{6}$
8. $1 \cdot 2 + 2 \cdot 3 + 3 \cdot 4 + \cdots + n(n + 1)$
   $= \dfrac{n(n + 1)(n + 2)}{3}$
9. $1^3 + 2^3 + 3^3 + \cdots + n^3 = \dfrac{n^2(n + 1)^2}{4}$
10. $1 + 5 + 9 + \cdots + (4n - 3) = n(2n - 1)$
11. $2 + 2^2 + 2^3 + \cdots + 2^n = 2^{n+1} - 2$
12. $a + ar + ar^2 + \cdots + ar^{n-1}$
    $= \dfrac{a(1 - r^n)}{1 - r} \quad (r \neq 1)$
13. $1 + 2^n < 3^n \quad (n \geq 2)$
14. $n < 2^n \quad (n > 1)$
15. $1 + 2 + 3 + \cdots + n < \frac{1}{8}(2n + 1)^2$
16. $\dfrac{1}{1 \cdot 2} + \dfrac{1}{2 \cdot 3} + \dfrac{1}{3 \cdot 4} + \cdots + \dfrac{1}{n(n + 1)} = \dfrac{n}{n + 1}$
17. $1 + a + a^2 + \cdots + a^{n-1} = \dfrac{a^n - 1}{a - 1}$
18. Show by mathematical induction that if a set A has n elements, then $P(A)$ has 2^n elements.
19. Show by mathematical induction that if $A_1, \ldots, A_n$ are any n subsets of a set U, then
    $$\overline{\left(\bigcap_{j=1}^{n} A_j \right)} = \bigcup_{j=1}^{n} \overline{A_j}$$

20. Show by mathematical induction that if $A_1, \ldots, A_n$, and B are any subsets of a set U, then
    $$\left(\bigcup_{j=1}^{n} A_j \right) \cap B = \bigcup_{j=1}^{n} (A_j \cap B)$$

21. Show by mathematical induction that if $A_1, A_2, \ldots, A_n$ and B are any subsets of a set U, then
    $$\left(\bigcap_{j=1}^{n} A_j \right) \cup B = \bigcap_{j=1}^{n} (A_j \cup B)$$

22. Show by mathematical induction that if $A_1, A_2, \ldots, A_n$ are any n subsets of a set U, then
    $$\overline{\left(\bigcup_{j=1}^{n} A_j \right)} = \bigcap_{j=1}^{n} \overline{A_j}$$

23. We define T-numbers recursively as follows:
    (1) 0 is a T-number
    (2) If X is a T-number, then $X + 3$ is a T-number.

    Write out a description of the set of T-numbers.

24. Define an S-number by:
    (1) 8 is an S-number.
    (2) If X is an S-number and Y is a multiple of X, then Y is an S-number.
    (3) If X is an S-number and X is a multiple of Y, then Y is an S-number.

    Describe the set of S-numbers.

25. Let $F(N)$ be a function defined for all nonegative integers by the following recursive definition.
    $$F(0) = 0$$
    $$F(1) = 1$$
    $$F(N + 2) = 2F(N) + F(N + 1), N \geq 0$$

    Write out the first six values of F.

26. Let $G(N)$ be a function defined for all nonnegative integers by the following recursive definition.

$$G(0) = 1$$

$$G(1) = 2$$

$$G(N + 2) = G(N)^2 \times G(N + 1), N \geq 0$$

Write out the first five values of G.

3.2 Division in the Integers

We shall now discuss some needed results on division and factoring in the nonnegative integers. If n and m are nonnegative integers, and n is not zero, we can plot the nonnegative integer multiples or n on a half line, and locate the point m as in Fig. 1. If m is a multiple of n, say $m = qn$, then we can write $m = qn + r$, where $r = 0$. On the other hand (as shown in Fig. 1), if m is not a multiple of n, we let qn be the nearest multiple of n to the left of m and let $r = m - qn$. Then r is the distance from qn to m, so clearly $0 < r < n$, and again we have $m = qn + r$. We state these observations as a theorem.

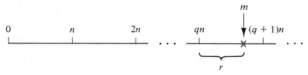

Figure 1

Theorem 1   If $n \neq 0$ and m are nonnegative integers, we can write $m = qn + r$ for some nonnegative integers q and r with $0 \leq r < n$. Moreover, there is just one way to do this.

Example 1   (a) If $n = 3$ and $m = 16$, then

$$16 = 5(3) + 1$$

so $q = 5$ and $r = 1$.

(b) If $n = 10$ and $m = 3$, then

$$3 = 0(10) + 3$$

so $q = 0$ and $r = 3$.

If r is zero in Theorem 1, so that m is a multiple of n, we write $n \mid m$, which is read "n divides m." If m is not a multiple of n, we write $n \nmid m$, which is read "n does not divide m." We now prove some simple properties of divisibility.

Theorem 2   Let a, b, and c be integers.
  (a) If $a \mid b$ and $a \mid c$, then $a \mid (b + c)$.
  (b) If $a \mid b$ and $a \mid c$, where $b > c$, then $a \mid (b - c)$.
  (c) If $a \mid b$ or $a \mid c$, then $a \mid bc$.
  (d) If $a \mid b$ and $b \mid c$, then $a \mid c$.

Proof. (a) If $a \mid b$ and $a \mid c$, then $b = k_1 a$ and $c = k_2 a$ for some nonnegative integers k_1 and k_2. Then

$$b + c = (k_1 + k_2)a$$

so $a \mid (b + c)$.
  (b) The proof is similar to the proof of part (a).
  (c) As in part (a), we have $b = k_1 a$ or $c = k_2 a$. Then either $bc = k_1 ac$ or $bc = k_2 ab$, so in either case $a \mid bc$.
  (d) If $a \mid b$ and $b \mid c$, we have $b = k_1 a$ and $c = k_2 b$, so

$$c = k_2 b = k_2(k_1 a) = (k_1 k_2)a$$

and hence $a \mid c$.

A number $p > 1$ in Z^+ is called **prime** if the only positive integers that divide p are p and 1.

Example 2   The numbers 2, 3, 5, 7, 11, and 13 are prime, while 4, 10, 16, and 21 are not prime. It is easy to write an algorithm to determine if a positive number $n > 1$ is a prime number. First we check whether $n = 2$. If $n > 2$ we could divide by every integer from 2 to $n - 1$, and if none of these is a divisor of n, then n is a prime. To make this process more efficient, we note that if $mk = n$, then either m or k is less than or equal to $\sqrt{n}$. This means that if n is not prime, it has a divisor k satisfying the inequality $1 < k \leq \sqrt{n}$, so we need only test for divisors in this range, Also, if n has any even number as a divisor, it must have 2 as a divisor. Thus, after checking for division by 2, we may skip all even integers.

Algorithm to test whether an integer $N > 1$ is prime.

Step 1. Check whether $N = 2$. If so, N is prime. If not, then go to
Step 2. Check whether $2 \mid N$. If so, N is not prime; otherwise, go to
Step 3. Compute the smallest integer K not exceeding $\sqrt{N}$. Then
Step 4. Check whether $D \mid N$, where D is any odd number less than or equal to K. If it does N is not prime; otherwise N is prime.

Theorem 3   Every positive integer $n > 1$ can be uniquely written as

$$n = p_1^{k_1} p_2^{k_2} \cdots p_s^{k_s}$$

where $p_1 < p_2 < \cdots < p_s$ are the distinct primes that divide n, and the k's are positive integers giving the number of times each prime occurs as a factor of n.

We omit the proof of Theorem 3, but we give several illustrations.

Example 3

$$9 = 3 \cdot 3 = 3^2$$
$$24 = 12 \cdot 2 = 2 \cdot 2 \cdot 2 \cdot 3$$
$$= 2^3 \cdot 3$$
$$30 = 2 \cdot 3 \cdot 5$$

Greatest Common Divisor

If a, b, and k are in Z^+, and $k \mid a$, $k \mid b$, we say that k is a common divisor of a and b. We say that the positive integer d is the **greatest common divisor,** or GCD, of a and b, and we write $d = \text{GCD }(a, b)$ if

1. d is a common divisor of a and b.
2. for any common divisor d' of a and b, $d' \mid d$.

Thus the greatest common divisor of a and b is a common divisor of a and b that is a multiple of every other common divisor of a and b.

The following theorem shows that GCD (a, b) exists and gives a method for computing it.

Theorem 4   Let $p_1, p_2, \ldots, p_k$ be all the prime numbers that are factors of either a or b. Then we can write

$$a = p_1^{a_1} p_2^{a_2} \cdots p_k^{a_k}$$

and

$$b = p_1^{b_1} p_2^{b_2} \cdots p_k^{b_k}$$

where some of the a_i and b_i may be zero.

It then follows that

$$\text{GCD}(a,\ b) = p_1^{\min(a_1,\,b_1)} p_2^{\min(a_2,\,b_2)} \cdots p_k^{\min(a_k,\,b_k)}$$

where $\min(a_i,\ b_i)$ is the smaller of the numbers a_i and b_i.

We omit the proof.

Example 4 Let

$$a = 540 \quad \text{and} \quad b = 504$$

Factoring a and b into primes, we obtain

$$a = 540 = 2^2 \cdot 3^3 \cdot 5$$
$$b = 540 = 2^3 \cdot 3^2 \cdot 7$$

Thus all the prime numbers that are factors of either a or b are

$$p_1 = 2, \qquad p_2 = 3, \qquad p_3 = 5, \qquad p_4 = 7$$

Then

$$a = 540 = 2^2 \cdot 3^3 \cdot 5^1 \cdot 7^0$$
$$b = 540 = 2^3 \cdot 3^2 \cdot 5^0 \cdot 7^1$$

We then have

$$\text{GCD}(540,\ 504) = 2^{\min(2,\,3)} \cdot 3^{\min(3,\,2)} \cdot 5^{\min(1,\,0)} \cdot 7^{\min(0,\,1)}$$
$$= 2^2 \cdot 3^2 \cdot 5^0 \cdot 7^0$$
$$= 2^2 \cdot 3^2$$
$$= 36$$

Theorem 4 requires us to know the prime factorization of a and b in order to compute $\text{GCD}(a,\ b)$. This is fine for small integers, but if a and b are not small, it may be inefficient or impossible for us to factor them into primes.

We now present an alternative method, called the **Euclidean algorithm,** for finding $\text{GCD}(a,\ b)$. This method works for all a and b, does not require finding prime factorizations, and uses only long division. Thus it is easily computerized, and in addition it will establish other interesting properties of the greatest common divisor.

Suppose that $a > b > 0$ (otherwise, interchange a and b). Then by Theorem 1 we may write

$$a = k_1 b + r_1 \tag{1}$$

where k_1 and r_1 are in Z^+ and $0 < r_1 < b$. Now Theorem 2 tells us that if n divides a and b, then it must divide r_1, since $r_1 = a - k_1 b$. Similarly, if n divides b and r_1, then it must divide a. We see that the common divisors of a and b are the same as the common divisors of b and r_1, so $GCD(a, b) = GCD(b, r_1)$.

We now continue using Theorem 1 as follows:

$$
\begin{array}{lll}
\text{divide } b \text{ by } r_1\text{:} & b = k_2 r_1 + r_2 & 0 \le r_2 < r_1 \\
\text{divide } r_1 \text{ by } r_2\text{:} & r_1 = k_3 r_2 + r_3 & 0 \le r_3 < r_2 \\
\text{divide } r_2 \text{ by } r_3\text{:} & r_2 = k_4 r_3 + r_4 & 0 \le r_4 < r_3 \\
& \quad\vdots & \quad\vdots \\
\text{divide } r_{n-3} \text{ by } r_{n-2}\text{:} & r_{n-3} = k_{n-1} r_{n-2} + r_{n-1} & 0 \le r_{n-1} < r_{n-2} \\
\text{divide } r_{n-2} \text{ by } r_{n-1}\text{:} & r_{n-2} = k_n r_{n-1} + r_n & 0 \le r_n < r_{n-1} \\
\text{divide } r_{n-1} \text{ by } r_n\text{:} & r_{n-1} = k_{n+1} r_n + r_{n+1} & 0 \le r_{n+1} < r_n
\end{array}
\tag{2}
$$

By the same reasoning as above, we have

$$GCD(a, b) = GCD(b, r_1) = GCD(r_1, r_2) = \cdots = GCD(r_{n-1}, r_n)$$

Since

$$a > b > r_1 > r_2 > r_3 > r_4, \ldots$$

the remainders will eventually become zero, so let us suppose that

$$r_{n+1} = 0$$

Then $r_{n-1} = k_{n+1} r_n$, so we see that $GCD(r_{n-1}, r_n) = r_n$. Hence $r_n = GCD(a, b)$.

Example 5   (a) Let $a = 190$ and $b = 34$. Then

$$
\begin{array}{ll}
\text{divide 190 by 34:} & 190 = 5 \cdot 34 + 20 \\
\text{divide 34 by 20:} & 34 = 1 \cdot 20 + 14 \\
\text{divide 20 by 14:} & 20 = 1 \cdot 14 + 6 \\
\text{divide 14 by 6:} & 14 = 2 \cdot 6 + 2 \\
\text{divide 6 by 2:} & 6 = 3 \cdot 2 + 0
\end{array}
$$

so

$$GCD(190, 34) = 2$$

(b) Let $a = 108$ and $b = 60$. Then

$$\begin{aligned}
\text{divide 108 by 60:} \quad & 108 = 1 \cdot 60 + 48 \\
\text{divide 60 by 48:} \quad & 60 = 1 \cdot 48 + 12 \\
\text{divide 48 by 12:} \quad & 48 = 4 \cdot 12
\end{aligned}$$

so

$$GCD(108, 60) = 12$$

Let us now manipulate the equations in (2) to obtain another result. Solve the next-to-last equation in (2) for r_n:

$$r_n = r_{n-2} - k_n r_{n-1} \tag{3}$$

Here we have expressed r_n as a multiple of r_{n-2} plus a multiple of r_{n-1}. Now solve the second-from-last equation in (2),

$$r_{n-3} = k_{n-1} r_{n-2} + r_{n-1}$$

for r_{n-1}:

$$r_{n-1} = r_{n-3} - k_{n-1} r_{n-2}$$

and substitute this expression for r_{n-1} in (3):

$$r_n = r_{n-2} - k_n [r_{n-3} - k_{n-1} r_{n-2}]$$

Now we have expressed r_n as a multiple of r_{n-3} plus a multiple of r_{n-2}. Continue up the equations in (2) and (1) replacing r_i by an expression involving r_{i-2} and r_{i-1}, finally arriving at an expression for r_n involving only a and b. Thus we have shown the following result.

Theorem 5   If $d = GCD(a, b)$, we can find integers s and t such that

$$d = sa + tb$$

Example 6   (a) Let $a = 190$ and $b = 34$ as in Example 5(a). Then

$$\begin{aligned}
\text{GCD}(190, 34) = 2 &= 14 - 2(6) \\
&= 14 - 2[20 - 1(14)] \qquad\qquad 6 = 20 - 1 \cdot 14 \\
&= 3(14) - 2 \cdot 20 \\
&= 3[34 - 1(20)] - 2(20) \qquad 14 = 34 - 1 \cdot 20 \\
&= 3(34) - 5(20) \\
&= 3(34) - 5(190 - 5 \cdot 34) \qquad 20 = 190 - 5 \cdot 34 \\
&= 28(34) - 5(190)
\end{aligned}$$

Hence $s = -5$ and $t = 28$.

(b) Let $a = 108$ and $b = 60$ as in Example 5(b). Then

$$\begin{aligned}
\text{GCD}(108, 60) = 12 &= 60 - 1(48) \\
&= 60 - 1[108 - 1(60)] \qquad 48 = 108 - 1 \cdot 60 \\
&= 2(60) - 108
\end{aligned}$$

Hence $s = -1$ and $t = 2$.

Least Common Multiple

If a, b, and k are in Z^+, and $a \mid k$, $b \mid k$, we say that k is a common multiple of a and b. An integer c is called the **least common multiple,** or LCM, of a and b, and we write $c = \text{LCM}(a, b)$ if

1. c is a common multiple of a and b.
2. For any common multiple c' of a and b, $c \mid c'$.

Thus the least common multiple of a and b is a common multiple of a and b that divides any other common multiple of a and b.

As in the case of GCD, we can see from the prime factorization of a and b that $\text{LCM}(a, b)$ exists, and we now give a formula for it.

Theorem 6　Let $p_1, p_2, \cdots p_k$ be all the prime numbers that are factors of *either* a or b. As before, we can write

$$a = p_1^{a_1} p_2^{a_2} \cdots p_k^{a_k}$$

and

$$b = p_1^{b_1} p_2^{b_2} \cdots p_k^{b_k}$$

where some of the a_i and b_i may be zero. Then

$$\mathrm{LCM}(a,\ b) = p_1^{\max(a_1,\,b_1)} p_2^{\max(a_2,\,b_2)} \cdots p_k^{\max(a_k,\,b_k)}$$

where $\max(a_i,\ b_i)$ is the larger of the numbers a_i and b_i
Again, we omit the proof.

Example 7   As in Example 4, let $a = 540$ and $b = 504$. Then

$$a = 540 = 2^2 \cdot 3^3 \cdot 5$$
$$b = 504 = 2^3 \cdot 3^2 \cdot 7$$

Then

$$\mathrm{LCM}(540,\ 504) = 2^{\max(2,\,3)} \cdot 3^{\max(3,\,2)} \cdot 5^{\max(1,\,0)} \cdot 7^{\max(0,\,1)}$$
$$= 2^3 \cdot 3^3 \cdot 5^1 \cdot 7^1$$
$$= 7560$$

The following result shows that we can obtain the least common multiple from the greatest common divisor, so we do not need a separate procedure for finding the least common multiple.

Theorem 7   If a and b are two positive integers, then

$$\mathrm{GCD}(a,\ b) \cdot \mathrm{LCM}(a, b) = ab$$

Proof. Using the notation and results from Theorems 4 and 6, we have

$$\mathrm{GCD}(a,\ b) = p_1^{\min(a_1,\,b_1)} p_2^{\min(a_2,\,b_2)} \cdots p_k^{\min(a_k,\,b_k)}$$
$$\mathrm{LCM}(a,\ b) = p_1^{\max(a_1,\,b_1)} p_2^{\max(a_2,\,b_2)} \cdots p_k^{\max(a_k,\,b_k)}$$

Hence

$$\mathrm{GCD}(a,\ b) \cdot \mathrm{LCM}(a,\ b) = p_1^{a_1+b_1} p_2^{a_1+b_2} \cdots p_k^{a_k+b_k}$$
$$= (p_1^{a_1} p_2^{a_2} \cdots p_k^{a_k}) \cdot (p_1^{b_1} p_2^{b_2} \cdots p_k^{b_k})$$
$$= ab$$

Example 8   Let $a = 540$ and $b = 504$. Then

$$\mathrm{GCD}(540,\ 504) \cdot \mathrm{LCM}(540,\ 504) = 36 \cdot 7560$$
$$= 272{,}160 = 540 \cdot 504$$

EXERCISE SET 3.2

In Exercises 1-6, for the given integers n and m, write m as $qn + r$, with $0 \le r < n$.

1. $m = 20$, $n = 3$ 2. $m = 64$, $n = 37$
3. $m = 3$, $n = 22$ 4. $m = 48$, $n = 12$
5. $m = 8$, $n = 3$ 6. $m = 12$, $n = 140$
7. Which of the following numbers are prime?
   (a) 22   (b) 29
   (c) 47   (d) 81
   (e) 527   (f) 247
   (g) 1180   (h) 1201
8. Write each integer as a product of powers of primes (as indicated by Theorem 3).
   (a) 60   (b) 350

   (c) 858   (d) 1666
   (e) 1125   (f) 210

In Exercises 9-14, find the greatest common divisor d of the integers a and b, and write d as $sa + tb$.

9. 32, 27   10. 45, 33
11. 40, 88   12. 60, 100
13. 34, 58   14. 77, 128

In Exercises 15-18, find the least common multiple of the integers.

15. 72, 108   16. 150, 70
17. 175, 245   18. 32, 27

THEORETICAL EXERCISES

T1. Let a and b be integers. Prove that if p is a prime and $p \mid ab$ then $p \mid a$ or $p \mid b$. (*Hint:* If $p \nmid a$, then $1 = GCD(a, p)$; use Theorem 4 to write $1 = sa + tp$.)

T2. Show that if $GCD(a, c) = 1$ and $c \mid ab$, then $c \mid b$.

T3. Show that if $GCD(a, c) = 1$, $a \mid m$, and $c \mid m$, then $ac \mid m$. (*Hint:* Use Exercise T2.)

T4. Show that if $d = GCD(a, c)$, $a \mid b$, and $c \mid b$, then $ac \mid bd$.

T5. Show that $GCD(ca, cb) = c \, GCD(a, b)$.

T6. Use induction to show that if p is a prime and $p \mid a^n \, (n \ge 1)$, then $p \mid a$.

T7. Prove that if $GCD(a, b) = 1$, then $GCD(a^n, b^n) = 1$ for all $n \ge 1$. (*Hint:* use Exercise T6.)

T8. Show that $LCM(a, ab) = ab$.

T9. Show that if $GCD(a, b) = 1$, then $LCM(a, b) = ab$.

3.3 Matrix Algebra

A **matrix** is a rectangular arrangement of numbers consisting of m horizontal **rows** and n vertical **columns:**

$$\mathbf{A} = \begin{bmatrix} a_{11} & a_{12} & \cdots & a_{1n} \\ a_{21} & a_{22} & \cdots & a_{2n} \\ \vdots & \vdots & & \vdots \\ a_{m1} & a_{m2} & \cdots & a_{mn} \end{bmatrix} \tag{1}$$

The *i*th **row** of $\mathbf{A}$ is

$$[a_{i1} \, a_{i2} \, \cdots \, a_{in}] \qquad (1 \le i \le m)$$

the *j*th column of **A** is

$$\begin{bmatrix} a_{1j} \\ a_{2j} \\ \vdots \\ a_{mj} \end{bmatrix} \quad (1 \le j \le n)$$

We shall say that **A** is *m* by *n*, written as $m \times n$. If $m = n$, we say that **A** is a **square matrix** of order n and that the numbers $a_{11}, a_{22}, \ldots, a_{nn}$ form the **main diagonal of A**. We refer to the number a_{ij}, which is in the *i*th row and *j*th column of **A**, as the *i, j*th **element** of **A** or as the (i, j) **entry** of **A**, and we often write (1) as

$$\mathbf{A} = [a_{ij}]$$

Example 1   Let

$$\mathbf{A} = \begin{bmatrix} 2 & 3 & 5 \\ 0 & -1 & 2 \end{bmatrix}, \quad \mathbf{B} = \begin{bmatrix} 2 & 3 \\ 4 & 6 \end{bmatrix}, \quad \mathbf{C} = \begin{bmatrix} -1 \\ 2 \\ 0 \end{bmatrix},$$

$$\mathbf{D} = \begin{bmatrix} 1 & 0 & -1 \\ -1 & 2 & 3 \\ 2 & 4 & 5 \end{bmatrix}, \quad \mathbf{E} = \begin{bmatrix} 1 & -1 & 3 & 4 \end{bmatrix}$$

Then **A** is 2×3 with $a_{12} = 3$ and $a_{23} = 2$, **B** is 2×2 with $b_{21} = 4$, **C** is 3×1, **D** is 3×3, and **E** is 1×4.

A square matrix $\mathbf{A} = [a_{ij}]$ for which every entry off the main diagonal is zero, that is, $a_{ij} = 0$ for $i \ne j$, is called a **diagonal matrix**.

Example 2   The following are diagonal matrices:

$$\mathbf{F} = \begin{bmatrix} 4 & 0 \\ 0 & 3 \end{bmatrix} \quad \text{and} \quad \mathbf{G} = \begin{bmatrix} 2 & 0 & 0 \\ 0 & -3 & 0 \\ 0 & 0 & 5 \end{bmatrix}$$

Matrices are used in many applications in computer science and we shall see them in our study of relations and graphs. At this point we present the following simple application showing that matrices can be used to display data in a tabular form.

Example 3 The following matrix gives the airline distances between the cities indicated (in statute miles).

$$
\begin{array}{c}
 \\
\text{London} \\
\text{Madrid} \\
\text{New York} \\
\text{Tokyo}
\end{array}
\begin{array}{cccc}
\text{London} & \text{Madrid} & \text{New York} & \text{Tokyo} \\
\left[\begin{array}{cccc}
0 & 785 & 3469 & 5959 \\
785 & 0 & 3593 & 6707 \\
3469 & 3593 & 0 & 6757 \\
5959 & 6707 & 6757 & 0
\end{array}\right]
\end{array}
$$

Two $m \times n$ matrices $\mathbf{A} = [a_{ij}]$ and $\mathbf{B} = [b_{ij}]$ are said to be **equal** if $a_{ij} = b_{ij}$, $1 \le i \le m$, $1 \le j \le n$, that is, if corresponding elements agree.

Example 4 If

$$
\mathbf{A} = \begin{bmatrix} 2 & -3 & -1 \\ 0 & 5 & 2 \\ 4 & -4 & 6 \end{bmatrix} \quad \text{and} \quad \mathbf{B} = \begin{bmatrix} 2 & x & -1 \\ y & 5 & 2 \\ 4 & -4 & z \end{bmatrix}
$$

Then $\mathbf{A} = \mathbf{B}$ if and only if $x = -3$, $y = 0$, and $z = 6$.

If $\mathbf{A} = [a_{ij}]$ and $\mathbf{B} = [b_{ij}]$ are $m \times n$ matrices, then the **sum** of $\mathbf{A}$ and $\mathbf{B}$ is the $m \times n$ matrix $\mathbf{C} = [c_{ij}]$ defined by

$$
c_{ij} = a_{ij} + b_{ij} \qquad (1 \le i \le m, 1 \le j \le n)
$$

That is, $\mathbf{C}$ is obtained by adding corresponding elements of $\mathbf{A}$ and $\mathbf{B}$.

Example 5 Let

$$
\mathbf{A} = \begin{bmatrix} 3 & 4 & -1 \\ 5 & 0 & -2 \end{bmatrix} \quad \text{and} \quad \mathbf{B} = \begin{bmatrix} 4 & 5 & 3 \\ 0 & -3 & 2 \end{bmatrix}
$$

Then

$$
\mathbf{A} + \mathbf{B} = \begin{bmatrix} 3+4 & 4+5 & -1+3 \\ 5+0 & 0+(-3) & -2+2 \end{bmatrix} = \begin{bmatrix} 7 & 9 & 2 \\ 5 & -3 & 0 \end{bmatrix}
$$

Observe that the sum of the matrices $\mathbf{A}$ and $\mathbf{B}$ is defined only when $\mathbf{A}$ and $\mathbf{B}$ have the same number of rows and the same number of columns. We agree to write $\mathbf{A} + \mathbf{B}$ only when the sum is defined.

A matrix all of whose entries are zero is called a **zero matrix** and is denoted by **0**.

Example 6 The following are all zero matrices:

$$\begin{bmatrix} 0 & 0 \\ 0 & 0 \end{bmatrix} \qquad \begin{bmatrix} 0 & 0 & 0 \\ 0 & 0 & 0 \end{bmatrix} \qquad \begin{bmatrix} 0 & 0 & 0 \\ 0 & 0 & 0 \\ 0 & 0 & 0 \end{bmatrix}$$

The following theorem gives some basic properties of matrix addition; the proofs are omitted.

Theorem 1
(a) $\mathbf{A} + \mathbf{B} = \mathbf{B} + \mathbf{A}$ **(Commutative Property for Addition)**
(b) $(\mathbf{A} + \mathbf{B}) + \mathbf{C} = \mathbf{A} + (\mathbf{B} + \mathbf{C})$ **(Associative Property for Addition)**
(c) $\mathbf{A} + \mathbf{0} = \mathbf{0} + \mathbf{A} = \mathbf{A}$ **(Zero Property for Addition)**

Matrix Multiplication

If $\mathbf{A} = [a_{ij}]$ is an $m \times p$ matrix and $\mathbf{B} = [b_{ij}]$ is a $p \times n$ matrix, then the **product** of $\mathbf{A}$ and $\mathbf{B}$, denoted $\mathbf{AB}$, is the $m \times n$ matrix $\mathbf{C} = [c_{ij}]$ defined by

$$c_{ij} = a_{i1}b_{1j} + a_{i2}b_{2j} + \cdots + a_{ip}b_{pj} \qquad (1 \le i \le m, \ 1 \le j \le n) \qquad (2)$$

Let us explain equation (2) in more detail. The elements $a_{i1}, a_{i2}, \ldots, a_{ip}$ form the ith row of $\mathbf{A}$, and the elements $b_{1j}, b_{2j}, \ldots, b_{pj}$ form the jth column of $\mathbf{B}$. Then equation (2) states that for any i and j, the element c_{ij} of $\mathbf{C} = \mathbf{AB}$ can be computed in the following way, as illustrated by Fig. 1.

1. Select row i of $\mathbf{A}$ and column j of $\mathbf{B}$, and arrange them side by side.
2. Multiply corresponding entries together and add the results.

Example 7 Let

$$\mathbf{A} = \begin{bmatrix} 2 & 3 & -4 \\ 1 & 2 & 3 \end{bmatrix} \qquad \text{and} \qquad \mathbf{B} = \begin{bmatrix} 3 & 1 \\ -2 & 2 \\ 5 & -3 \end{bmatrix}$$

Then

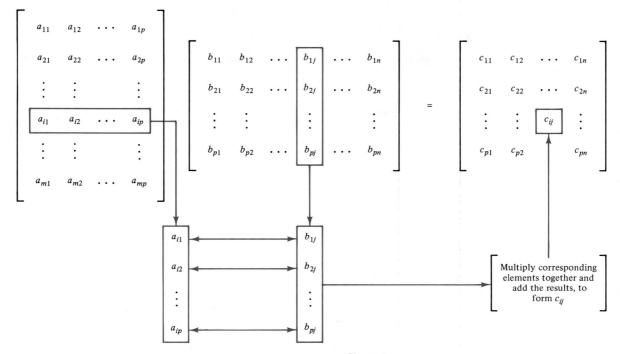

Figure 1

$$\mathbf{AB} = \begin{bmatrix} (2)(3) + (3)(-2) + (-4)(5) & (2)(1) + (3)(2) + (-4)(-3) \\ (1)(3) + (2)(-2) + (3)(5) & (1)(1) + (2)(2) + (3)(-3) \end{bmatrix}$$

$$= \begin{bmatrix} -20 & 20 \\ 14 & -4 \end{bmatrix}$$

An **array of dimension two** is a modification of the idea of a matrix, in the same way that a linear array is a modification of the idea of a sequence (see Section 1.3). By an $m \times n$ **array A** we will mean an $m \times n$ matrix **A** of positions. We may assign numbers to these positions later, and make future changes in these assignments, and we will still refer to the array as **A**. This is a model for two-dimensional storage of information in a computer. The number assigned to row i and column j of an array **A** will be denoted $\mathbf{A}[i, j]$.

As we have seen in Theorem 1, the properties of matrix addition resemble familiar properties for the addition of real numbers. However, most properties of matrix multiplication do not resemble those of real number multiplication. First, observe that if **A** is an $m \times p$ matrix and **B** is a $p \times n$ matrix, then **AB** can be computed and is an $m \times n$ matrix. As for **BA**, we have any of the following four possibilities:

1. **BA** may not be defined (we may have $n \neq m$).
2. **BA** may be defined ($m = n$) and then **BA** is $p \times p$, while **AB** is $m \times m$, $m \neq p$. Thus **AB** and **BA** are not equal.
3. **AB** and **BA** may both be of the same sizes, but unequal as matrices.
4. **AB** = **BA**

We now agree to write **AB** only when the product is defined.

Example 8   Let

$$A = \begin{bmatrix} 2 & 1 \\ 3 & -2 \end{bmatrix} \quad \text{and} \quad B = \begin{bmatrix} 1 & -1 \\ 2 & -3 \end{bmatrix}$$

Then

$$AB = \begin{bmatrix} 4 & -5 \\ -1 & 3 \end{bmatrix} \quad \text{and} \quad BA = \begin{bmatrix} -1 & 3 \\ -5 & 8 \end{bmatrix}$$

The basic properties of matrix multiplication are given by the following theorem.

Theorem 2   (a) **A(BC) = (AB)C**   (**Associative Property for Multiplication**)
(b) **A(B + C) = AB + AC**   (**Distributive Property**)
(c) **(A + B)C = AC + BC**   (**Distributive Property**)

The $n \times n$ diagonal matrix

$$I_n = \begin{bmatrix} 1 & 0 & \cdot & \cdot & \cdot & 0 \\ 0 & 1 & \cdot & \cdot & \cdot & 0 \\ \cdot & \cdot & & & & \cdot \\ \cdot & \cdot & & & & \cdot \\ \cdot & \cdot & & & & \cdot \\ 0 & 0 & \cdot & \cdot & \cdot & 1 \end{bmatrix}$$

all of whose diagonal elements are 1, is called the **identity matrix** of order n. If **A** is an $m \times n$ matrix, it is easy to verify (Exercise T1) that

$$I_m A = A I_n = A$$

If **A** is an $n \times n$ matrix and p is a positive integer, we define

$$A^p = \underbrace{A \cdot A \cdots A}_{p \text{ factors}}$$

and

$$\mathbf{A}^0 = \mathbf{I}_n$$

If p and q are nonnegative integers, we can prove the following laws of exponents:

$$\mathbf{A}^p \mathbf{A}^q = \mathbf{A}^{p+q}$$

and

$$(\mathbf{A}^p)^q = \mathbf{A}^{pq}$$

Observe that the rule

$$(\mathbf{AB})^p = \mathbf{A}^p \mathbf{B}^p$$

does not hold for square matrices. However, if $\mathbf{AB} = \mathbf{BA}$, then $(\mathbf{AB})^p = \mathbf{A}^p \mathbf{B}^p$ (Exercise T 7).

If $\mathbf{A} = [a_{ij}]$ is an $m \times n$ matrix, then the $n \times m$ matrix $\mathbf{A}^T = [a_{ij}^T]$, where

$$a_{ij}^T = a_{ji} \qquad (1 \leq i \leq m, \, 1 \leq j \leq n)$$

is called the **transpose** of $\mathbf{A}$. Thus the transpose of $\mathbf{A}$ is obtained by interchanging the rows and columns of $\mathbf{A}$.

Example 9   Let

$$\mathbf{A} = \begin{bmatrix} 2 & -3 & 5 \\ 6 & 1 & 3 \end{bmatrix} \quad \text{and} \quad \mathbf{B} = \begin{bmatrix} 3 & 4 & 5 \\ 2 & -1 & 0 \\ 1 & 6 & -2 \end{bmatrix}$$

Then

$$\mathbf{A}^T = \begin{bmatrix} 2 & 6 \\ -3 & 1 \\ 5 & 3 \end{bmatrix} \quad \text{and} \quad \mathbf{B}^T = \begin{bmatrix} 3 & 2 & 1 \\ 4 & -1 & 6 \\ 5 & 0 & -2 \end{bmatrix}$$

The following theorem gives the basic properties of transpose.

Theorem 3   If $\mathbf{A}$ and $\mathbf{B}$ are matrices, then
(a) $(\mathbf{A}^T)^T = \mathbf{A}$
(b) $(\mathbf{A} + \mathbf{B})^T = \mathbf{A}^T + \mathbf{B}^T$
(c) $(\mathbf{AB})^T = \mathbf{B}^T \mathbf{A}^T$

A matrix $\mathbf{A} = [a_{ij}]$ is called **symmetric** if

$$\mathbf{A}^T = \mathbf{A}$$

Thus, if $\mathbf{A}$ is symmetric, it must be a square matrix. It is easy to show (Exercise T8) that $\mathbf{A}$ is symmetric if and only if

$$a_{ij} = a_{ji}$$

That is, $\mathbf{A}$ is symmetric if and only if the entries of $\mathbf{A}$ are symmetric with respect to the main diagonal of $\mathbf{A}$.

Example 10 If

$$\mathbf{A} = \begin{bmatrix} 1 & 2 & -3 \\ 2 & 4 & 5 \\ -3 & 5 & 6 \end{bmatrix} \quad \text{and} \quad \mathbf{B} = \begin{bmatrix} 1 & 2 & -3 \\ 2 & 4 & 0 \\ 3 & 2 & 1 \end{bmatrix}$$

Then $\mathbf{A}$ is symmetric and $\mathbf{B}$ is not symmetric.

EXERCISE SET 3.3

1. Let

$$\mathbf{A} = \begin{bmatrix} 3 & -2 & 5 \\ 4 & 1 & 2 \end{bmatrix}, \quad \mathbf{B} = \begin{bmatrix} 3 \\ -2 \\ 4 \end{bmatrix},$$

$$\mathbf{C} = \begin{bmatrix} 2 & 3 & 4 \\ 5 & 6 & -1 \\ 2 & 0 & 8 \end{bmatrix}$$

(a) What is a_{12}, a_{22}, a_{23}?
(b) What is b_{11}, b_{31}?
(c) What is c_{13}, c_{21}, c_{33}?
(d) List the elements on the main diagonal of $\mathbf{C}$.

2. Which of the following are diagonal matrices?

$$\mathbf{A} = \begin{bmatrix} 2 & 3 \\ 0 & 0 \end{bmatrix}, \quad \mathbf{B} = \begin{bmatrix} 3 & 0 & 0 \\ 0 & -2 & 0 \\ 0 & 0 & 5 \end{bmatrix},$$

$$\mathbf{C} = \begin{bmatrix} 0 & 0 & 0 \\ 0 & 0 & 0 \\ 0 & 0 & 0 \end{bmatrix}, \quad \mathbf{D} = \begin{bmatrix} 2 & 6 & -2 \\ 0 & -1 & 0 \\ 0 & 0 & 3 \end{bmatrix},$$

$$\mathbf{E} = \begin{bmatrix} 4 & 0 & 0 \\ 0 & 4 & 0 \\ 0 & 0 & 4 \end{bmatrix}$$

3. If $\begin{bmatrix} a+b & c+d \\ c-d & a-b \end{bmatrix} = \begin{bmatrix} 4 & 6 \\ 10 & 2 \end{bmatrix}$, find a, b, c, and d.

4. If $\begin{bmatrix} a+2b & 2a-b \\ 2c+d & c-2d \end{bmatrix} = \begin{bmatrix} 4 & -2 \\ 4 & -3 \end{bmatrix}$, find a, b, c, and d.

In Exercises 5–11, let

$$\mathbf{A} = \begin{bmatrix} 2 & 1 & 3 \\ 4 & 1 & -2 \end{bmatrix}, \quad \mathbf{B} = \begin{bmatrix} 0 & 1 \\ 1 & 2 \\ 2 & 3 \end{bmatrix},$$

$$C = \begin{bmatrix} 1 & -2 & 3 \\ 4 & 2 & 5 \\ 3 & 1 & 2 \end{bmatrix}, \quad D = \begin{bmatrix} -3 & 2 \\ 4 & 1 \end{bmatrix},$$

$$E = \begin{bmatrix} 3 & 2 & -1 \\ 5 & 4 & -3 \\ 0 & 1 & 2 \end{bmatrix}, \quad F = \begin{bmatrix} -2 & 3 \\ 4 & 5 \end{bmatrix}$$

5. If possible, compute.
   (a) $C + E$
   (b) AB and BA
   (c) $CB + F$
   (d) $AB + DF$
6. If possible, compute.
   (a) $A(BD)$ and $(AB)D$
   (b) $A(C + E)$ and $AC + AE$
   (c) $FD + AB$
7. If possible, compute.
   (a) $EB + FA$
   (b) $A(B + D)$ and $AB + AD$
   (c) $(F + D)A$
   (d) $AC + DE$
8. If possible, compute.
   (a) A^T and $(A^T)^T$
   (b) $(C + E)^T$ and $C^T + E^T$
   (c) $(AB)^T$ and $B^T A^T$
   (d) $(B^T C) + A$
9. If possible, compute.
   (a) $A^T(D + F)$

(b) $(BC)^T$ and $C^T B^T$
(c) $(B^T + A)C$
(d) $(D^T + E)F$
10. Compute D^3.
11. Compute BD^2.
12. Let
$$A = \begin{bmatrix} 2 & 1 \\ 3 & -2 \end{bmatrix} \quad \text{and} \quad B = \begin{bmatrix} -1 & 2 \\ 3 & 4 \end{bmatrix}$$
Show that $AB \neq BA$.
13. Let
$$A = \begin{bmatrix} 3 & 0 & 0 \\ 0 & -2 & 0 \\ 0 & 0 & 4 \end{bmatrix}$$
(a) Compute A^3.
(b) What is A^k?
14. If $A = \begin{bmatrix} 0 & 1 \\ 1 & 0 \end{bmatrix}$, show that $A^2 = I_2$.
15. If 0 is the 2×2 zero matrix, find two 2×2 matrices A and B, $A \neq 0$, $B \neq 0$, such that $AB = 0$.
16. Determine all 2×2 matrices $A = \begin{bmatrix} 0 & a \\ b & c \end{bmatrix}$ such that $A^2 = I_2$.

THEORETICAL EXERCISES

T1. Let A be an $m \times n$ matrix. Show that $I_m A = A$ and $AI_n = A$.
T2. Show that $A0 = 0$ for any matrix A.
T3. Show that $I_n^T = I_n$.
T4. (a) Prove that if A has a row of zeros, then AB has a corresponding row of zeros.
    (b) Prove that if B has a column of zeros, then AB has a corresponding column of zeros.
T5. Prove that the sum and product of diagonal matrices is diagonal.
T6. Show that the jth column of the matrix product

AB is equal to the matrix product AB_j, where B_j is the jth column of B.
T7. Let A and B be square matrices. Use mathematical induction to show that if $AB = BA$, then $(AB)^n = A^n B^n$, for all integers $n \geq 1$.
T8. Show that $A = [a_{ij}]$ is symmetric if and only if $a_{ij} = a_{ji}$.
T9. Let A and B be symmetric matrices.
    (a) Show that $A + B$ is symmetric.
    (b) Show that AB is symmetric if and only if $AB = BA$.

T10. Let A be an $n \times n$ matrix.
   (a) Show that AA^T and A^TA are symmetric.
   (b) Show that $A + A^T$ is symmetric.

T11. Prove Theorem 3. [*Hint for part (c):* Show that the ijth element of $(AB)^T$ equals the ijth element of B^TA^T.]

   In Exercises T12–T15 prove the indicated result by mathematical induction.

T12. $(A_1 + A_2 + \cdots + A_n)^T = A_1^T + A_2^T + \cdots + A_n^T$

T13. $(A_1 A_2 \cdots A_n)^T = A_n^T A_{n-1}^T \cdots A_2^T A_1^T$

T14. $A^2 A^n = A^{2+n}$

T15. $(A^2)^n = A^{2n}$

T16. Let A and B be square matrices. Show that $(A + B)^2 = A^2 + 2AB + B^2$ if and only if $AB = BA$.

3.4 Boolean Matrix Algebra

A **Boolean matrix** is a matrix whose entries are either all zero or all 1. Although these matrices may be subjected to the usual algebraic operations, we can also define new operations on them. These new operations always result in Boolean matrices. We define three such operations in this section. These operations will be used later to study relations.

Join and Meet of Boolean Matrices

Let $A = [a_{ij}]$ and $B = [b_{ij}]$ be $m \times n$ Boolean matrices. We define $A \vee B = C = [c_{ij}]$, the **join of A and B,** by

$$c_{ij} = \begin{cases} 1 & \text{if } a_{ij} = 1 \text{ or } b_{ij} = 1 \\ 0 & \text{if } a_{ij} \text{ and } b_{ij} \text{ are both } 0 \end{cases}$$

and $A \wedge B = D = [d_{ij}]$, the **meet of A and B,** by

$$d_{ij} = \begin{cases} 1 & \text{if } a_{ij} \text{ and } b_{ij} \text{ are both } 1 \\ 0 & \text{if } a_{ij} = 0 \text{ or } b_{ij} = 0 \end{cases}$$

Note that the operations above are possible only when A and B have the same size, just as in the case of matrix addition. Instead of adding corresponding elements in A and B to compute the elements of the answer, we simply examine corresponding elements for particular patterns.

Example 1   Let

$$A = \begin{bmatrix} 1 & 0 & 1 \\ 0 & 1 & 1 \\ 1 & 1 & 0 \\ 0 & 0 & 0 \end{bmatrix}, \qquad B = \begin{bmatrix} 1 & 1 & 0 \\ 1 & 0 & 1 \\ 0 & 0 & 1 \\ 1 & 1 & 0 \end{bmatrix}$$

Compute (a) $A \vee B$; (b) $A \wedge B$.

Solution. (a) Let $\mathbf{A} \vee \mathbf{B} = \mathbf{C} = [c_{ij}]$. Then since a_{43} and b_{43} are both 0, we see that $c_{43} = 0$. In all other cases, either a_{ij} or b_{ij} is 1, so c_{ij} is also 1. Thus

$$\mathbf{A} \vee \mathbf{B} = \begin{bmatrix} 1 & 1 & 1 \\ 1 & 1 & 1 \\ 1 & 1 & 1 \\ 1 & 1 & 0 \end{bmatrix}$$

(b) Let $\mathbf{A} \wedge \mathbf{B} = \mathbf{D} = [d_{ij}]$. Then since a_{11} and b_{11} are both 1, $d_{11} = 1$, and since a_{23} and b_{23} are both 1, $d_{23} = 1$. In all other cases, either a_{ij} or b_{ij} is 0, so $d_{ij} = 0$. Thus

$$\mathbf{A} \wedge \mathbf{B} = \begin{bmatrix} 1 & 0 & 0 \\ 0 & 0 & 1 \\ 0 & 0 & 0 \\ 0 & 0 & 0 \end{bmatrix}$$

Boolean Product

Finally, suppose that $\mathbf{A} = [a_{ij}]$ is an $m \times p$ Boolean matrix and $\mathbf{B} = [b_{ij}]$ is a $p \times n$ Boolean matrix. Notice that the conditions on the sizes of $\mathbf{A}$ and $\mathbf{B}$ are exactly the conditions that we need in order to form the matrix product $\mathbf{AB}$. We now define another kind of product.

The **Boolean product** of $\mathbf{A}$ and $\mathbf{B}$, denoted $\mathbf{A} \odot \mathbf{B}$, is the $m \times n$ Boolean matrix $\mathbf{E} = [e_{ij}]$ defined by

$$e_{ij} = \begin{cases} 1 & \text{if } a_{ik} = 1 \text{ and } b_{kj} = 1 \text{ for some } k, \ 1 \le k \le p \\ 0 & \text{otherwise} \end{cases}$$

This multiplication is quite similar to ordinary multiplication. The formula above states that for any i and j, the element e_{ij} of $\mathbf{E} = \mathbf{A} \odot \mathbf{B}$ can be computed in the following way, as illustrated in Fig. 1. (Compare this figure with Fig. 1 of Section 3.3.)

1. Select row i of $\mathbf{A}$ and column j of $\mathbf{B}$, and arrange them side by side.
2. Compare corresponding entries. If even a single pair of corresponding entries consists of two 1's, then $e_{ij} = 1$. If this is not the case, then $e_{ij} = 0$.

One can easily perform the indicated comparisons and checks for each position of the Boolean product. Thus, at least for human beings, the computation of elements in $\mathbf{A} \odot \mathbf{B}$ is considerably easier than the computation of elements in $\mathbf{AB}$.

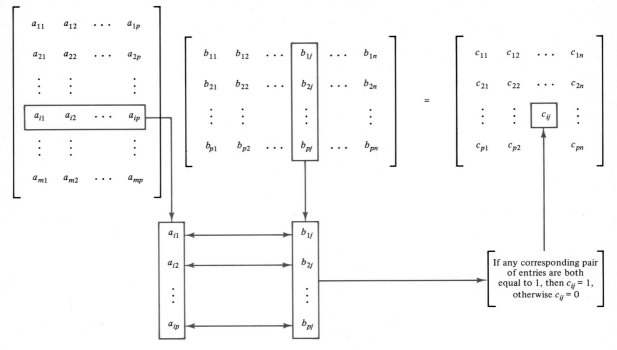

Figure 1

Example 2 Let

$$\mathbf{A} = \begin{bmatrix} 1 & 1 & 0 \\ 0 & 1 & 0 \\ 1 & 1 & 0 \\ 0 & 0 & 1 \end{bmatrix}, \qquad \mathbf{B} = \begin{bmatrix} 1 & 0 & 0 & 0 \\ 0 & 1 & 1 & 0 \\ 1 & 0 & 1 & 1 \end{bmatrix}.$$

Compute $\mathbf{A} \odot \mathbf{B}$.

Solution. let $\mathbf{A} \odot \mathbf{B} = [e_{ij}]$. Then $e_{11} = 1$, since row 1 of $\mathbf{A}$ and column 1 of $\mathbf{B}$ each have a 1 as the first entry. Similarly, $e_{12} = 1$ since $a_{12} = 1$ and $b_{22} = 1$; that is, the first row of $\mathbf{A}$ and the second column of $\mathbf{B}$ each have a 1 in the second position. In a similar way we see that $e_{13} = 1$. On the other hand, $e_{14} = 0$, since row 1 of $\mathbf{A}$ and column 4 of $\mathbf{B}$ do not have common 1's in any position. Proceeding in this way, we obtain (verify)

$$\mathbf{A} \odot \mathbf{B} = \begin{bmatrix} 1 & 1 & 1 & 0 \\ 0 & 1 & 1 & 0 \\ 1 & 1 & 1 & 0 \\ 1 & 0 & 1 & 1 \end{bmatrix}.$$

The following theorem, whose proof is left as an exercise, gives the basic properties of the Boolean matrix operations defined above.

Theorem 1   If **A**, **B**, and **C** are Boolean matrices, then

1. (a) $\mathbf{A} \vee \mathbf{B} = \mathbf{B} \vee \mathbf{A}$;
   (b) $\mathbf{A} \wedge \mathbf{B} = \mathbf{B} \wedge \mathbf{A}$ **(Commutative Properties)**

2. (a) $(\mathbf{A} \vee \mathbf{B}) \vee \mathbf{C} = \mathbf{A} \vee (\mathbf{B} \vee \mathbf{C})$;
   (b) $(\mathbf{A} \wedge \mathbf{B}) \wedge \mathbf{C} = \mathbf{A} \wedge (\mathbf{B} \wedge \mathbf{C})$ **(Associative Properties)**

3. (a) $\mathbf{A} \wedge (\mathbf{B} \vee \mathbf{C}) = (\mathbf{A} \wedge \mathbf{B}) \vee (\mathbf{A} \wedge \mathbf{C})$
   (b) $\mathbf{A} \vee (\mathbf{B} \wedge \mathbf{C}) = (\mathbf{A} \vee \mathbf{B}) \wedge (\mathbf{A} \vee \mathbf{C})$ **(Distributive Properties)**

4. $\mathbf{A} \odot (\mathbf{B} \odot \mathbf{C}) = (\mathbf{A} \odot \mathbf{B}) \odot \mathbf{C}$ **(Associative Property for $\odot$)**

EXERCISE SET 3.4

In Exercises 1–3 compute $\mathbf{A} \vee \mathbf{B}$, $\mathbf{A} \wedge \mathbf{B}$, and $\mathbf{A} \odot \mathbf{B}$ for the given matrices **A** and **B**.

1. (a) $\mathbf{A} = \begin{bmatrix} 1 & 0 \\ 0 & 1 \end{bmatrix}$, $\mathbf{B} = \begin{bmatrix} 1 & 1 \\ 0 & 1 \end{bmatrix}$

   (b) $\mathbf{A} = \begin{bmatrix} 1 & 0 & 0 \\ 0 & 1 & 1 \\ 1 & 0 & 0 \end{bmatrix}$, $\mathbf{B} = \begin{bmatrix} 1 & 1 & 1 \\ 0 & 0 & 1 \\ 1 & 0 & 1 \end{bmatrix}$

2. (a) $\mathbf{A} = \begin{bmatrix} 1 & 1 \\ 0 & 1 \end{bmatrix}$, $\mathbf{B} = \begin{bmatrix} 0 & 0 \\ 1 & 1 \end{bmatrix}$

   (b) $\mathbf{A} = \begin{bmatrix} 0 & 0 & 1 \\ 1 & 1 & 0 \\ 1 & 0 & 0 \end{bmatrix}$, $\mathbf{B} = \begin{bmatrix} 0 & 1 & 1 \\ 1 & 1 & 0 \\ 1 & 0 & 1 \end{bmatrix}$

3. (a) $\mathbf{A} = \begin{bmatrix} 1 & 1 \\ 1 & 1 \end{bmatrix}$, $\mathbf{B} = \begin{bmatrix} 0 & 0 \\ 1 & 0 \end{bmatrix}$

   (b) $\mathbf{A} = \begin{bmatrix} 1 & 0 & 0 \\ 0 & 0 & 1 \\ 1 & 0 & 1 \end{bmatrix}$, $\mathbf{B} = \begin{bmatrix} 1 & 1 & 1 \\ 1 & 1 & 1 \\ 1 & 0 & 0 \end{bmatrix}$

4. Compute $\mathbf{A} \odot \mathbf{B}$ for the given matrices **A** and **B**.

   (a) $\mathbf{A} = \begin{bmatrix} 1 & 0 & 1 \\ 0 & 0 & 1 \end{bmatrix}$, $\mathbf{B} = \begin{bmatrix} 1 & 1 \\ 1 & 0 \\ 0 & 1 \end{bmatrix}$

   (b) $\mathbf{A} = \begin{bmatrix} 1 & 1 & 0 & 1 \\ 0 & 1 & 1 & 1 \\ 1 & 0 & 0 & 1 \\ 0 & 1 & 1 & 1 \end{bmatrix}$, $\mathbf{B} = \begin{bmatrix} 1 & 1 \\ 0 & 1 \\ 0 & 1 \\ 1 & 0 \end{bmatrix}$

THEORETICAL EXERCISES

T1. (a) Show that $\mathbf{A} \vee \mathbf{A} = \mathbf{A}$.
   (b) Show that $\mathbf{A} \wedge \mathbf{A} = \mathbf{A}$.

T2. (a) Show that $\mathbf{A} \vee \mathbf{B} = \mathbf{B} \vee \mathbf{A}$.
   (b) Show that $\mathbf{A} \wedge \mathbf{B} = \mathbf{B} \wedge \mathbf{A}$.

T3. (a) Show that $\mathbf{A} \vee (\mathbf{B} \vee \mathbf{C}) = (\mathbf{A} \vee \mathbf{B}) \vee \mathbf{C}$.
   (b) Show that $\mathbf{A} \wedge (\mathbf{B} \wedge \mathbf{C}) = (\mathbf{A} \wedge \mathbf{B}) \wedge \mathbf{C}$.
   (c) Show that $\mathbf{A} \odot (\mathbf{B} \odot \mathbf{C}) = (\mathbf{A} \odot \mathbf{B}) \odot \mathbf{C}$.

T4. (a) Show that $\mathbf{A} \wedge (\mathbf{B} \vee \mathbf{C}) = (\mathbf{A} \wedge \mathbf{B}) \vee (\mathbf{A} \wedge \mathbf{C})$.
   (b) Show that $\mathbf{A} \vee (\mathbf{B} \wedge \mathbf{C}) = (\mathbf{A} \vee \mathbf{B}) \wedge (\mathbf{A} \vee \mathbf{C})$.

KEY IDEAS FOR REVIEW

☐ Principle of Mathematical Induction: Let n_0 be a fixed integer. Suppose that for each integer $n \geq n_0$ we have a proposition $P(n)$. Suppose that
(a) $P(n_0)$ is true.

(b) If $P(k)$ is true for some $k \geq n_0$, then $P(k + 1)$ must also be true.

Then the principle of mathematical induction states that $P(n)$ is true for all $n \geq n_0$.

☐ Recursion: An algorithm is recursive if it makes reference to earlier or simpler cases of itself.

☐ Theorem: If $n \neq 0$ and m are nonnegative integers, we can write $m = qn + r$, for some nonnegative integers q and r with $0 \leq r < n$ in one and only one way.

☐ GCD(a, b): $d = $ GCD(a, b) if $d \mid a$, $d \mid b$, and whenever $c \mid a$ and $c \mid b$, then $c \mid d$.

☐ If $d = $ GCD(a, b), then $d = sa + tb$ for some integers s and t.

☐ Euclidean algorithm: method used to find GCD(a, b): see page 52.

☐ LCM(a, b): $c = $ LCM(a, b) if $a \mid c$, $b \mid c$, and whenever $a \mid d$ and $b \mid d$, then $c \mid d$.

☐ GCD$(a, b) \cdot$ LCM$(a, b) = ab$.

☐ Matrix: a rectangular array of numbers.

☐ Entries of a matrix: the numbers in a matrix.

☐ Size of a matrix: $\mathbf{A}$ is $m \times n$ if it has m rows and n columns.

☐ Diagonal matrix: a square matrix with zero entries off the main diagonal.

☐ Equal matrices: matrices of the same size whose corresponding entries are equal.

☐ $\mathbf{A} + \mathbf{B}$: the matrix obtained by adding corresponding entries in $\mathbf{A}$ and $\mathbf{B}$.

☐ Zero matrix: a matrix all of whose entries are zero.

☐ $\mathbf{AB}$: see page 60.

☐ $\mathbf{I}_n$ (identity matrix): a square matrix with ones on the main diagonal and zeros elsewhere.

☐ $\mathbf{A}^T$: The matrix obtained from $\mathbf{A}$ by interchanging the rows and columns of $\mathbf{A}$.

☐ Symmetric matrix: $\mathbf{A}^T = \mathbf{A}$.

☐ Array of dimension two: see page 61.

☐ Boolean matrix: a matrix whose entries are either zero or one.

☐ $\mathbf{A} \vee \mathbf{B}$: see page 66.

☐ $\mathbf{A} \wedge \mathbf{B}$: see page 66.

☐ $\mathbf{A} \odot \mathbf{B}$: see page 67.

☐ Properties of Boolean matrix operations: see page 69.

REVIEW EXERCISES

1. Prove by mathematical induction that

$$3 + 7 + 11 + 15 + (4n - 1) = 2n^2 + n$$

2. Prove by mathematical induction that

$$\begin{bmatrix} a & 0 \\ 0 & b \end{bmatrix}^n = \begin{bmatrix} a^n & 0 \\ 0 & b^n \end{bmatrix}$$

where a and b are real numbers.

3. Consider the sequence of matrices defined recursively as follows:

$$\mathbf{A}_0 = \begin{bmatrix} 1 & 0 \\ 0 & 1 \end{bmatrix}, \qquad \mathbf{A}_1 = \begin{bmatrix} 1 & 0 \\ -1 & 1 \end{bmatrix}$$

$$\mathbf{A}_{n+2} = \mathbf{A}_n^2 + \mathbf{A}_{n+1}$$

Compute $\mathbf{A}_2$, $\mathbf{A}_3$, and $\mathbf{A}_4$.

4. Write 750 as a product of prime powers.
5. Find GCD (180, 150) and LCM (180, 150) by factoring into primes and using Theorem 4.
6. Find GCD (476, 440) by the Euclidean algorithm, and write it as $476a + 440b$ (find a and b).

In Exercises 7–9, let

$$\mathbf{A} = \begin{bmatrix} 1 & -1 & 2 \\ 3 & 4 & 5 \end{bmatrix}, \qquad \mathbf{B} = \begin{bmatrix} 1 & 2 \\ 2 & -1 \\ 3 & 2 \end{bmatrix}$$

$$\mathbf{C} = \begin{bmatrix} -2 & 3 \\ 4 & 2 \end{bmatrix}, \qquad \mathbf{D} = \begin{bmatrix} 0 & 1 & 0 \\ -1 & 3 & 2 \\ 4 & 1 & 0 \end{bmatrix}$$

7. If possible, compute
   (a) $(\mathbf{A} + \mathbf{B})\mathbf{C}$ (b) $\mathbf{AB} + \mathbf{C}$ (c) $\mathbf{B} + \mathbf{A}^2$
8. If possible, compute
   (a) $(\mathbf{A}^T + \mathbf{B})^T$ (b) $(\mathbf{AB})^T + \mathbf{C}$ (c) $\mathbf{D}(\mathbf{A}^T + \mathbf{B})$
9. Compute $\mathbf{ABC}$.

In Exercises 10 and 11, compute $\mathbf{A} \vee \mathbf{B}$, $\mathbf{A} \wedge \mathbf{B}$, and $\mathbf{A} \odot \mathbf{B}$ for the given matrices $\mathbf{A}$ and $\mathbf{B}$.

10. $\mathbf{A} = \begin{bmatrix} 0 & 0 \\ 1 & 1 \end{bmatrix}$, $\mathbf{B} = \begin{bmatrix} 1 & 1 \\ 1 & 0 \end{bmatrix}$

11. $\mathbf{A} = \begin{bmatrix} 1 & 0 & 1 \\ 0 & 1 & 1 \\ 1 & 1 & 0 \end{bmatrix}$, $\mathbf{B} = \begin{bmatrix} 1 & 1 & 0 \\ 1 & 0 & 1 \\ 0 & 0 & 1 \end{bmatrix}$

CHAPTER TEST

1. Prove by mathematical induction that $9 + 11 + 13 + 11 + 15 + \cdots + (2n + 7) = n^2 + 8n$.
2. Find GCD (357, 140) and LCM (357, 140).
3. Find GCD (525, 616) and write this number in the form $525a + 616b$.
4. Compute

$$\begin{bmatrix} 1 & 0 & -1 \\ 3 & 1 & 5 \end{bmatrix} \begin{bmatrix} 0 & -1 \\ -1 & 3 \\ 2 & 0 \end{bmatrix} + \begin{bmatrix} 2 & 3 \\ 1 & -2 \end{bmatrix}$$

5. Compute

$$\begin{bmatrix} 1 & 0 & -2 \\ 3 & 2 & 1 \end{bmatrix}^T \left(\begin{bmatrix} 1 & 0 \\ 2 & 3 \end{bmatrix} + \begin{bmatrix} 4 & -1 \\ -2 & 1 \end{bmatrix} \right)$$

6. Let

$$\mathbf{A} = \begin{bmatrix} 1 & 0 & 1 \\ 1 & 1 & 0 \\ 1 & 0 & 1 \end{bmatrix}, \qquad \mathbf{B} = \begin{bmatrix} 1 & 1 & 0 \\ 1 & 1 & 0 \\ 0 & 0 & 1 \end{bmatrix}$$

Compute $\mathbf{A} \vee \mathbf{B}$, $\mathbf{A} \wedge \mathbf{B}$, and $\mathbf{A} \odot \mathbf{B}$.

7. Answer each of the following as true (T) or false (F).
   (a) If $c = \text{GCD}(a, b)$ and $d = \text{LCM}(a, b)$, where a and b are in Z^+, then $a + b = c + d$.
   (b) If p and q are prime numbers, and $p \neq q$, then $\text{GCD}(p, q) = 1$.
   (c) If $\mathbf{A}$ and $\mathbf{B}$ are $n \times n$ matrices, then $\mathbf{AB} = \mathbf{BA}$.
   (d) The product $\mathbf{A}^T \mathbf{A}$ is defined if and only if $\mathbf{A}$ is an $n \times n$ matrix, for some $n \geq 1$.
   (e) If $\mathbf{A}$ and $\mathbf{B}$ are 2×2 Boolean matrices, then $\mathbf{AB}$ and $\mathbf{A} \odot \mathbf{B}$ are equal.

Relations

Relationships between people, numbers, sets, and many other entities can be formalized in the idea of a binary relation. In this chapter we develop the concept of binary relation, and we give several geometric and algebraic methods of representing such objects.

4.1 Product Sets and Partitions

Product Sets

An **ordered pair** (a, b) is a listing of the objects a and b in a prescribed order with a appearing first and b appearing second. Thus an ordered pair is merely a sequence of length 2 from our earlier discussion of sequences (see Section 1.2), it follows that the ordered pairs (a_1, b_1) and (a_2, b_2) are equal if and only if $a_1 = a_2$ and $b_1 = b_2$.

If A and B are two nonempty sets, we define the **product set** or **Cartesian product** $A \times B$ as the set of all ordered pairs (a, b) with $a \in A$ and $b \in B$. Thus

$$A \times B = \{(a, b) \mid a \in A \quad \text{and} \quad b \in B\}$$

Example 1   Let

$$A = \{1, 2, 3\} \qquad \text{and} \qquad B = \{r, s\}$$

then

$$A \times B = \{(1, r), (1, s), (2, r), (2, s), (3, r), (3, s)\}$$

Observe that the elements of $A \times B$ can be arranged in a convenient tabular representation as shown in Fig. 1.

^{B} $_A$	r	s
1	$(1, r)$	$(1, s)$
2	$(2, r)$	$(2, s)$
3	$(3, r)$	$(3, s)$

Figure 1

Example 2   If A and B are as in Example 1, then

$$B \times A = \{(r, 1), (s, 1), (r, 2)\ (s, 2), (r, 3), (s, 3)\}$$

From Examples 1 and 2, we see that $A \times B$ need not equal $B \times A$. On the other hand, the following result shows that $A \times B$ and $B \times A$ have the same number of elements.

Theorem 1   For any two finite, nonempty sets A and B, $|A \times B| = |A||B|$. (See Section 1.1 for a definition of $|A|$.)

   Proof. Suppose that $|A| = m$ and $|B| = n$. In order to form an ordered pair (a, b), $a \in A$ and $b \in B$, we must perform two successive tasks. Task 1 is to choose a first element from A, and task 2 is to choose a second element from B. There are m ways to perform task 1 and, for each of these ways, there are n ways to perform task 2, so by the multiplication principle (see Section 2.2), there are $m \cdot n$ ways to form an ordered pair (a, b). In other words, $|A \times B| = m \cdot n = |A||B|$.

Example 3   If $A = B = \mathbb{R}$, the set of all real numbers, then $\mathbb{R} \times \mathbb{R}$ also denoted by $\mathbb{R}^2$, is the set of all points in the plane. The ordered pair (a, b) is the point (a, b) in the plane.

Example 4   A marketing research firm classifies a person according to the following two criteria:

   Sex: m = male; f = female

   Highest level of education completed: e = elementary school; h = high school;

   c = college; g = graduate school

Let $S = \{m, f\}$ and $L = \{e, h, c, g\}$. Then the product set $S \times L$ contains all the categories into which the population is classified. Thus the classification (f, g) represents a female who has completed graduate school. There are eight categories in this classification scheme.

We now define the Cartesian product of three or more nonempty sets by generalizing the earlier definition of the Cartesian product of two sets. That is, the **Cartesian product** $A_1 \times A_2 \times \cdots \times A_m$ of the nonempty sets $A_1, A_2, \ldots, A_m$ is the set of all ordered m-tuples $(a_1, a_2, \ldots, a_m)$, where $a_i \in A_i$, $i = 1, 2, \ldots, m$. Thus

$$A_1 \times A_2 \times \cdots \times A_m = \{(a_1, a_2, \ldots, a_m) \mid a_i \in A_i, i = 1, 2, \ldots, m\}$$

Example 5 A software firm provides the following three characteristics for each program that it sells:

Language:	f = FORTRAN, p = PASCAL, l = LISP
Memory:	48 = 48,000, 64 = 64,000, 128 = 128,000
Operating system:	u = UNIX, c = CP/M

Let $L = \{f, p, l\}$, $M = \{48, 64, 128\}$, $O = \{u, c\}$. Then the Cartesian product $L \times M \times O$ contains all the categories that describe a program. There are $3 \cdot 3 \cdot 2 = 18$ categories in this classification scheme.

Proceeding in a manner similar to that used to prove Theorem 1, we can show that if A_1 has n_1 elements, A_2 has n_2 elements, . . . , and A_m has n_m elements, then $A_1 \times A_2 \times \cdots \times A_m$ has $n_1 \cdot n_2 \cdots n_m$ elements.

Partitions

A **partition** or **quotient set** of a nonempty set A is a collection $\mathcal{P}$ of nonempty subsets of A such that

1. Each element of A belongs to one of the sets in $\mathcal{P}$.
2. If A_1 and A_2 are distinct elements of $\mathcal{P}$, then $A_1 \cap A_2 = \varnothing$.

The sets in $\mathcal{P}$ are called the **blocks** or **cells** of the partition. Figure 2 shows diagrammatically a partition $\mathcal{P} = \{A_1, A_2, A_3, A_4, A_5, A_6, A_7\}$ of A into seven blocks.

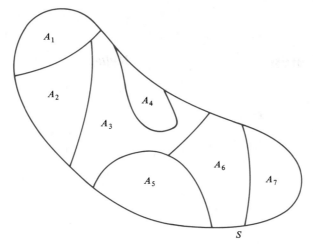

Figure 2

Example 6   Let

$$A = \{a, b, c, d, e, f, g, h\}$$

Consider the following subsets of A:

$$A_1 = \{a, b, c, d\} \qquad A_2 = \{a, c, e, f, g, h\}, \qquad A_3 = \{a, c, e, g\},$$
$$A_4 = \{b, d\}, \qquad A_5 = \{f, h\}$$

Then $\{A_1, A_2\}$ is not a partition since $A_1 \cap A_2 \neq \varnothing$. Also, $\{A_1, A_5\}$ is not a partition since $e \notin A_1$ and $e \notin A_5$. The collection $\mathcal{P} = \{A_3, A_4, A_5\}$ is a partition of A.

Example 7   Consider the set A of all employees of General Motors. If we form subsets of A by grouping in one subset all employees who make exactly the same salary, we obtain a partition of A.

Example 8   Let

$$A = \text{set of all integers}$$
$$A_1 = \text{set of all even integers}$$
$$A_2 = \text{set of all odd integers}$$

Then $\{A_1, A_2\}$ is a partition of A.

Since the members of a partition of a set A are subsets of A, we see that a partition is a subset of $P(A)$. That is, partitions can be considered as particular kinds of subsets of $P(A)$.

EXERCISE SET 4.1

1. In each part, find x or y so that the ordered pairs are equal.
   - (a) $(x, 3) = (4, 3)$
   - (b) $(a, 3y) = (a, 9)$
   - (c) $(3x + 1, 2) = (7, 2)$
   - (d) $(\text{ALGOL, PASCAL}) = (y, \text{PASCAL})$

2. In each part find x and y so that the ordered pairs are equal.
   - (a) $(4x, 6) = (16, 6)$
   - (b) $(2x - 3, 3y - 1) = (5, 5)$
   - (c) $(x^2, 25) = (49, 25)$
   - (d) $(x, y) = (x^2, y^2)$

3. Let $A = \{a, b\}$ and $B = \{4, 5, 6\}$.
   - (a) List the elements in $A \times B$.
   - (b) List the elements in $B \times A$.
   - (c) List the elements in $A \times A$.
   - (d) List the elements in $B \times B$.

4. Let $A = \{\text{Fine, Smith}\}$ and $B = \{\text{president, vice-president, secretary, treasurer}\}$.
   - (a) List the elements in $A \times B$.
   - (b) List the elements in $B \times A$.
   - (c) List the elements in $A \times A$.

5. A genetics experiment classifies fruit flies according to the following two criteria.

   Sex: m = male, f = female

   Wing span: s = short-winged, m = medium-winged, l = long-winged

   - (a) How many categories are there in this classification scheme?
   - (b) List all the categories.

6. A car manufacturer makes three different types of frames and two types of engines:

   Frames: s = sedan, c = coupe, w = station wagon

   Engines: g = gas, d = diesel

   List all the possible models.

7. How many two-letter abbreviations can be formed from the 26-letter English alphabet if a letter can be repeated?

8. A medical experiment classifies each subject according to two criteria:

   Smoking pattern: s = smoker, n = nonsmoker

   Weight: u = underweight, a = average weight, o = overweight

   List all possible classifications in this scheme.

9. If $A = \{a, b, c\}$, $B = \{1, 2\}$, and $C = \{r, s\}$, list all elements in $A \times B \times C$.

10. If $A = \{x, y, z, u\}$ and $B = \{1, 9\}$, list all elements in $A \times A \times B \times B$.

11. If $A = \{x \mid x \text{ is real and } -2 \leq x \leq 3\}$ and $Y = \{y \mid y \text{ is real and } 1 \leq y \leq 5\}$, sketch the set $A \times B$ in the Cartesian plane.

12. Let $A = \{1, 2, 3, 4, 5, 6, 7, 8, 9, 10\}$ and let

    $A_1 = \{1, 2, 3, 4\}$ $A_2 = \{5, 6, 7\}$

    $A_3 = \{4, 5, 7, 9\}$ $A_4 = \{4, 8, 10\}$

    $A_5 = \{8, 9, 10\}$ $A_6 = \{1, 2, 3, 6, 8, 10\}$

    Which of the following are partitions of A?
    - (a) $\{A_1, A_2, A_5\}$
    - (b) $\{A_1, A_3, A_5\}$
    - (c) $\{A_3, A_6\}$
    - (d) $\{A_2, A_3, A_4\}$

13. If A_1 = the set of all positive integers and A_2 = the set of all negative integers, is $\{A_1, A_2\}$ a partition of Z?

14. List all partitions of the set $A = \{1, 2, 3\}$.

15. List all partitions of the set $A = \{a, b, c, d\}$.

THEORETICAL EXERCISES

T1. If A has three elements and B has $n \geq 1$ elements, show by mathematical induction that $A \times B$ has $3n$ elements.

T2. Show that if A_1 has n_1 elements, A_2 has n_2 elements, and A_3 has n_3 elements, then $A_1 \times A_2 \times A_3$ has $n_1 \cdot n_2 \cdot n_3$ elements.

T3. If $A \subseteq C$ and $B \subseteq D$, show that $A \times B \subseteq C \times D$.

4.2 Relations

The notion of a relation between two sets of objects is quite common and intuitively clear (a precise definition will be given below). If A is the set of all living human males and B is the set of all living human females, then the relation F (father) can be defined from A to B. Thus, if $x \in A$ and $y \in B$, then x is related to y by the relation F if x is the father of y, and we write $x \, F \, y$. We could also consider the relations S and H from A to B by letting $x \, S \, y$ mean that x is a son of y and $x \, H \, y$ means that x is the husband of y.

If A is the set of all real numbers, there are many commonly used relations from A to A. Examples are the relation "x is less than y," denoted by $x < y$, and the other order relations $>$, $\leq$, $\geq$ = , and $\neq$. We see that a relation is often described in English and is denoted by a familiar name or symbol. The problem with this approach is that we will need to discuss *any possible* relation from one abstract set to another. Most of these relations have no simple verbal description and no familiar name or symbol to remind us of their nature or properties. Furthermore, it is usually awkward, and sometimes nearly impossible, to give any precise proofs of the properties that a relation satisfies if we must deal with a verbal description of it.

To get around this problem, observe that the only thing that really matters about a relation is that we know precisely which elements in A are related to which elements in B. Thus suppose that $A = \{1, 2, 3, 4\}$ and R is a relation from A to A. If we know that $1 \, R \, 2$, $1 \, R \, 3$, $1 \, R \, 4$, $2 \, R \, 3$, $2 \, R \, 4$, $3 \, R \, 4$, then we know everything we need to know about R. Actually, R is the familiar relation $<$, "less than," but we need not know this. It would be enough to be given the foregoing list of related pairs. Thus we may say that R is completely known if we know all R-related pairs. We could then write $R = \{(1, 2), (1, 3), (1, 4), (2, 3), (2, 4), (3, 4)\}$, since R is essentially equal to or completely specified by this list of ordered pairs. Each ordered pair specifies that its first element is related to its second element, and all possible related pairs are assumed to be given, at least in principle. This method of specifying a relation does not require any special symbol or description, and so is suitable for any relation between any two sets. Note that from this point of view, a relation from A to B is simply a subset of $A \times B$ (giving the related pairs) and conversely, any subset of

$A \times B$ can be considered a relation, even if it is an unfamiliar relation for which we have no name or alternative description. We choose this approach for defining relations.

Let A and B be nonempty sets. **A relation R from A to B** is a subset of $A \times B$. If $R \subseteq A \times B$ and $(a, b) \in R$, we say that a is related to b by R, and we also write $a \, R \, b$. If a is not related to b by R, we write $a \, \not{R} \, b$. Frequently, A and B are equal. In this case, we often say that $R \subseteq A \times A$ is a **relation on A,** instead of a relation from A to A.

Relations are extremely important in mathematics and its applications. It is not an exaggeration to say that 90 percent of what will be discussed in the remainder of this text will concern some type of object which may be considered a relation. We now give a number of examples.

Example 1   Let

$$A = \{1, 2, 3\} \qquad \text{and} \qquad B = \{r, s\}$$

Then

$$R = \{(1, r), (2, s), (3, r)\}$$

is a relation from A to B.

Example 2   Let A and B be sets of real numbers. We define the following relation R (equals) from A to B:

$$a \, R \, b \qquad \text{if and only if} \qquad a = b$$

Example 3   Let

$$A = \{1, 2, 3, 4, 5\} = B$$

Define the following relation R (less than) on A:

$$a \, R \, b \qquad \text{if and only if} \qquad a < b$$

Then

$$R = \{(1, 2), (1, 3), (1, 4), (1, 5), (2, 3), (2, 4), (2, 5), (3, 4), (3, 5), (4, 5)\}$$

Example 4   Let $A = B = Z^+$, the set of all positive integers. Define the following relation R on Z^+:

$$a \; R \; b \qquad \text{if and only if} \qquad a \text{ divides } b$$

Then $4 \; R \; 12$, but $5 \not{R} \; 7$.

Example 5   Let A be the set of all people in the world. We define the following relation R on A:

$a \; R \; b$ if and only if there is a sequence $a_0, a_1, \ldots, a_n$ of people such that $a_0 = a$, $a_n = b$ and a_{i-1} knows a_i, $i = 1, 2, \ldots, n$ (n will depend on a and b)

Example 6   Let $A = B = \mathbb{R}$, the set of all real numbers. We define the following relation R on A:

$$x \; R \; y \quad \text{if and only if} \quad x \text{ and } y \text{ satisfy the equation } \quad \frac{x^2}{4} + \frac{y^2}{9} = 1$$

The set R consists of all points on the ellipse shown in Fig. 1.

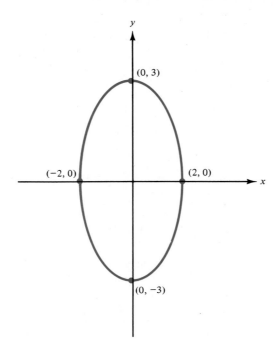

Figure 1

Example 7   Let A be the set of all possible inputs to a given computer program, and let B be the set of all possible outputs from the same program. Define the following relation R from A to B:

$a \ R \ b$ if and only if b is the output produced by the program when input a is used

Example 8   Let

$$A = B = \text{the set of all lines in the plane}$$

Define the following relation R on A:

$$l_1 \ R \ l_2 \qquad \text{if and only if} \qquad l_1 \text{ is parallel to } l_2$$

where l_1 and l_2 are lines in the plane.

Example 9   An airline services the five cities c_1, c_2, c_3, c_4, and c_5. The following table gives the cost (in dollars) of going from c_i to c_j. Thus the cost of going from c_1 to c_3 is \$100, while the cost of going from c_4 to c_2 is \$200.

To From	c_1	c_2	c_3	c_4	c_5
c_1		140	100	150	200
c_2	190		200	160	220
c_3	110	180		190	250
c_4	190	200	120		150
c_5	200	100	200	150	

We now define the following relation R on the set of cities $A = \{c_1, c_2, c_3, c_4, c_5\}$:

$c_i \ R \ c_j$ if and only if   the cost of going from c_i to c_j is less than or equal to \$180

Find R.

   Solution. The relation R is the subset of $A \times A$ consisting of all cities (c_i, c_j), where the cost of going from c_i to c_j is less than or equal to \$180. Hence

$$R = \{(c_1, c_2), (c_1, c_3), (c_1, c_4), (c_2, c_4), (c_3, c_1), (c_3, c_2,),$$
$$(c_4, c_3), (c_4, c_5), (c_5, c_2), (c_5, c_4)\}$$

Sets Arising from Relations

Let $R \subseteq A \times B$ be a relation from A to B. We now define various important and useful sets connected with R.

The **domain** of R, denoted by Dom (R), is the set of elements in A which are related to some element in B. In other words, Dom (R), a subset of A, is the set of all first elements in the pairs that make up R. Similarly, we define the **range** of R, denoted by Ran (R), to be the set of elements in B which are second elements of pairs in R, that is, all elements in B which are related to some element in A.

Elements of A that are not in Dom (R) are not involved in the relation R in any way. This is also true for elements of B that are not in Ran (R).

Example 10   If R is the relation defined in Example 1 or 2, then Dom $(R) = A$ and Ran $(R) = B$.

Example 11   If R is the relation given in Example 3, then Dom $(R) = \{1, 2, 3, 4\}$ and Ran $(R) = \{2, 3, 4, 5\}$.

Example 12   Let R be the relation of Example 6. Then Dom $(R) = (-2, 2)$ and Ran $(R) = (-3, 3)$.

If R is a relation from A to B and $x \in A$, we define $R(x)$, the **R-relative set of x**, to be the set of all y in B with the property that x is R-related to y. Thus in symbols,

$$R(x) = \{y \in B \,|\, x \, R \, y\}$$

Similarly, if $A_1 \subseteq A$, then $R(A_1)$, the **R-relative set of A_1**, is the set of all y in B with the property that x is R-related to y for some x in A_1. That is,

$$R(A_1) = \{y \in B \,|\, x \, R \, y \text{ for some } x \text{ in } A_1\}$$

From the definitions above we see that $R(A_1)$ is the union of the sets $R(x)$, where $x \in A_1$. The sets $R(x)$ play an important role in the study of many types of relations.

Example 13   Let $A = B = \{a, b, c, d\}$ and let $R = \{(a, a), (a, b), (b, c), (c, a), (d, c), (c, b)\}$. Then

$$R(a) = \{a, b\}, \, R(b) = \{c\}$$

and if $A_1 = \{c, d\}$, then

$$R(A_1) = \{a, b, c\}$$

Example 14   Let R be the relation of Example 6, and let $x \in \mathbb{R}$. If $x R y$ for some y, then $x^2/4 + y^2/9 = 1$. We see that if x is not in the interval $(-2, 2)$, then no y can satisfy the equation above, since $x^2/4 > 1$. Thus, in this case, $R(x) = \varnothing$. If $x = -2$, then $x^2/4 = 1$, so x can only be related to 0. Hence $R(-2) = \{0\}$. Similarly, $R(2) = \{0\}$. Finally, if $-2 < x < 2$ and $x R y$, then we must have $y = \sqrt{9 - (9x^2/4)}$ or $y = -\sqrt{9 - (9x^2/4)}$, as we see by solving the equation above, so $R(x) = \{\sqrt{9 - (9x^2/4)}, -\sqrt{9 - (9x^2/4)}\}$. For example, $R(1) = \{(3\sqrt{3})/2, -(3\sqrt{3})/2\}$.

It is a useful and easily seen fact that the sets $R(a)$, for all a in A, completely determine a relation R. We state this fact precisely in the following theorem.

Theorem 1   Let R and S be relations from A to B. If $R(a) = S(a)$ for all a in A, then $R = S$.

Proof. If $a R b$, then $b \in R(a)$. Therefore, $b \in S(a)$ and $a S b$. A completely similar argument shows that if $a S b$, then $a R b$. Thus $R = S$.

EXERCISE SET 4.2

1. Consider the relation defined in Example 4. Which of the following ordered pairs belong to R?
   - (a) $(2, 3)$
   - (b) $(-6, 24)$
   - (c) $(0, 8)$
   - (d) $(8, 0)$
   - (e) $(1, 3)$
   - (f) $(6, 18)$

2. Consider the relation defined in Example 6. Which of the following ordered pairs belong to R?
   - (a) $(2, 0)$
   - (b) $(0, 2)$
   - (c) $(0, 3)$
   - (d) $(1, 3/(2\sqrt{3}))$
   - (e) $(0, 0)$
   - (f) $(-2, 0)$

In Exercises 3–13, find the domain and range of the relation R.

3. $A = \{a, b, c, d\}$, $B = \{1, 2, 3\}$,
   $R = \{(a, 1), (a, 2), (b, 1), (c, 2), (d, 1)\}$

4. $A = \{4, 5, 6, 7\}$, $B = \{r, s, t, u, v\}$,
   $R = \{(4, r), (4, u), (5, r), (5, v), (7, u)\}$

5. $A = \{\text{IBM, Univac, Commodore, Atari, Xerox}\}$,
   $B = \{370, 1110, 4000, 8000, 400, 800, 820\}$,
   $R = \{(\text{IBM}, 370), (\text{Univac}, 1110), (\text{Commodore}, 4000), (\text{Atari}, 800)\}$.

6. $A = \{1, 2, 3, 4\}$, $B = \{1, 4, 6, 8, 9\}$; $a R b$ if and only if $b = a^2$.

7. $A = \{1, 2, 3, 4, 8\} = B$; $a R b$ if and only if $a = b$.

8. $A = \{1, 2, 3, 4, 8\}$, $B = \{1, 4, 6, 9\}$; $a R b$ if and only if $a \mid b$.

9. $A = \{1, 2, 3, 4, 6\} = B$; $a R b$ if and only if a is a multiple of b.

10. $A = \{1, 2, 3, 4, 5\} = B$; $a R b$ if and only if $a \leq b$.

11. $A = \{1, 3, 5, 7, 9\}$, $B = \{2, 4, 6, 8\}$; $a R b$ if and only if $b < a$.

12. $A = \{1, 2, 3, 4, 8\} = B$; $a R b$ if and only if $a + b \leq 9$.

13. $A = \{1, 2, 3, 4, 5, 8\} = B$; $a R b$ if and only if $b = a + 1$.

14. Let $A = B = \mathbb{R}$. Consider the following relation R on A; $a R b$ if and only if $2a + 3b = 6$. Find Dom (R) and Ran (R).

15. Let $A = B = \mathbb{R}$. Consider the following relation R on A: $a\, R\, b$ if and only if $a^2 + b^2 = 25$. Find Dom (R) and Ran (R).

16. Let R be the relation defined in Example 6.
    (a) Find $R(6)$.
    (b) If $A_1 = [0, 1]$, find $R(A_1)$.

17. Let R be the relation defined in Example 6. Find $R(A_1)$ for each of the following sets A_1.
    (a) $A_1 = \{1, 8\}$
    (b) $A_1 = \{3, 4, 5\}$
    (c) $A_1 = \varnothing$

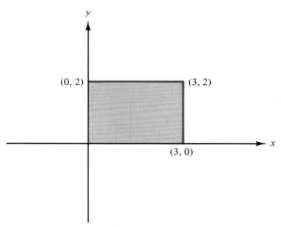

y

$(0, 2)$ $(3, 2)$

$(3, 0)$

x

Figure 2

18. Let R be the relation defined in Exercise 9. Find
    (a) $R(3)$
    (b) $R(6)$
    (c) $R(\{2, 4, 6\})$

19. Let R be the relation defined in Exercise 4. Find
    (a) $R\{(4, 6)\}$
    (b) $R(4)$
    (c) $R(\{5, 7\})$

20. Let $A = B = \mathbb{R}$. Find the relation specified by the shaded region in Fig. 2.

21. A manufacturer of automobiles has five factories $\{a_1, a_2, a_3, a_4, a_5\}$ and six distribution centers $\{b_1, b_2, b_3, b_4, b_5, b_6\}$. The following table gives the distance (in miles) from a_i to b_j.

	b_1	b_2	b_3	b_4	b_5	b_6
a_1	1200	1100	400	600	1800	700
a_2	800	700	1200	450	400	500
a_3	1000	600	1000	650	600	600
a_4	250	400	500	350	900	600
a_5	800	280	300	400	1300	2400

We define the following relation: $a_i\, R\, b_j$ if and only if the distance from a_i to b_j is at least 800 miles. List the elements in R.

THEORETICAL EXERCISE

T1. If A has n elements and B has m elements, how many different relations are there from A to B?

4.3 Representing Relations

In this section we present two ways to represent relations, one algebraic and one geometric. Each method is useful, but certain examples lend themselves more naturally to one method than to the other.

The Matrix of a Relation

If $A = \{a_1, a_2, \ldots, a_m\}$ and $B = \{b_1, b_2, \ldots, b_n\}$ are finite sets, containing m and n elements, respectively, and R is a relation from A to B, we can represent R by the

$m \times n$ matrix $\mathbf{M}_R = [m_{ij}]$, which is defined as follows:

$$m_{ij} = \begin{cases} 1 & \text{if } (a_i, b_j) \in R \\ 0 & \text{if } (a_i, b_j) \notin R \end{cases}$$

The matrix $\mathbf{M}_R$ is called the **matrix of** R.

Example 1   Let $A = \{1, 2, 3\}$ and $B = \{r, s\}$ and consider the relation

$$R = \{(1, r), (2, s), (3, r)\}$$

from A to B. Find $\mathbf{M}_R$.

Solution. If we think of the elements of A as labels on the rows of $\mathbf{M}_R$ and the elements of B as labels on the columns of $\mathbf{M}_R$, then

$$\mathbf{M}_R = \begin{array}{c} \\ 1 \\ 2 \\ 3 \end{array} \begin{array}{cc} r & s \\ \begin{bmatrix} 1 & 0 \\ 0 & 1 \\ 1 & 0 \end{bmatrix} \end{array}$$

Conversely, given sets A and B with $|A| = m$ and $|B| = n$, an $m \times n$ matrix whose entries are zeros and ones determines a relation, as is illustrated in the following example.

Example 2   Consider the matrix

$$\mathbf{M} = \begin{bmatrix} 1 & 0 & 0 & 1 \\ 0 & 1 & 1 & 0 \\ 1 & 0 & 1 & 0 \end{bmatrix}$$

Since $\mathbf{M}$ is 3×4, we let

$$A = \{a_1, a_2, a_3\} \quad \text{and} \quad B = \{b_1, b_2, b_3, b_4\}$$

Then $(a_i, b_j) \in R$ if and only if $m_{ij} = 1$. Thus

$$R = \{(a_1, b_1), (a_1, b_4), (a_2, b_2), (a_2, b_3), (a_3, b_1), (a_3, b_3)\}$$

Digraphs

If A is a finite set and R is a relation on A, we can represent R pictorially as follows. Draw a small circle for each element of A and label the circle with the corresponding element of A. These circles are called **vertices.** Draw a directed line, called an **edge,** from vertex a_i to vertex a_j if and only if $a_i \, R \, a_j$. The resulting pictorial representation of R is called a **directed graph** or **digraph** of R.

Thus if R is a relation on A, the edges in the digraph of R correspond exactly to the pairs in R, and the vertices correspond exactly to the elements of the set A. Sometimes, when we want to emphasize the geometric nature of some property of R, we may refer to the pairs of R themselves as edges, and the elements of A as vertices.

Example 3   Let

$$A = \{1, 2, 3, 4\}$$
$$R = \{(1, 1), (1, 2), (2, 1), (2, 2), (2, 3), (2, 4), (3, 4), (4, 1)\}$$

Then the digraph of R is shown in Fig. 1.

A collection of vertices with edges between some of the vertices determines a relation in a natural manner.

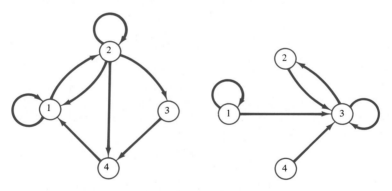

Figure 1   Figure 2

Example 4   Find the relation determined by Fig. 2.

Solution. Since $a_i \, R \, a_j$ if and only if there is an edge from a_i to a_j, we have

$$R = \{(1, 1), (1, 3), (2, 3), (3, 2), (3, 3), (4, 3)\}$$

In this book, digraphs are nothing but geometrical representations of relations, and any statement made about a digraph is actually a statement about the corresponding relation. This is especially important for theorems and their proofs. In some cases, it is easier or clearer to state a result in graphical terms, but a proof will always refer to the underlying relation. The reader should be aware that some authors allow more general objects as digraphs, for example by permitting several edges between the same vertices. We do not need, and will not again refer to, these generalizations, except to say that they could be covered by the idea of a function defined on a digraph (see Section 6.1).

An important concept for relations is inspired by the visual form of digraphs. If R is a relation on a set A, and $a \in A$, then the **in-degree** of a (relative to the relation R) is the number of $b \in A$ such that $(b, a) \in R$. The **out-degree** of a is the number of $b \in A$ such that $(a, b) \in R$.

What this means, in terms of the digraph of R, is that the in-degree of a vertex is the number of edges terminating at the vertex. The out-degree of a vertex is the number of edges leaving the vertex. Note that the out-degree of a is $|R(a)|$, the number of elements to which a is related.

Example 5 Consider the digraph of Fig. 1. Vertex 1 has in-degree 3 and out-degree 2. Also consider the digraph shown in Fig. 2. Vertex 3 has in-degree 4 and out-degree 2, while vertex 4 has in-degree 0 and out-degree 1.

Example 6 Let $A = \{a, b, c, d\}$, and let R be the relation on A that has the matrix

$$\mathbf{M}_R = \begin{bmatrix} 1 & 0 & 0 & 0 \\ 0 & 1 & 0 & 0 \\ 1 & 1 & 1 & 0 \\ 0 & 1 & 0 & 1 \end{bmatrix}$$

Construct the digraph of R, and list in-degrees and out-degrees of all vertices.

Solution. The digraph of R is shown in Fig. 3. The following table gives the in-degrees and out-degrees of all vertices.

	a	b	c	d
In-degree	2	3	1	1
Out-degree	1	1	3	2

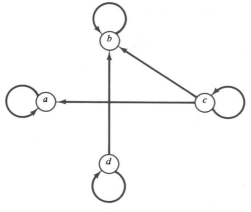

Figure 3

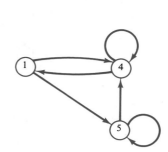

Figure 4

Example 7   Let $A = \{1, 4, 5\}$ and let R be given by the digraph shown in Fig. 4. Find $\mathbf{M}_R$ and R.

Solution

$$\mathbf{M}_R = \begin{bmatrix} 0 & 1 & 1 \\ 1 & 1 & 0 \\ 0 & 1 & 1 \end{bmatrix}, \quad R = \{(1, 4), (1, 5), (4, 1), (4, 4), (5, 4), (5, 5)\}$$

EXERCISE SET 4.3

In Exercises 1 and 2, let $A = \{a, b, c, d, e\}$. Find the matrix $\mathbf{M}_R$ that represents the given relation R on A.

1. $R = \{(a, a), (a, e), (b, a), (b, c), (b, d), (d, c), (c, b), (e, d)\}$

2. $R = \{(a, a), (b, b), (c, c), (d, d), (e, e)\}$

In Exercises 3 and 4, let $A = \{x, y, z, w\}$. Find the digraph of the given relations R on A.

3. $R = \{(x, x), (x, y), (y, y), (y, z), (z, z), (z, w), (w, w), (w, x)\}$

4. $R = \{(x, y), (x, z), (x, w), (w, x), (w, y), (w, z), (y, z)\}$

In Exercises 5 and 6, let $A = \{1, 2, 3, 4\}$. Find the relation R on A determined by the given matrix.

5. $\mathbf{M}_R = \begin{bmatrix} 1 & 0 & 1 & 0 \\ 0 & 0 & 1 & 0 \\ 1 & 0 & 0 & 0 \\ 1 & 1 & 0 & 1 \end{bmatrix}$

6. $\mathbf{M}_R = \begin{bmatrix} 1 & 1 & 0 & 1 \\ 0 & 1 & 1 & 0 \\ 0 & 0 & 1 & 1 \\ 1 & 0 & 0 & 0 \end{bmatrix}$

7. Find the relation determined by Fig. 5.

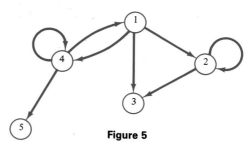

Figure 5

8. Find the relation determined by Fig. 6.

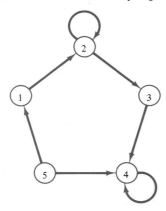

Figure 6

9. If A has n elements and B has m elements, how many different relations are there from A to B?

In Exercises 10 and 11, let $A = \{a, b, c, d, e\}$.

Find the digraph of the relation R on A whose matrix representation $\mathbf{M}_R$ is given.

10. $\mathbf{M}_R = \begin{bmatrix} 1 & 1 & 0 & 0 & 0 \\ 0 & 0 & 1 & 1 & 0 \\ 0 & 0 & 0 & 1 & 1 \\ 0 & 1 & 1 & 0 & 0 \\ 1 & 0 & 0 & 0 & 0 \end{bmatrix}$

11. $\mathbf{M}_R = \begin{bmatrix} 0 & 0 & 0 & 0 & 1 \\ 0 & 1 & 0 & 0 & 0 \\ 0 & 0 & 1 & 0 & 0 \\ 0 & 0 & 0 & 1 & 0 \\ 1 & 0 & 0 & 0 & 0 \end{bmatrix}$

12. Compute the matrix $\mathbf{M}_R$ of the relation whose digraph is shown in Fig. 5.

13. Compute the matrix $\mathbf{M}_R$ of the relation whose digraph is shown in Fig. 6.

14. Consider the digraph of Fig. 5. Compute the in-degree and out-degree of each vertex.

15. Consider the digraph of Fig. 6. Compute the in-degree and out-degree for each vertex.

4.4 Paths

Suppose that R is a relation on a set A. A **path of length n** in R from a to b is a finite sequence $\pi = a, x_1, x_2, \ldots, x_{n-1}, b$, beginning with a and ending with b, such that

$$a \, R \, x_1, \qquad x_1 \, R \, x_2, \qquad \ldots, \qquad x_{n-1} \, R \, b$$

Note that a path of length n involves $n + 1$ elements of A, although they are not necessarily distinct.

A path is most easily visualized with the aid of the digraph of the relation. It appears as a geometric "path" or succession of edges in such a digraph, where the indicated directions of the edges are followed, and in fact a path derives its name from this representation. Thus the length of a path is the number of edges in the path, where the vertices need not all be distinct, and an edge is counted as often as it is traversed.

Example 1 Consider the digraph in Fig. 1. Then $\pi_1 = 1, 2, 5, 4, 3$ is a path of length 4 from vertex 1 to vertex 3, $\pi_2 = 1, 2, 5, 1$ is a path of length 3 from vertex 1 to itself, and $\pi_3 = 2, 2$ is a path of length 1 from vertex 2 to itself.

A path that begins and ends at the same vertex is called a **cycle.** In Example 1, π_2 and π_3 are cycles of length 3 and 1, respectively. It is clear that the paths of length

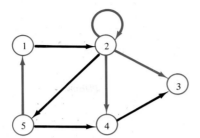

Figure 1

1 can be identified with the ordered pairs (x, y) that belong to R. Paths in a relation R can be used to define new relations which are quite useful. If n is a fixed positive integer we define a relation R^n on A as follows: $x\ R^n\ y$ means that there is a path of length n from x to y in R. We may also define a relation R^∞ on A, by letting $x\ R^\infty\ y$ mean that there is some path in R from x to y. The length of such a path will depend, in general, on x and y. R^∞ is sometimes called the **connectivity relation** for R.

Note that $R^n(x)$ consists of all vertices that can be reached from x by means of a path in R of length n. The set $R^\infty(x)$ consists of all vertices that can be reached from x by some path in R.

Example 2 Let A be the set of all living human beings, and let R be the relation of mutual acquaintance. That is, $a\ R\ b$ means that a and b know one another. Then $a\ R^2\ b$ means that a and b have an acquaintance in common. In general, $a\ R^n\ b$ if a knows someone x_1, who knows $x_2, \ldots$, who knows x_{n-1}, who knows b. Finally, $a\ R^\infty\ b$ means that some chain of acquaintances exists which begins at a and ends at b. It is interesting (and unknown) whether every two Americans, say, are related by R^∞.

Example 3 Let A be a set of U.S. cities, and let $x\ R\ y$ if there is a direct flight from x to y on at least one airline. Then x and y are related by R^n if one can book a flight from x to y having exactly $n - 1$ intermediate stops, and $x\ R^\infty\ y$ if one can get away from x to y by plane.

Example 4 Let $A = \{1, 2, 3, 4, 5, 6\}$. Let R be the relation whose digraph is shown in Fig. 2.

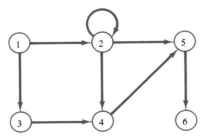

Figure 2

Figure 3 shows the digraph of the relation R^2 on A. A line connects two vertices in Fig. 3 if and only if they are R^2-related, that is, if and only if there is a path of length two connecting those vertices in Fig. 2. Thus

$$1 \; R^2 \; 2 \quad \text{since} \quad 1 \, R \, 2 \quad \text{and} \quad 2 \, R \, 2$$
$$1 \; R^2 \; 4 \quad \text{since} \quad 1 \, R \, 2 \quad \text{and} \quad 2 \, R \, 4$$
$$1 \; R^2 \; 5 \quad \text{since} \quad 1 \, R \, 2 \quad \text{and} \quad 2 \, R \, 5$$
$$2 \; R^2 \; 2 \quad \text{since} \quad 2 \, R \, 2 \quad \text{and} \quad 2 \, R \, 2$$
$$2 \; R^2 \; 4 \quad \text{since} \quad 2 \, R \, 2 \quad \text{and} \quad 2 \, R \, 4$$
$$2 \; R^2 \; 5 \quad \text{since} \quad 2 \, R \, 2 \quad \text{and} \quad 2 \, R \, 5$$
$$2 \; R^2 \; 6 \quad \text{since} \quad 2 \, R \, 5 \quad \text{and} \quad 5 \, R \, 6$$
$$3 \; R^2 \; 5 \quad \text{since} \quad 3 \, R \, 4 \quad \text{and} \quad 4 \, R \, 5$$
$$4 \; R^2 \; 6 \quad \text{since} \quad 4 \, R \, 5 \quad \text{and} \quad 5 \, R \, 6$$

In a similar way, we can construct the digraph of R^n for any n.

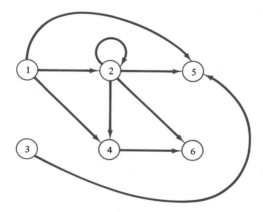

Figure 3

Example 5   Let

$$A = \{a, b, c, d, e\}$$
$$R = \{(a, a), (a, b), (b, c), (c, e), (c, d), (d, e)\}$$

Compute (a) R^2; (b) R^∞.

Solution. (a) The digraph of R is shown in Fig. 4. Then

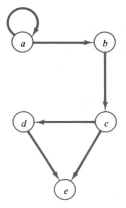

Figure 4

$$a \; R^2 \; a \quad \text{since} \quad a \; R \; a \quad \text{and} \quad a \; R \; a$$
$$a \; R^2 \; b \quad \text{since} \quad a \; R \; a \quad \text{and} \quad a \; R \; b$$
$$a \; R^2 \; c \quad \text{since} \quad a \; R \; b \quad \text{and} \quad b \; R \; c$$
$$b \; R^2 \; e \quad \text{since} \quad b \; R \; c \quad \text{and} \quad c \; R \; e$$
$$b \; R^2 \; d \quad \text{since} \quad b \; R \; c \quad \text{and} \quad c \; R \; d$$
$$c \; R^2 \; e \quad \text{since} \quad c \; R \; d \quad \text{and} \quad d \; R \; e$$

Hence

$$R^2 = \{(a, a), (a, b), (a, c), (b, e), (b, d), (c, e)\}$$

(b) To compute R^∞, we need all ordered pairs of vertices for which there is a path of any length from the first vertex to the second one. From Fig. 4 we see that

$$R^\infty = \{(a, a), (a, b), (a, c), (a, d), (a, e), (b, c), (b, d), (b, e), (c, d), (c, e), (d, e)\}$$

For example, $(a, d) \in R^\infty$, since there is a path of length 3 from a to d: a, b, c, d. Similarly, $(a, e) \in R^\infty$ since there is a path of length 3 from a to e: a, b, c, e as well as a path of length 4 from a to e: a, b, c, d, e.

Let R be a relation on a finite set $A = \{a_1, a_2, \ldots, a_n\}$, and let $\mathbf{M}_R$ be the $n \times n$ matrix representing R. We will show how the matrix $\mathbf{M}_{R^2}$, of R^2, can be computed from $\mathbf{M}_R$.

Theorem 1   If R is a relation on $A = \{a_1, a_2, \ldots, a_n\}$, then $\mathbf{M}_{R^2} = \mathbf{M}_R \odot \mathbf{M}_R$ (see Section 1.8).

Proof. Let $\mathbf{M}_R = [m_{ij}]$ and $\mathbf{M}_{R^2} = [n_{ij}]$. By definition, the i, jth element of

$\mathbf{M}_R \odot \mathbf{M}_R$ is equal to 1 if and only if row i of $\mathbf{M}_R$ and column j of $\mathbf{M}_R$ have a 1 in the same relative position, say position k. This means that $m_{ik} = 1$ and $m_{kj} = 1$ for some k, $1 \le k \le n$. By definition of the matrix $\mathbf{M}_R$, the conditions above mean that $a_i \, R \, a_k$ and $a_k \, R \, a_j$. Thus $a_i \, R^2 \, a_j$, so $n_{ij} = 1$. We have therefore shown that position i, j of $\mathbf{M}_R \odot \mathbf{M}_R$ is equal to 1 if and only if $n_{ij} = 1$. This means that $\mathbf{M}_R \odot \mathbf{M}_R = \mathbf{M}_{R^2}$.

For brevity, we will usually denote $\mathbf{M}_R \odot \mathbf{M}_R$ simply as $(\mathbf{M}_R)^2_{\odot}$ (the symbol $\odot$ reminds us that this is not the usual matrix product).

Example 6   Let A and R be as in Example 5. Then

$$\mathbf{M}_R = \begin{bmatrix} 1 & 1 & 0 & 0 & 0 \\ 0 & 0 & 1 & 0 & 0 \\ 0 & 0 & 0 & 1 & 1 \\ 0 & 0 & 0 & 0 & 1 \\ 0 & 0 & 0 & 0 & 0 \end{bmatrix}$$

From the discussion above,

$$\mathbf{M}_{R^2} = \mathbf{M}_R \odot \mathbf{M}_R = \begin{bmatrix} 1 & 1 & 0 & 0 & 0 \\ 0 & 0 & 1 & 0 & 0 \\ 0 & 0 & 0 & 1 & 1 \\ 0 & 0 & 0 & 0 & 1 \\ 0 & 0 & 0 & 0 & 0 \end{bmatrix} \odot \begin{bmatrix} 1 & 1 & 0 & 0 & 0 \\ 0 & 0 & 1 & 0 & 0 \\ 0 & 0 & 0 & 1 & 1 \\ 0 & 0 & 0 & 0 & 1 \\ 0 & 0 & 0 & 0 & 0 \end{bmatrix}$$

$$= \begin{bmatrix} 1 & 1 & 1 & 0 & 0 \\ 0 & 0 & 0 & 1 & 1 \\ 0 & 0 & 0 & 0 & 1 \\ 0 & 0 & 0 & 0 & 0 \\ 0 & 0 & 0 & 0 & 0 \end{bmatrix}$$

Computing $\mathbf{M}_{R^2}$ directly from R^2, we obtain the same result.

We can see from Examples 5 and 6 that it is often easier to compute R^2 by computing $\mathbf{M}_{R^2} = \mathbf{M}_R \odot \mathbf{M}_R$ instead of searching the digraph of R for all vertices that can be joined by a path of length 2. Similarly, we can show that $\mathbf{M}_{R^3} = \mathbf{M}_R \odot (\mathbf{M}_R \odot \mathbf{M}_R) = (\mathbf{M}_R)^3_{\odot}$. In fact, the following result shows how these results can be generalized.

Theorem 2   For $n \ge 2$, and R a relation on a finite set A, we have

$$\mathbf{M}_{R^n} = \mathbf{M}_R \odot \mathbf{M}_R \odot \cdots \odot \mathbf{M}_R \quad (n \text{ factors})$$

We omit the proof.

Let $\pi_1 = a, x_1, x_2, \ldots, x_{n-1}, b$ be a path in a relation R of length n from a to b, and let $\pi_2 = b, y_1, y_2, \ldots, y_{m-1}, c$ be a path in R of length m from b to c. Then the **composition of** π_1 **and** π_2 is the path $a, x_1, x_2, \ldots, b, y_1, y_2, \ldots, y_{m-1}, c$ of length $n + m$, which is denoted by $\pi_2 \circ \pi_1$. This is a path from a to c.

Example 7   Consider the relation whose digraph is given in Fig. 5, and the paths

$$\pi_1 = 1, 2, 3 \qquad \text{and} \qquad \pi_2 = 3, 5, 6, 2, 4$$

Then the composition of π_1 and π_2 is the path

$$\pi_2 \circ \pi_1 = 1, 2, 3, 5, 6, 2, 4$$

from 1 to 4 of length 6.

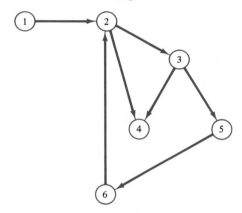

Figure 5

EXERCISE SET 4.4

For Exercises 1–11, let R be the relation whose digraph is given in Fig. 6.

1. List all paths of length 1.
2. List all paths of length 2 starting from vertex 2.
3. List all paths of length 3 starting from vertex 3.
4. Find a cycle starting at vertex 2.
5. Find a cycle starting at vertex 6.
6. List all paths of length 2.
7. List all paths of length 3.
8. Draw the digraph of R^2.
9. Find $\mathbf{M}_{R^2}$.
10. Find R^∞.
11. Find $\mathbf{M}_{R^\infty}$.

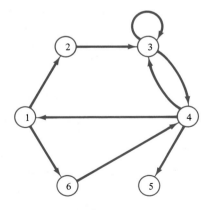

Figure 6

For Exercises 12–22, let R be the relation whose digraph is given in Fig. 7.

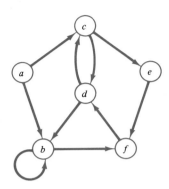

Figure 7

12. List all paths of length 1.
13. List all paths of length 2 starting from vertex c.
14. List all paths of length 3 starting from vertex a.
15. Find a cycle starting at vertex c.
16. Find a cycle starting at vertex d.
17. Find all paths of length 2.
18. Find all paths of length 3.

19. Draw the digraph of R^2.
20. Find $\mathbf{M}_{R^2}$.
21. Find R^∞.
22. Find $\mathbf{M}_{R^\infty}$.

In Exercises 23 and 24, let R be the relation whose digraph is given in Fig. 8.

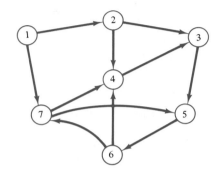

Figure 8

23. If $\pi_1 = 1, 2, 4, 3$ and $\pi_2 = 3, 5, 6, 4$, find the composition $\pi_2 \circ \pi_1$.
24. If $\pi_1 = 1, 7, 5$ and $\pi_2 = 5, 6, 7, 4, 3$, find the composition $\pi_2 \circ \pi_1$.

KEY IDEAS FOR REVIEW

☐ $A \times B$ (product set or Cartesian product): $\{(a, b) \,|\, a \in A$ and $b \in B\}$.
☐ $|A \times B| = |A| \, |B|$.
☐ Partition or quotient set: see page 000.
☐ Relation from A to B: subset of $A \times B$.
☐ Domain and range of a relation: see pages 000 and 000.
☐ Relative sets $R(a)$, a in A, and $R(A_1)$, A_1 a subset of A, see page 000.
☐ Matrix of a relation: see page 000.
☐ Digraph of a relation: pictorial representation of a relation: see page 000.
☐ Path of length n from a to b in a relation R: finite sequence $a, x_1, x_2, \ldots, x_{n-1}, b$ such that $a \, R \, x_1, x_1 \, R \, x_2, \ldots, x_{n-1} \, R \, b$.
☐ $x \, R^n \, y$ (R a relation on A): There is a path of length n from x to y in R.
☐ $x \, R^\infty \, y$ (connectivity relation for R): Some path exists in R from x to y.
☐ $\mathbf{M}_{R^n} = \mathbf{M}_R \odot \mathbf{M}_R \odot \cdots \odot \mathbf{M}_R$: see page 94.

REVIEW EXERCISES

1. Let $A = \{2, 5\}$ and $B = \{0, 1, 4\}$.
   (a) List the elements in $A \times B$.
   (b) List the elements in $B \times B$.
   (c) List the elements in $A \times A$.
   (d) List the elements in $B \times A$.

2. List $A = \{\alpha, \beta, \gamma\}$ and $B = \{1, 2, 3, 4\}$.
   (a) List the elements in $A \times B$.
   (b) List the elements in $B \times B$.
   (c) List the elements in $A \times A$.
   (d) List the elements in $B \times A$.

3. If $A = \{1, 3, 5\}$, $B = \{e, f\}$, and $C = \{\alpha, \beta\}$, list all elements in $A \times B \times C$.

4. If $A = \{1, 2, 3, 4\}$ and $B = \{2, 5\}$, sketch the set $A \times B$ in the Cartesian plane.

5. Write the shaded set of points shown in this figure as a Cartesian product of two sets A and B.

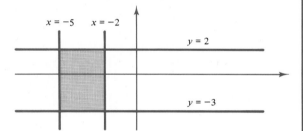

6. A computer dealer carries the following inventory:

   Computers: IBM PC, Compac, IBM XT

   Operating systems: DOS 2.0, DOS 1.0, DOS 3.0

   Printers: Epson, Okidata

   A configuration consists of a computer, an operating system, and a printer. How many configurations are possible?

7. Give four different partitions of $A = \{a, b, c, d, e, f, g\}$.

8. Give three different partitions of the set of all integers.

9. Let $A = \{a, b, c\}$ and $B = \{2, 3, 5, 7\}$. Let R be the relation from A to B defined by

$R = \{(a, 2), (a, 3), (a, 5), (b, 2), (c, 5), (c, 7)\}$

Find the domain, matrix, and range of R.

10. Let $A = \{1, 2, 3, 5\}$. Let R be the relation on A defined by $a\,R\,b$ if and only if $a \leq b - 2$. Find the domain, range, matrix, and digraph of R.

11. Let $A = B = \mathbb{R}$. Consider the following relation R on A: $a\,R\,b$ if and only if $3a - 2b = 6$.
    (a) Find $R(1)$.
    (b) If $A_1 = \{-2, 2, 3, 5\}$, find $R(A_1)$.

12. Let $A = \{1, 2, 3, 4, 5\}$ and $B = \{3, 4, 5, 6, 7\}$. Let R be the relation from A to B defined by $a\,R\,b$ if and only if $a < b$. Find $R(A_1)$ for each of the following sets.
    (a) $A_1 = \{1, 3, 4\}$
    (b) $A_1 = \{5\}$
    (c) $A_1 = \{4, 5\}$

13. Let $A = Z = B$ and let R be the relation from A to B defined by $a\,R\,b$ if and only if $2 \mid (a - b)$. Find (a) $R(2)$, (b) $R(3)$, and (c) $R(A_1)$, where A_1 is the set of all even integers.

14. Let $R = \{(a, a), (a, b), (b, a), (b, d), (c, d), (d, a), (d, c), (d, d)\}$ be a relation on the set $A = \{a, b, c, d\}$. Find
    (a) the matrix of R.
    (b) the digraph of R.

15. Let $A = \{a, b, c, d\}$. Find the relation R on A determined by the matrix

$$\mathbf{M}_R = \begin{bmatrix} 1 & 0 & 0 & 1 \\ 1 & 0 & 1 & 0 \\ 0 & 0 & 1 & 1 \\ 1 & 0 & 1 & 1 \end{bmatrix}$$

16. Find the relation determined by the following figure.

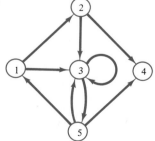

17. Compute the matrix $\mathbf{M}_R$ on the relation whose digraph is shown in

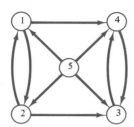

18. Compute the in-degree and out-degree for each vertex in the digraph shown in

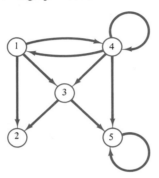

19. Let R be the relation whose digraph is shown in

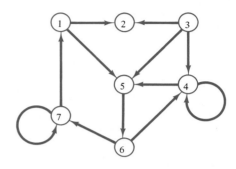

(a) List all paths of length 1.
(b) List all paths of length 2 starting from vertex 2.
(c) List all paths of length 3 starting from vertex 3.
(d) Find a cycle starting at vertex 5.
(e) Find a cycle starting at vertex 4.

CHAPTER TEST

1. If $A = \{a, b, c\}$ and $B = \{1, 2\}$ list all elements in $A \times B$.

2. Let $A = \{1, 2, 3, 4\}$ and $B = \{a, b, c\}$. Let R be the relation from A to B defined by

$R = \{(1, a), (1, b), (1, c), (2, a), (3, b), (4, a), (4, c)\}$

Find the domain, matrix, and range of R.

3. Let $A = \{1, 2, 3, 4, 5\}$ and $B = \{4, 5, 6, 7, 8\}$. Let R be the relation from A to B defined by $a\ R\ b$ if and only if $b = a + 1$. Find (a) $R(3)$, (b) $R(2)$, and (c) $R(A_1)$, where $A_1 = \{3, 4, 5\}$.

4. Find the digraph of the relation on the set $A = \{a, b, c, d\}$ whose matrix is

$$\begin{bmatrix} 1 & 0 & 0 & 1 & 0 \\ 0 & 1 & 1 & 1 & 0 \\ 1 & 0 & 1 & 1 & 0 \\ 0 & 0 & 1 & 1 & 1 \\ 1 & 0 & 1 & 0 & 1 \end{bmatrix}$$

5. Let R be the relation whose digraph is shown below

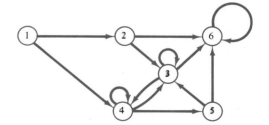

(a) Find a path of length 2.

(b) Find a cycle starting at vertex 4.

(c) Find R^∞.

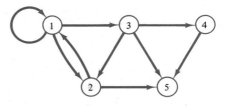

6. Answer each of the following as true (T) or false (F).

(a) If A has five elements and $A \times B$ has 30 elements, then B has 150 elements.

(b) $\{\{a, b, c\}, \{c, d, e\}, \{f, g\}\}$ is a partition of the set $\{a, b, c, d, e, f, g\}$

(c) Let R be a relation from A to B. If A_1 and A_2 are subsets of A and $A_1 \subseteq A_2$, then $R(A_2) \subseteq R(A_1)$.

(d) The in-degree of vertex 1 in the following digraph is 2.

(e) If R is the relation on $A = \{1, 2, 3, 4\}$ whose digraph is shown below

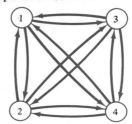

then $R^\infty = \{(1, 1),(2, 2),(3, 3), (4, 4)\}$.

5

Properties of Relations

Prerequisites: Chapters 1, 3, and 4.

In this chapter we study several properties that a relation can possess, and investigate some of the relationships between these properties. We also learn to combine relations in various algebraic ways, and we discuss the basic features of this matrix algebra.

5.1 Some Fundamental Properties

In many of the applications of relations to computer science and applied mathematics, we deal with relations on a set A rather than relations from A to B. Moreover, these relations often satisfy certain properties that will be discussed in this section.

Reflexive and Irreflexive Relations

A relation R on a set A is **reflexive** if $(a, a) \in R$ for every $a \in A$, that is, if $a\,R\,a$ for all $a \in A$. A relation R on a set A is **irreflexive** if $a\,\not{R}\,a$ for every $a \in A$.

Thus R is reflexive if every element $a \in A$ is related to itself and it is irreflexive if no element is related to itself.

Example 1   (a) Let $\Delta = \{(a, a) \,|\, a \in A\}$, so that Δ is the relation of **equality** on the set A. Then Δ is reflexive, since $(a, a) \in \Delta$ for all $a \in A$.

(b) Let $R = \{(a, b) \in A \times A \,|\, a \neq b\}$, so that R is the relation of **inequality** on the set A. Then R is irreflexive, since $(a, a) \notin R$ for all $a \in A$.

(c) Let $A = \{1, 2, 3\}$, and let $R = \{(1, 1), (1, 2)\}$. Then R is not reflexive since $(2, 2) \notin R$ and $(3, 3) \notin R$. Also, R is not irreflexive, since $(1, 1) \in R$.

(d) Let A be a nonempty set. Let $R = \varnothing$, the empty subset of $A \times A$, which is called the **empty relation**. Then R is not reflexive, since $(a, a) \notin R$ for all $a \in A$ (the empty set has no elements). However, R is irreflexive.

We can characterize a reflexive or irreflexive relation by its matrix as follows. The matrix of a reflexive relation must have all 1's on its main diagonal, while the matrix of an irreflexive relation must have all 0's on its main diagonal.

Similarly, we can characterize a reflexive or irreflexive relation by its digraph as follows. A reflexive relation has a cycle of length 1 at every vertex, while an irreflexive relation has no cycles of length 1. Another useful way of saying the same thing uses the equality relation Δ on a set A : R is reflexive if and only if $\Delta \subseteq R$, and R is irreflexive if and only if $\Delta \cap R = \varnothing$.

Finally, we may note that if R is reflexive on a set A, then Dom $(R) = $ Ran $(R) = A$.

Symmetric, Asymmetric, and Antisymmetric Relations

A relation R on a set A is **symmetric** if whenever $a \; R \; b$, then $b \; R \; a$. It then follows that R is not symmetric if we have some a and $b \in A$ with $a \; R \; b$, but $b \; \cancel{R} \; a$. A relation R on a set A is **asymmetric** if whenever $a \; R \; b$, then $b \; \cancel{R} \; a$. It then follows that R is not asymmetric if we have some a and $b \in A$ with both $a \; R \; b$ and $b \; R \; a$. A relation R on a set A is **antisymmetric** if whenever $a \; R \; b$ and $b \; R \; a$, then $a = b$. A restatement of this definition is that R is antisymmetric if whenever $a \neq b$, we have $a \; \cancel{R} \; b$ or $b \; \cancel{R} \; a$. It then follows that R is not antisymmetric if we have a and b in A, $a \neq b$, and both $a \; R \; b$ and $b \; R \; a$.

Given a relation R, we shall want to determine which properties hold for R. Keep in mind the following remarks. A property fails to hold in general if we can find one situation where the property does not hold. If there is no situation where the property fails to hold, we must conclude that the property holds at all times.

Example 2   Let $A = Z$, the set of integers and let

$$R = \{(a, b) \in A \times A \,|\, a < b\}$$

so that R is the relation **less than**. Is R symmetric, asymmetric, or antisymmetric?

Solution.

Symmetry: If $a < b$, then it is not true that $b < a$, so R is not symmetric.
Asymmetry: If $a < b$, then $b \not< a$ (b is not less than a), so R is asymmetric.
Antisymmetry: If $a \neq b$, then either $a \not< b$ or $b \not< a$, so R is antisymmetric.

Example 3   Let A be a set of people that let

$$R = \{(x, y) \in A \times A \,|\, x \text{ is a cousin of } y\}$$

Then R is a symmetric relation (verify).

Example 4   Let $A = \{1, 2, 3, 4\}$ and let

$$R = \{(1, 2), (2, 2), (3, 4), (4, 1)\}$$

Then R is not symmetric, since $(1, 2) \in R$ but $(2, 1) \notin R$. Also, R is not asymmetric, since $(2, 2) \in R$. Finally, R is antisymmetric, since if $a \neq b$, either $(a, b) \notin R$ or $(b, a) \notin R$.

Example 5   Let $A = Z^+$, the set of positive integers, and let

$$R = \{(a, b) \in A \times A \,|\, a \text{ divides } b\}$$

Is R symmetric, asymmetric, or antisymmetric?

Solution.

If $a \,|\, b$, it does not follow that $b \,|\, a$, so R is not symmetric.
If $a = b = 3$, say, then $a \, R \, b$ and $b \, R \, a$, so R is not asymmetric.
If $a \,|\, b$ and $b \,|\, a$, then $a = b$, so R is antisymmetric.

We now characterize symmetric, asymmetric, or antisymmetric properties of a relation by properties of its matrix. The matrix $\mathbf{M}_R = [m_{ij}]$ of a symmetric relation satisfies the property that

$$\text{if } m_{ij} = 1, \quad \text{then } m_{ji} = 1$$

Moreover, if $m_{ji} = 0$, then $m_{ij} = 0$. Thus $\mathbf{M}_R$ is a matrix such that each pair of locations, symmetrically placed about the main diagonal, are either both 0 or both 1. It follows that $\mathbf{M}_R = \mathbf{M}_R^T$, so that $\mathbf{M}_R$ is a symmetric matrix (see Section 3.3).

The matrix $\mathbf{M}_R = [m_{ij}]$ of an asymmetric relation R satisfies the property that

$$\text{if } m_{ij} = 1, \quad \text{then} \quad m_{ji} = 0.$$

If R is asymmetric, it follows that $m_{ii} = 0$ for all i; that is, the main diagonal of the matrix $\mathbf{M}_R$ consists entirely of 0's. This must be true since the asymmetric property implies that if $m_{ii} = 1$, then $m_{ii} = 0$, which is a contradiction.

Finally, the matrix $\mathbf{M}_R = [m_{ij}]$ of an antisymmetric relation R satisfies the property that if $i \neq j$, then $m_{ij} = 0$ or $m_{ji} = 0$.

Example 6   Consider the matrices in Fig. 1, each of which is the matrix of a relation, as indicated.

Relations R_1 and R_2 are symmetric since the matrices $\mathbf{M}_{R_1}$ and $\mathbf{M}_{R_2}$ are symmetric matrices. Relation R_3 is antisymmetric, since no symmetrically situated, off-diagonal positions of $\mathbf{M}_{R_3}$ can both contain 1's. Such positions may both have 0's, however, and the diagonal elements are unrestricted. The relation R_3 is not asymmetric because $\mathbf{M}_{R_3}$ has 1's on the main diagonal.

Relation R_4 has none of the three properties: $\mathbf{M}_{R_4}$ is not symmetric. The presence of the 1's in positions 4, 1 and 1, 4 of $\mathbf{M}_{R_4}$ violates both asymmetry and antisymmetry.

Finally, R_5 is antisymmetric but not asymmetric, and R_6 is both asymmetric and antisymmetric.

We now consider the digraphs of these three types of relations. If R is an asymmetric relation, then the digraph of R cannot simultaneously have an edge from vertex i to vertex j, and an edge from vertex j to vertex i. This is true for any i and j, and in particular if i equals j. Thus there can be no cycles of length 1, and all paths are "one-way streets."

$$\begin{bmatrix} 1 & 0 & 1 \\ 0 & 0 & 1 \\ 1 & 1 & 1 \end{bmatrix} = M_{R_1}$$

(a)

$$\begin{bmatrix} 0 & 1 & 1 & 0 \\ 1 & 1 & 0 & 0 \\ 1 & 0 & 1 & 1 \\ 0 & 0 & 1 & 1 \end{bmatrix} = M_{R_2}$$

(b)

$$\begin{bmatrix} 1 & 1 & 1 \\ 0 & 1 & 0 \\ 0 & 0 & 0 \end{bmatrix} = M_{R_3}$$

(c)

$$\begin{bmatrix} 0 & 0 & 1 & 1 \\ 0 & 0 & 1 & 0 \\ 0 & 0 & 0 & 1 \\ 1 & 0 & 0 & 0 \end{bmatrix} = M_{R_4}$$

(d)

$$\begin{bmatrix} 1 & 0 & 0 & 1 \\ 0 & 1 & 1 & 1 \\ 0 & 0 & 1 & 0 \\ 0 & 0 & 0 & 1 \end{bmatrix} = M_{R_5}$$

(e)

$$\begin{bmatrix} 0 & 1 & 1 & 1 \\ 0 & 0 & 1 & 0 \\ 0 & 0 & 0 & 1 \\ 0 & 0 & 0 & 0 \end{bmatrix} = M_{R_6}$$

(f)

Figure 1

If R is an antisymmetric relation, then for different vertices i and j, there cannot be an edge from vertex i to vertex j, and an edge from vertex j to vertex i. When $i = j$, no condition is imposed. Thus there may be cycles of length 1, but again all paths are "one way."

We consider the digraphs of symmetric relations in more detail.

The digraph of a symmetric relation R has the property that if there is an edge from vertex i to vertex j, then there is an edge from vertex j to vertex i. Thus if two vertices are connected by an edge, they must always be connected in both directions. Because of this, it is possible, and quite useful, to give a different representation of a symmetric relation. We keep the vertices as they appear in the digraph, but if two vertices a and b are connected by edges in each direction, we replace these two edges with one undirected edge, or a "two-way street." This undirected edge is just a single line without arrows, and connects a and b. The resulting diagram will be called the **graph** of the symmetric relation.

Example 7   Let $A = \{a, b, c, d, e\}$ and let R be the symmetric relation given by

$$R = \{(a, b), (b, a), (a, c), (c, a), (b, c), (c, b), (b, e), (e, b), (e, d), (d, e),$$
$$(c, d), (d, c)\}$$

The usual digraph of R is shown in Fig. 2(a), while Fig. 2(b) shows the graph of R. Note that each undirected edge corresponds to two ordered pairs in the relation R.

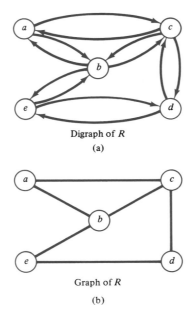

Digraph of R
(a)

Graph of R

(b)   **Figure 2**

An undirected edge between a and b, in the graph of a symmetric relation R, corresponds to a set $\{a, b\}$ such that $(a, b) \in R$ and $(b, a) \in R$. Sometimes we will also refer to such a set $\{a, b\}$ as an **undirected edge** of the relation R.

A symmetric relation R on a set A will be called **connected** if there is a path from any element of A to any other element of A. This simply means that the graph of R is all in one piece. In Fig. 3 we show the graphs of two symmetric relations. The graph in Fig. 3(a) is connected, whereas that in Fig. 3(b) is not connected.

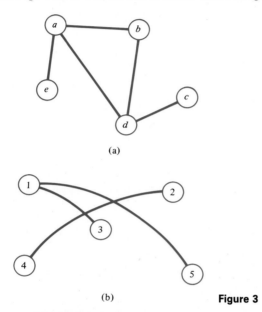

(a)

(b)   **Figure 3**

Transitive Relations

We say that a relation R on a set A is **transitive** if whenever $a\ R\ b$ and $b\ R\ c$, then $a\ R\ c$. It is often convenient to say what it means for a relation to be not transitive. A relation R on A is not transitive if there exist a, b, and c in A so that $a\ R\ b$ and $b\ R\ c$, but $a\ \not{R}\ c$. If such a, b, and c do not exist, then R is transitive.

Example 8   Let $A = Z$, the set of integers, and let R be the relation considered in Example 2. To see whether R is transitive, we assume that $a\ R\ b$ and $b\ R\ c$. Thus $a < b$ and $b < c$. It then follows that $a < c$, so $a\ R\ c$. Hence R is transitive.

Example 9   Let $A = Z^+$ and let R be the relation considered in Example 5. Is R transitive?

Solution. Suppose that $a\ R\ b$ and $b\ R\ c$, so that $a\,|\,b$ and $b\,|\,c$. It then follows that $a\,|\,c$. [See Theorem 2(c) of Section 3.2.] Thus R is transitive.

Example 10 Let $A = \{1, 2, 3, 4\}$ and let

$$R = \{(1, 2), (1, 3), (4, 2)\}$$

Is R transitive?

Solution. Since we cannot find elements a, b, and c in A such that $a \, R \, b$ and $b \, R \, c$, but $a \, \not{R} \, c$, we conclude that R is transitive.

We can characterize a transitive relation by its matrix $\mathbf{M}_R = [m_{ij}]$ as follows:

$$\text{if } m_{ij} = 1 \text{ and } m_{jk} = 1, \quad \text{then } m_{ik} = 1$$

The left-hand side of the statement above simply means that $(\mathbf{M}_R)^2_\odot$ has a 1 in position i, k. Thus the transitivity of R means that if $(\mathbf{M}_R)^2_\odot$ has a 1 in any position, then $\mathbf{M}_R$ must have a 1 in the same position. In other words, each element of $(\mathbf{M}_R)^2_\odot$ is less than or equal to the corresponding element of $\mathbf{M}_R$, using the usual meaning of $\leq$ for 0 and 1. When this condition is true, we write $(\mathbf{M}_R)^2_\odot \leq \mathbf{M}_R$. In particular, if $(\mathbf{M}_R)^2_\odot = \mathbf{M}_R$ then R is transitive.

Example 11 Let $A = \{1, 2, 3\}$ and let R be the relation on A whose matrix is

$$\mathbf{M}_R = \begin{bmatrix} 1 & 1 & 1 \\ 0 & 0 & 1 \\ 0 & 0 & 1 \end{bmatrix}$$

Show that R is transitive.

Solution. By direct computation, $(\mathbf{M}_R)^2_\odot = \mathbf{M}_R$; therefore, R is transitive.

To see what transitivity means for the digraph of a relation, we translate the definition of transitivity into geometric terms.

If we consider particular vertices a and c of the digraph of R, the conditions $a \, R \, b$ and $b \, R \, c$ mean that there is a path of length 2 in R from a to c. In other words, $a \, R^2 \, c$. Therefore, we may rephrase the definition of transitivity as follows: If $a \, R^2 \, c$, then $a \, R \, c$; that is, $R^2 \subseteq R$ (as subsets of $A \times A$). In other words, if a and c are connected by a path of length 2 in R, then they must be connected by a path of length 1.

We can slightly generalize the foregoing geometric characterization of transitivity as follows.

Theorem 1 A relation R is transitive if and only if it satisfies the following property: If there is a path of length greater than 1 from vertex a to vertex b, then there is a path of length 1 from a to b (that is, a is related to b). Algebraically stated, R is transitive if and only if $R^n \subseteq R$ for all $n \geq 1$.

Proof. The proof is left as an exercise.

It will be convenient to have a restatement of some of the properties above in terms of relative sets. We list these statements below without proof.

Theorem 2   Let R be a relation on a set A. Then

(a) Reflexivity of R means that $a \in R(a)$ for all a in A.
(b) Symmetry of R means that $a \in R(b)$ if and only if $b \in R(a)$.
(c) Transitivity of R means that if $b \in R(a)$ and $c \in R(b)$, then $c \in R(a)$.

EXERCISE SET 5.1

In Exercises 1–8, let $A = \{1, 2, 3, 4\}$. Determine whether the relation is reflexive, irreflexive, symmetric, asymmetric, antisymmetric, or transitive.

1. $R = \{(1, 1), (1, 2), (2, 1), (2, 2), (3, 3), (3, 4), (4, 3), (4, 4)\}$
2. $R = \{(1, 2), (1, 3), (1, 4), (2, 3), (2, 4), (3, 4)\}$
3. $R = \{(1, 3), (1, 1), (3, 1), (1, 2), (3, 3), (4, 4)\}$
4. $R = \{(1, 1), (2, 2), (3, 3)\}$
5. $R = \emptyset$
6. $R = A \times A$
7. $R = \{(1, 2), (1, 3), (3, 1), (1, 1), (3, 3), (3, 2), (1, 4), (4, 2), (3, 4)\}$
8. $R = \{(1, 3), (4, 2), (2, 4), (3, 1), (2, 2)\}$

In Exercises 9 and 10, let $A = \{1, 2, 3, 4, 5\}$. Determine whether the relation R whose digraph is given is reflexive, irreflexive, symmetric, asymmetric, antisymmetric, or transitive.

9.

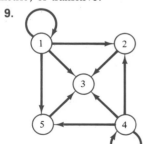

10.

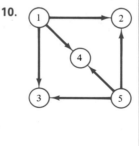

In Exercises 11 and 12, let $A = \{1, 2, 3, 4\}$. Determine whether the relation R whose matrix $\mathbf{M}_R$ is given is reflexive, irreflexive, symmetric, asymmetric, antisymmetric, or transitive.

11. $\begin{bmatrix} 0 & 1 & 0 & 1 \\ 1 & 0 & 1 & 1 \\ 0 & 1 & 0 & 0 \\ 1 & 1 & 0 & 0 \end{bmatrix}$ 12. $\begin{bmatrix} 1 & 1 & 0 & 0 \\ 1 & 1 & 0 & 0 \\ 0 & 0 & 1 & 0 \\ 0 & 0 & 0 & 1 \end{bmatrix}$

In Exercises 13–20, determine whether the relation R on the set A is reflexive, irreflexive, symmetric, asymmetric, antisymmetric, or transitive.

13. $A = Z$; $a R b$ if and only if $a \le b + 1$.
14. $A = Z$; $a R b$ if and only if $|a - b| \le 2$.
15. $A = Z$; $a R b$ if and only if $a = b^k$ for some $k \in Z^+$.
16. $A = Z$; $a R b$ if and only if $a + b$ is even.
17. $A = Z$; $a R b$ if and only if $|a - b| = 2$.
18. $A = $ the set of all real numbers; $a R b$ if and only if $a^2 + b^2 = 4$.
19. $A = Z^+$; $a R b$ if and only if GCD $(a, b) = 1$, that is, if and only if 1 is the only factor that a and b have in common. In this case, we say that a and b are **relatively prime.**
20. $A = $ the set of all ordered pairs of real numbers; $(a, b) R (c, d)$ if and only if $a = c$.
21. Let R be the following symmetric relation on the set $A = \{1, 2, 3, 4, 5\}$.

$$R = \{(1, 2), (2, 1), (3, 4), (4, 3), (3, 5), (5, 3),$$
$$(4, 5), (5, 4), (5, 5)\}$$

Draw the graph of R.

22. Let $A = \{a, b, c, d\}$ and let R be the symmetric relation

$$R = \{(a, b), (b, a), (a, c), (c, a), (a, d), (d, a)\}$$

Draw the graph of R.

23. Consider the graph of a symmetric relation R on $A = \{1, 2, 3, 4, 5, 6, 7\}$ shown in Fig. 6. Determine R (list all pairs).

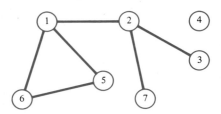

Figure 6

24. Consider the graph of a symmetric relation R on $A = \{a, b, c, d, e\}$ shown in Fig. 7. Determine R (list all pairs).

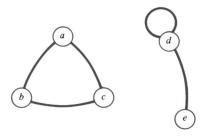

Figure 7

THEORETICAL EXERCISES

T1. Show that if a relation on a set A is transitive and irreflexive, then it is asymmetric.

T2. Prove by induction that if a relation R on a set A is symmetric, then R^n is symmetric, for $n \geq 1$.

T3. Let R be a nonempty relation on a set A. Suppose R is symmetric and transitive. Show that R is not irreflexive.

5.2 Equivalence Relations

A relation R on a set A is called an **equivalence relation** if it is reflexive, symmetric, and transitive.

Example 1   Let A be the set of all triangles in the plane and let R be the relation on A defined as follows.

$$R = \{(a, b) \in A \times A \,|\, a \text{ is congruent to } b\}$$

It is easy to see that R is an equivalence relation.

Example 2   Let $A = \{1, 2, 3, 4\}$ and let

$$R = \{(1, 1), (1, 2), (2, 1), (2, 2), (3, 4), (4, 3), (3, 3), (4, 4)\}$$

It is easy to verify that R is an equivalence relation.

Example 3   Let $A = Z$, the set of integers, and let

$$R = \{(a, b) \in A \times A \,|\, a \leq b\}$$

Is R an equivalence relation?

Solution. Since $a \leq a$, R is reflexive. If $a \leq b$, it need not follow that $b \leq a$, so R is not symmetric. Incidentally, R is transitive, since $a \leq b$ and $b \leq c$ imply that $a \leq c$. We see that R is not an equivalence relation.

Example 4   Let $A = Z$ and let

$$R = \{(a, b) \in A \times A \,|\, 2 \text{ divides } a - b\}$$

We write $a\,R\,b$ as

$$a \equiv b \pmod 2$$

read "a is congruent to b mod 2." Show that congruence mod 2 is an equivalence relation.

Solution. First,

$$a \equiv a \pmod 2$$

since $2\,|\,(a - a)$.

Second, if $a \equiv b \pmod 2$, then $2\,|\,(a - b)$, or $a - b = 2k$ for some $k \in Z$. Then

$$b - a = 2(-k)$$

so $2\,|\,(b - a)$ and $b \equiv a \pmod 2$.

Finally, suppose that $a \equiv b \pmod 2$ and $b \equiv c \pmod 2$. Then $2\,|\,(a - b)$, so

$$a - b = 2k_1 \tag{1}$$

where $k_1 \in Z$. Also, $2 \mid (b - c)$, so

$$b - c = 2k_2 \tag{2}$$

where $k_2 \in Z$. Adding equations (1) and (2), we obtain

$$a - c = 2(k_1 + k_2)$$

so $2 \mid (a - c)$, which means that

$$a \equiv c \pmod{2}$$

Hence congruence mod 2 is an equivalence relation.

Example 5   Let $A = Z$ and let $n \in Z^+$. We generalize the relation defined in Example 4 as follows. Let

$$R = \{(a, b) \in A \times A \mid n \text{ divides } a - b\}$$

and we write $a \, R \, b$ as

$$a \equiv b \pmod{n}$$

read "a is congruent to b mod n." Proceeding exactly as in Example 4, we can show that congruence mod n is an equivalence relation.

Equivalence Relations and Partitions

The following result shows that if $\mathcal{P}$ is a partition of a set A (see Section 4.1), then $\mathcal{P}$ can be used to construct an equivalence relation on A.

Theorem 1   Let $\mathcal{P}$ be a partition of a set A. Recall that the sets in $\mathcal{P}$ are called the "blocks" of $\mathcal{P}$. Define the relation R on A as follows:

$a \, R \, b$ if and only if a and b are members of the same block

Then R is an equivalence relation on A.

Proof.

(a) If $a \in A$, then clearly a is in the same block as itself, so $a \, R \, a$.

(b) If $a \, R \, b$, then a and b are in the same block, so $b \, R \, a$.

(c) If $a \, R \, b$ and $b \, R \, c$, then a, b, and c must all lie in the same block of $\mathcal{P}$. Thus $a \, R \, c$.

Since R is reflexive, symmetric, and transitive, R is an equivalence relation. R will be called the **equivalence relation determined by $\mathcal{P}$**.

Example 6   Let $A = \{1, 2, 3, 4\}$, and consider the partition $\mathscr{P} = \{\{1, 2, 3\}, \{4\}\}$ of A. Find the equivalence relation R of A, determined by $\mathscr{P}$.

Solution. The blocks of $\mathscr{P}$ are $\{1, 2, 3\}$ and $\{4\}$. Each element in a block is related to every other element in the same block, and only to those elements. Thus in this case

$$R = \{(1, 1), (1, 2), (1, 3), (2, 1), (2, 2), (2, 3), (3, 1), (3, 2), (3, 3), (4, 4)\}$$

If $\mathscr{P}$ is a partition of A and R is the equivalence relation determined by $\mathscr{P}$, then the blocks of $\mathscr{P}$ can easily be described in terms of R. If A_1 is a block of $\mathscr{P}$ and $a \in A_1$, we see by definition that A_1 consists of all elements x of A with $a \mathrel{R} x$. That is, $A_1 = R(a)$. Thus the partition $\mathscr{P} = \{R(a) \mid a \in A\}$. In words, $\mathscr{P}$ consists of all distinct R-relative sets that arise from elements of A. For instance, in Example 6, the blocks $\{1, 2, 3\}$ and $\{4\}$ can be described, respectively, as $R(1)$ and $R(4)$. Of course, $\{1, 2, 3\}$ could also be described as $R(2)$ or $R(3)$, so this way of representing the blocks is not unique.

The foregoing construction of equivalence relations from partitions is quite simple. One might be tempted to believe that few equivalence relations could be produced in this way. The fact is, as we will now show, that all equivalence relations on A can be produced from partitions.

We begin with the following result. Since its proof uses Theorem 2 of Section 5.1, the reader might first want to review that theorem.

Lemma 1[†]   Let R be an equivalence relation on a set A, and let $a \in A$ and $b \in A$. Then

$$a \mathrel{R} b \quad \text{if and only if} \quad R(a) = R(b)$$

Proof. First suppose that $R(a) = R(b)$. Since R is reflexive, $b \in R(b)$; therefore, $b \in R(a)$, so $a \mathrel{R} b$.

Conversely, suppose that $a \mathrel{R} b$. Then note that

(1) $b \in R(a)$ by definition; therefore, since R is symmetric

(2) $a \in R(b)$ by Theorem 2(b) of Section 5.1

We must show that $R(a) = R(b)$. First let $x \in R(b)$. Since R is transitive, the fact that $x \in R(b)$, together with (1) above imply by Theorem 2(c) of Section 5.1 that $x \in R(a)$. Thus $R(b) \subseteq R(a)$. Now let $x \in R(a)$. This fact and (2) above imply, as before, that $x \in R(b)$. Thus $R(b) \subseteq R(a)$, so we must have $R(a) = R(b)$.

We now prove our main result.

[†]A lemma is a theorem whose main purpose is to aid in proving some other theorem.

Theorem 2   Let R be an equivalence relation on A, and let $\mathscr{P}$ be the collection of all distinct relative sets $R(a)$, for a in A. Then $\mathscr{P}$ is a partition of A, and R is the equivalence relation determined by $\mathscr{P}$.

Proof. According to the definition of a partition, we must show the following two properties:

(a) Every element of A belongs to some relative set.

(b) If $R(a)$ and $R(b)$ are not identical, then $R(a) \cap R(b) = \varnothing$.

Now part (a) is true, since $a \in R(a)$ by reflexivity of R. To show part (b) we prove the following equivalent statement:

$$\text{If } R(a) \cap R(b) \neq \varnothing \qquad \text{then } R(a) = R(b)$$

To prove this, we assume that $c \in R(a) \cap R(b)$. Then $a \, R \, c$ and $b \, R \, c$.

Since R is symmetric, we have $c \, R \, b$. Then $a \, R \, c$ and $c \, R \, b$, so by transitivity of R, $a \, R \, b$. Lemma 1 then tells us that $R(a) = R(b)$. We have now proved that $\mathscr{P}$ is a partition. By Lemma 1, we see that $a \, R \, b$ if and only if a and b belong to the same block of $\mathscr{P}$. Thus $\mathscr{P}$ determines R, and the theorem is proved.

If R is an equivalence relation on A, then the sets $R(a)$ are traditionally called **equivalence classes** of R. Many authors denote the equivalence class $R(a)$ by $[a]$. The partition $\mathscr{P}$ constructed in Theorem 2 therefore consists of all equivalence classes of R, and this partition will be denoted by A/R. Recall that partitions of A are also called **quotient sets** of A, and the notation A/R reminds us that $\mathscr{P}$ is the quotient set of A that is constructed from, and determines, R.

Example 7   Let R be the relation defined in Example 2. Determine A/R.

Solution. From Example 2 we have $R(1) = \{1, 2\} = R(2)$. Also $R(3) = \{3, 4\} = R(4)$. Hence $A/R = \{\{1, 2\}, \{3, 4\}\}$.

Example 8   Let R be the equivalence relation defined in Example 4. Determine A/R.

Solution. First,

$$R(0) = \{\ldots, -6, -4, -2, 0, 2, 4, 6, 8, \ldots\}$$

the set of even integers, since 2 divides the difference between two even integers.

$$R(1) = \{\ldots, -5, -3, -1, 1, 3, 5, 7, \ldots\}$$

the set of odd integers, since 2 divides the difference between two odd integers. Hence A/R consists of the set of even integers and the set of odd integers.

From Examples 7 and 8 we can extract a general procedure for determining partitions A/R. The procedure is as follows:

> **To compute the partition A/R**
>
> *Step 1.* Choose any element of A and compute the equivalence class $R(a)$.
>
> *Step 2.* If $R(a) \neq A$, choose an element b, not included in $R(a)$, and compute the equivalence class $R(b)$.
>
> *Step 3.* If A is not the union of previously computed equivalence classes, then choose an element x of A that is not in any of those equivalence classes and compute $R(x)$.
>
> *Step 4.* Repeat step 3 until all elements of A are included in the computed equivalence classes. If A is infinite, this process could continue indefinitely. In that case continue until a pattern emerges that allows us to describe or give a formula for all equivalence classes.

EXERCISE SET 5.2

In Exercises 1 and 2, let $A = \{a, b, c\}$. Determine whether the relation R whose matrix $\mathbf{M}_R$ is given is an equivalence relation.

1. $\mathbf{M}_R = \begin{bmatrix} 1 & 0 & 0 \\ 0 & 1 & 1 \\ 0 & 1 & 1 \end{bmatrix}$ 2. $\mathbf{M}_R = \begin{bmatrix} 1 & 0 & 1 \\ 0 & 1 & 0 \\ 0 & 0 & 1 \end{bmatrix}$

In Exercises 3 and 4, determine whether the relation R whose digraph is given is an equivalence relation.

3.

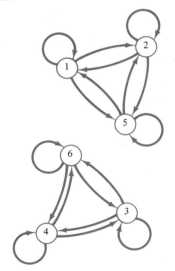

Figure 1

4.

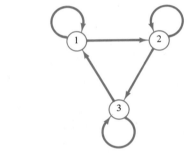

Figure 2

In Exercises 5–11, determine whether the relation R on the set A is an equivalence relation.

5. $A = \{a, b, c, d\}$,
   $R = \{(a, a), (b, a), (b, b), (c, c), (d, d), (d, c)\}$.

6. $A = \{1, 2, 3, 4, 5\}$, $R = \{(1, 1), (1, 2), (1, 3), (2, 1), (2, 2), (3, 1), (2, 3), (3, 3), (4, 4), (3, 2), (5, 5)\}$

7. $A = \{1, 2, 3, 4\}$, $R = \{(1, 1), (1, 2), (2, 1), (2, 2), (3, 1), (3, 3), (1, 3), (4, 1), (4, 4)\}$

8. $A =$ the set of all members of the Software-of-the-Month Club: $a \, R \, b$ if and only if a and b buy the same number of programs.

9. $A =$ the set of all members of the Software-of-the-Month Club; $a \, R \, b$ if and only if a and b buy the same programs.

10. A = the set of all people in the Social Security data base; $a\,R\,b$ if and only if a and b have the same last name.

11. A = the set of all triangles in the plane; $a\,R\,b$ if and only if a is similar to b.

12. If $\{\{a, c, e\}, \{b, d, f\}\}$ is a partition of the set $A = \{a, b, c, d, e, f\}$, determine the corresponding equivalence relation R.

13. Let $S = \{1, 2, 3, 4, 5\}$ and let $A = S \times S$. Define the following relation R on A: $(a, b)\,R\,(a', b')$ if and only if $ab' = a'b$.

(a) Show that R is an equivalence relation.
(b) Compute A/R.

14. Let $S = \{1, 2, 3, 4\}$ and let $A = S \times S$. Define the following relation R on A: $(a, b)\,R\,(a', b')$ if and only if $a + b = a' + b'$.

(a) Show that R is an equivalence relation.
(b) Compute A/R.

15. If $\{\{1, 3, 5\}, \{2, 4\}\}$ is a partition of the set $A = \{1, 2, 3, 4, 5\}$, determine the corresponding equivalence relation R.

THEORETICAL EXERCISES

T1. A relation R on a set A is called **circular** if $a\,R\,b$ and $b\,R\,c$ imply $c\,R\,a$. Show that R is reflexive and circular if and only if it is an equivalence relation.

T2. Show that if R_1 and R_2 are equivalence relations on A, then $R_1 \cap R_2$ is an equivalence relation on A.

5.3 The Algebra of Relations

Just as we can manipulate numbers and formulas using the rules of algebra, we can also define operations that allow us to manipulate relations. With these operations we can change, combine, and refine existing relations to produce new ones.

Let R and S be relations from a set A to a set B. Then if we remember that R and S are simply subsets of $A \times B$, we can use set operations on R and S. For example, the complement of R, $\overline{R}$ is referred to as the **complementary relation**. It is, of course, a relation from A to B which can be expressed simply in terms of R:

$$a\,\overline{R}\,b \quad \text{if and only if} \quad a\,\not{R}\,b$$

We can also form the intersection $R \cap S$ and the union $R \cup S$ of the relations R and S. In relational terms, we see that $a\,(R \cap S)\,b$ means that $a\,R\,b$ and $a\,S\,b$. Similarly, $a\,(R \cup S)\,b$ means that $a\,R\,b$ or $a\,S\,b$. All of our set-theoretic operations can be used in this way to produce new relations. The reader should try to give a relational description of the relation $R \oplus S$ (see Section 1.3).

A different type of operation on a relation R from A to B is the formation of the **inverse**, usually written R^{-1}. The relation R^{-1} is a relation from B to A (reverse order from R), defined by

$$b\,R^{-1}\,a \quad \text{if and only if} \quad a\,R\,b$$

It is not hard to see that Dom (R^{-1}) = Ran (R) and Ran (R^{-1}) = Dom (R). We leave these simple facts as exercises.

Observe that the relation R^{-1} consists of all ordered pairs in R, written in reverse order. It is clear from this observation that $(R^{-1})^{-1} = R$.

Example 1   Let $A = \{1, 2, 3, 4\}$ and $B = \{a, b, c\}$. Let

$$R = \{(1, a), (1, b), (2, b), (2, c), (3, b), (4, a)\}$$
$$S = \{(1, b), (2, c), (3, b), (4, b)\}$$

Compute: (a) $\bar{R}$; (b) $R \cap S$; (c) $R \cup S$; (d) R^{-1}.

Solution. (a) We first find

$$A \times B = \{(1, a), (1, b), (1, c), (2, a), (2, b), (2, c), (3, a),$$
$$(3, b), (3, c), (4, a), (4, b), (4, c)\}$$

Then the complement of R in $A \times B$ is

$$\bar{R} = \{(1, c), (2, a), (3, a), (3, c), (4, b), (4, c)\}$$

(b) We have

$$R \cap S = \{(1, b), (3, b), (2, c)\}$$

(c) We have

$$R \cup S = \{(1, a), (1, b), (2, b), (2, c), (3, b), (4, a), (4, b)\}$$

(d) Since $(x, y) \in R^{-1}$ if and only if $(y, x) \in R$, we have

$$R^{-1} = \{(a, 1), (b, 1), (b, 2), (c, 2), (b, 3), (a, 4)\}$$

Example 2   Let $A = B = \mathbb{R}$. Let R be the relation "$\leq$" and let S be "$\geq$." Then the complement of R is the relation "$>$," since $a \not\leq b$ means that $a > b$. Similarly, the complement of S is "$<$." On the other hand, $R^{-1} = S$, since for any numbers a and b,

$a\ R^{-1}\ b$ if and only if $b\ R\ a$ if and only if $b \leq a$ if and only if $a \geq b$

Similarly, we have $S^{-1} = R$. Also, we note that $R \cap S$ is the relation of equality, since $a\ (R \cap S)\ b$ if and only if $a \leq b$ and $b \leq a$ if and only if $a = b$. Since, for any a and b, $a \leq b$ or $b \leq a$ must hold, we see that $R \cup S = A \times B$, that is $R \cup S$ is the "universal" relation in which any a is related to any b.

Example 3   Let $A = \{a, b, c, d, e\}$ and let R and S be two relations on A whose corresponding digraphs are shown in Fig. 1. Then the reader can verify the following facts:

$$R = \{(a, b), (a, d), (a, e), (b, c), (b, d), (b, e), (c, c), (d, d), (d, c), (e, e)\}$$

$$S = \{(a, a), (a, b), (b, b), (c, b), (c, c), (b, e), (c, e), (d, b), (e, a), (e, d)\}$$

$$\overline{R} = \{(a, a), (b, b), (a, c), (b, a), (c, b), (c, d), (c, e), (c, a), (d, b), (d, a),$$
$$(d, e), (e, b), (e, a), (e, d), (e, c)\}$$

$$R^{-1} = \{(b, a), (e, b), (c, c), (c, d), (d, d), (d, b), (c, b), (d, a), (e, e), (e, a)\}$$

$$R \cap S = \{(a, b), (b, e), (c, c)\}$$

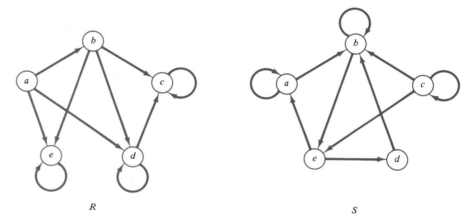

R S

Figure 1

Example 4   Let $A = \{1, 2, 3\}$ and let R and S be relations on A. Suppose that the matrices of R and S are

$$\mathbf{M}_R = \begin{bmatrix} 1 & 0 & 1 \\ 0 & 1 & 1 \\ 0 & 0 & 0 \end{bmatrix} \quad \text{and} \quad \mathbf{M}_S = \begin{bmatrix} 0 & 1 & 1 \\ 1 & 1 & 0 \\ 0 & 1 & 0 \end{bmatrix}$$

Then we can verify that

$$\mathbf{M}_{\overline{R}} = \begin{bmatrix} 0 & 1 & 0 \\ 1 & 0 & 0 \\ 1 & 1 & 1 \end{bmatrix}, \quad \mathbf{M}_{R^{-1}} = \begin{bmatrix} 1 & 0 & 0 \\ 0 & 1 & 0 \\ 1 & 1 & 0 \end{bmatrix}$$

$$\mathbf{M}_{R \cap S} = \begin{bmatrix} 0 & 0 & 1 \\ 0 & 1 & 0 \\ 0 & 0 & 0 \end{bmatrix}, \quad \mathbf{M}_{R \cup S} = \begin{bmatrix} 1 & 1 & 1 \\ 1 & 1 & 1 \\ 0 & 1 & 0 \end{bmatrix}$$

Example 4 illustrates some general facts. Recalling the operations on Boolean matrices from Section 3.4, we can show (Exercise T7) that if R and S are relations on set A, then

$$\mathbf{M}_{R \cap S} = \mathbf{M}_R \wedge \mathbf{M}_S$$
$$\mathbf{M}_{R \cup S} = \mathbf{M}_R \vee \mathbf{M}_S$$
$$\mathbf{M}_{R^{-1}} = (\mathbf{M}_R)^T$$

Moreover, if $\mathbf{M}$ is a Boolean matrix, we define the **complement** $\overline{\mathbf{M}}$ of $\mathbf{M}$ as the matrix obtained from $\mathbf{M}$ by replacing every 1 in $\mathbf{M}$ by a 0, and every 0 by a 1. Thus if

$$\mathbf{M} = \begin{bmatrix} 1 & 0 & 0 \\ 0 & 1 & 1 \\ 1 & 0 & 0 \end{bmatrix}$$

then

$$\overline{\mathbf{M}} = \begin{bmatrix} 0 & 1 & 1 \\ 1 & 0 & 0 \\ 0 & 1 & 1 \end{bmatrix}$$

We can also show (Exercise T7) that if R is a relation on a set A, then

$$\mathbf{M}_{\overline{R}} = \overline{\mathbf{M}}_R$$

We know that a symmetric relation is a relation R such that $\mathbf{M}_R = (\mathbf{M}_R)^T$, and since $(\mathbf{M}_R)^T = \mathbf{M}_{R^{-1}}$, we see that R is symmetric if and only if $R = R^{-1}$.

We will now see what effect our relational algebra has on some of the properties of relations which we presented in Section 5.1.

Theorem 1   Let R and S be relations on a set A.

(a) If R is reflexive, so is R^{-1}.
(b) R is reflexive if and only if $\overline{R}$ is irreflexive.
(c) If R and S are reflexive, then so are $R \cap S$ and $R \cup S$.

Proof. (a) If R is reflexive, then $(a, a) \in R$ for all $a \in A$. Thus $(a, a) \in R^{-1}$ for all $a \in A$, so R^{-1} is also reflexive.
(b) If R is reflexive, then $(a, a) \in R$ for all $a \in A$. Therefore we never have $(a, a) \in \overline{R}$ for any a, so $\overline{R}$ is irreflexive. Conversely, if $(a, a) \in \overline{R}$ is false for every $a \in A$, then $(a, a) \in R$ for all $a \in A$ so R is reflexive.
(c) If R and S are reflexive, then $(a, a) \in R$ and $(a, a) \in S$, for all a in A. Thus, for all a, $(a, a) \in R \cap S$ and $(a, a) \in R \cup S$. This means that $R \cap S$ and $R \cup S$ are reflexive.

Example 5   Let $A = \{1, 2, 3\}$ and let

$$R = \{(1, 1), (1, 2), (1, 3), (2, 2), (3, 3)\}$$
$$S = \{(1, 1), (1, 2), (2, 2), (3, 2), (3, 3)\}$$

Then

(a) $R^{-1} = \{(1, 1), (2, 1), (3, 1), (2, 2), (3, 3)\}$, so R and R^{-1} are both reflexive.
(b) $\bar{R} = \{(2, 1), (2, 3), (3, 1), (3, 2)\}$, which is irreflexive while R is reflexive.
(c) $R \cap S = \{(1, 1), (1, 2), (2, 2), (3, 3)\}$ and $R \cup S = \{(1, 1), (1, 2), (1, 3), (2, 2), (3, 2), (3, 3)\}$, which are both reflexive.

Theorem 2   Let R and S be relations on A.

(a) If R is symmetric, so are R^{-1} and $\bar{R}$.
(b) If R and S are symmetric, so are $R \cap S$ and $R \cup S$.

Proof. (a) If R is symmetric, $R = R^{-1}$ and thus $(R^{-1})^{-1} = R = R^{-1}$, which means that R^{-1} is also symmetric. Also, $(a, b) \in (\bar{R})^{-1}$ if and only if $(b, a) \in \bar{R}$ if and only if $(b, a) \notin R$ if and only if $(a, b) \notin R^{-1} = R$ if and only if $(a, b) \in \bar{R}$, so $\bar{R}$ is symmetric and part (a) is proved.

(b) If $(a, b) \in R \cap S$, then $(a, b) \in R$ and $(a, b) \in S$. Since R and S are symmetric, we must have $(b, a) \in R$ and $(b, a) \in S$. Thus $(b, a) \in R \cap S$. We have therefore shown that whenever $(a, b) \in R \cap S$, we must also have $(b, a) \in R \cap S$. This means that $R \cap S$ is symmetric. The proof for $R \cup S$ is quite similar, and we omit the details.

Example 6   Let $A = \{1, 2, 3\}$ and consider the symmetric relations

$$R = \{(1, 1), (1, 2), (2, 1), (1, 3), (3, 1)\}$$
$$S = \{(1, 1), (1, 2), (2, 1), (2, 2), (3, 3)\}$$

Then

(a) $R^{-1} = \{(1, 1), (2, 1), (1, 2), (3, 1), (1, 3)\}$, $\bar{R} = \{(2, 2), (2, 3), (3, 2), (3, 3)\}$, so R^{-1} and $\bar{R}$ are symmetric.
(b) $R \cap S = \{(1, 1), (1, 2), (2, 1)\}$ and $R \cup S = \{(1, 1), (1, 2), (1, 3), (2, 1), (2, 2), (3, 1), (3, 3)\}$, which are both symmetric.

Theorem 3   Let R and S be relations on A.
(a) If R is transitive, so is R^{-1}.
(b) If R and S are transitive, so is $R \cap S$.
(c) If R and S are equivalence relations, so is $R \cap S$.

Proof. (a) Suppose that $a \, R^{-1} \, b$ and $b \, R^{-1} \, c$. Then $c \, R \, b$ and $b \, R \, a$. Since R is transitive, this means that $c \, R \, a$, and therefore $a \, R^{-1} \, c$. We have thus shown that R^{-1} is transitive.

(b) Suppose that $a \, (R \cap S) b$ and $b \, (R \cap S) c$. Then, by definition, the following four statements must hold:

(1) $a \, R \, b$, (2) $a \, S \, b$, (3) $b \, R \, c$, and (4) $b \, S \, c$.

Since R is transitive, (1) and (3) imply that $a \, R \, c$. Since S is transitive, (2) and (4) imply that $a \, S \, c$. Thus $a \, (R \cap S) c$ and we see that $R \cap S$ is transitive.

(c) This follows from Theorems 1(c), 2(b), and 3(b).

Example 7 Let $A = \{1, 2, 3, 4, 5\}$, and let R and S be the relations on A given, respectively, by the matrices

$$\mathbf{M}_R = \begin{bmatrix} 1 & 1 & 1 & 0 & 0 \\ 1 & 1 & 1 & 0 & 0 \\ 1 & 1 & 1 & 0 & 0 \\ 0 & 0 & 0 & 1 & 1 \\ 0 & 0 & 0 & 1 & 1 \end{bmatrix} \quad \text{and} \quad \mathbf{M}_S = \begin{bmatrix} 1 & 1 & 0 & 0 & 0 \\ 1 & 1 & 0 & 0 & 0 \\ 0 & 0 & 1 & 1 & 1 \\ 0 & 0 & 1 & 1 & 1 \\ 0 & 0 & 1 & 1 & 1 \end{bmatrix}$$

An examination of these matrices shows that each is symmetric, all diagonal elements are 1, and each satisfies the equation $\mathbf{M} \odot \mathbf{M} = \mathbf{M}$. Thus the relations R and S are equivalence relations. Then

$$\mathbf{M}_{R \cap S} = \mathbf{M}_R \wedge \mathbf{M}_S = \begin{bmatrix} 1 & 1 & 0 & 0 & 0 \\ 1 & 1 & 0 & 0 & 0 \\ 0 & 0 & 1 & 0 & 0 \\ 0 & 0 & 0 & 1 & 1 \\ 0 & 0 & 0 & 1 & 1 \end{bmatrix}$$

and this is easily seen to be the matrix of an equivalence relation.

In terms of partitions, note first that the partition of A corresponding to R is $\mathcal{P}_R = \{\{1, 2, 3\}, \{4, 5\}\}$. The partition of A corresponding to S is $\mathcal{P}_S = \{\{1, 2\}, \{3, 4, 5\}\}$. Finally, the partition corresponding to $R \cap S$ is $\mathcal{P}_{R \cap S} = \{\{1, 2\}, \{3\}, \{4, 5\}\}$.

The observation about partitions given in the example above is typical in the following sense. For any two equivalence relations R and S on a set A, the partition corresponding to the equivalence relation $R \wedge S$ can be computed by taking all possible nonempty intersections of an R-equivalence class with an S-equivalence class.

Closures

If R is a relation on a set A, it may well happen that R lacks some of the important relational properties discussed in Section 5.1, especially reflexivity, symmetry, and transitivity. If R does not possess a particular property, we may wish to add related pairs to R until we get a relation that *does* have the required property. Naturally we want to add as few new pairs as possible, so what we need to find is the *smallest* relation R_1 on A that contains R and possesses the property we desire. Sometimes R_1 does not exist. If a relation such as R_1 does exist, we call it the **closure** of R with respect to the property in question.

Example 8 Suppose that R is a relation on a set A, and R is not reflexive. This can only occur because some pairs of the diagonal relation Δ are not in R. Thus $R_1 = R \cup \Delta$ is the smallest reflexive relation on A, containing R; that is, the **reflexive closure** of R is $R \cup \Delta$.

Example 9 Suppose now that R is a relation on A which is not symmetric. Then there must exist pairs (x, y) in R such that (y, x) is not in R. Of course, $(y, x) \in R^{-1}$, so if R is to be symmetric we must add all pairs from R^{-1}; that is, we must enlarge R to $R \cup R^{-1}$. Clearly, $(R \cup R^{-1})^{-1} = R^{-1} \cup R = R \cup R^{-1}$, so $R \cup R^{-1}$ is the smallest symmetric relation containing R; that is, $R \cup R^{-1}$ is the **symmetric closure** of R.

If $A = \{a, b, c, d\}$ and $R = \{(a, b), (b, c), (a, c), (c, d)\}$, then $R^{-1} = \{(b, a), (c, b), (c, a), (d, c)\}$, so the symmetric closure of R is

$$R \cup R^{-1} = \{(a, b), (b, a), (b, c), (c, b), (a, c)\ (c, a), (c, d), (d, c)\}$$

The symmetric closure of a relation R is very easy to visualize geometrically. All edges in the digraph of R become "two-way streets" in $R \cup R^{-1}$. Thus the graph of the symmetric closure of R is simply the digraph of R with all edges made bidirectional. We show in Fig. 2(a) the digraph of the relation R of Example 9. Figure 2(b) shows the graph of the symmetric closure $R \cup R^{-1}$.

The **transitive closure** of a relation R is the smallest transitive relation containing R. We will discuss the transitive closure in Section 5.4.

Composition

Now suppose that A, B, and C are sets, R is a relation from A to B, and S is a relation from B to C. We can then define a new relation, the **composition** of R and S, written $S \circ R$. The relation $S \circ R$ is a relation from A to C, and is defined as follows. If a is in A and c is in C, then $a(S \circ R)c$ if and only if for some b in B, we have $a R b$ and $b S c$. In other words, a is related to c by $S \circ R$ if we can get from a to c in two stages; first to an intermediate vertex b by relation R, and then from b to c by relation S.

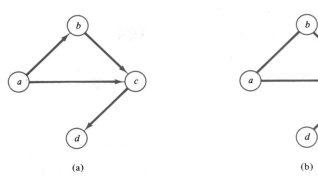

(a)   (b)

Figure 2

The relation $S \circ R$ might be thought of as "S following R" since it represents the combined effect of two relations, first R then S.

Example 10   Let $A = \{1, 2, 3, 4\}$, $R = \{(1, 2), (1, 1), (1, 3), (2, 4), (3, 2)\}$. and $S = \{(1, 4), (1, 3), (2, 3), (3, 1), (4, 1)\}$. Since $(1, 2) \in R$ and $(2, 3) \in S$, we must have $(1, 3) \in S \circ R$. Similarly, since $(1, 1) \in R$ and $(1, 4) \in S$, we see that $(1, 4) \in S \circ R$. Proceeding in this way, we find that $S \circ R = \{(1, 4), (1, 3), (1, 1), (2, 1), (3, 3)\}$.

The following result shows how to compute relative sets for the composition of two relations.

Theorem 4   Let R be a relation from A to B and let S be a relation from B to C. Then if A_1 is any subset of A, we have

$$(S \circ R)(A_1) = S(R(A_1))$$

Proof. If an element $z \in C$ is in $(S \circ R)(A_1)$, then $x(S \circ R)z$ for some x in A_1. By the definition of composition, this means that $x \, R \, y$ and $y \, S \, z$ for some y in B. Thus $y \in R(x)$, so $z \in S(R(x))$. Since $\{x\} \subseteq A_1$, we see that $S(R(x)) \subseteq S(R(A_1))$. Hence $z \in S(R(A_1))$, so $(S \circ R)(A_1) \subseteq S(R(A_1))$.

Conversely, suppose that $z \in S(R(A_1))$. Then $z \in S(y)$ for some y in $R(A_1)$, and similarly $y \in R(x)$ for some x in A_1. This means that $x \, R \, y$ and $y \, S \, z$, so $x(S \circ R)z$. Thus $z \in (S \circ R)(A_1)$, so $S(R(A_1)) \subseteq (S \circ R)(A_1)$. This proves the theorem.

Example 11   Let $A = \{a, b, c\}$, and let R and S be relations on A whose matrices are

$$\mathbf{M}_R = \begin{bmatrix} 1 & 0 & 1 \\ 1 & 1 & 1 \\ 0 & 1 & 0 \end{bmatrix}, \qquad \mathbf{M}_S = \begin{bmatrix} 1 & 0 & 0 \\ 0 & 1 & 1 \\ 1 & 0 & 1 \end{bmatrix}$$

We see from the matrices that

$$
\begin{array}{llll}
(a,\,a) \in R & \text{and} & (a,\,a) \in S, & \text{so} & (a,\,a) \in S \circ R \\
(a,\,c) \in R & \text{and} & (c,\,a) \in S, & \text{so} & (a,\,a) \in S \circ R \\
(a,\,c) \in R & \text{and} & (c,\,c) \in S, & \text{so} & (a,\,c) \in S \circ R
\end{array}
$$

It is easily seen that $(a,\,b) \notin S \circ R$, since if we had $(a,\,x) \in R$ and $(x,\,b) \in S$, matrix $\mathbf{M}_R$ tells us that x would have to be a or c, but matrix $\mathbf{M}_S$ tells us that neither $(a,\,b)$ nor $(c,\,b)$ is an element of S.

We see that the first row of $\mathbf{M}_{S \circ R}$ is 1 0 1. The reader may show by similar analysis that

$$
\mathbf{M}_{S \circ R} = \begin{bmatrix} 1 & 0 & 1 \\ 1 & 1 & 1 \\ 0 & 1 & 1 \end{bmatrix}
$$

We note that $\mathbf{M}_{S \circ R} = \mathbf{M}_R \odot \mathbf{M}_S$ (verify).

Example 11 illustrates a general and useful fact. let A, B, and C be finite sets with n, p, and m elements, respectively, let R be a relation from A to B, and let S be a relation from B to C. Then R and S have Boolean matrices $\mathbf{M}_R$ and $\mathbf{M}_S$ with respective sizes $n \times p$ and $p \times m$. Then $\mathbf{M}_R \odot \mathbf{M}_S$ can be computed, and it equals $\mathbf{M}_{S \circ R}$

In the special case where R and S are equal, we have $S \circ R = R^2$ and $\mathbf{M}_{S \circ R} = \mathbf{M}_{R^2} = \mathbf{M}_R \odot \mathbf{M}_R$, as was shown in Section 4.4.

Example 12   Let us redo Example 10 using matrices. We see that

$$
\mathbf{M}_R = \begin{bmatrix} 1 & 1 & 1 & 0 \\ 0 & 0 & 0 & 1 \\ 0 & 1 & 0 & 0 \\ 0 & 0 & 0 & 0 \end{bmatrix} \quad \text{and} \quad \mathbf{M}_S = \begin{bmatrix} 0 & 0 & 1 & 1 \\ 0 & 0 & 1 & 0 \\ 1 & 0 & 0 & 0 \\ 1 & 0 & 0 & 0 \end{bmatrix}
$$

Then

$$
\mathbf{M}_R \odot \mathbf{M}_S = \begin{bmatrix} 1 & 0 & 1 & 1 \\ 1 & 0 & 0 & 0 \\ 0 & 0 & 1 & 0 \\ 0 & 0 & 0 & 0 \end{bmatrix}
$$

so

$$S \circ R = \{(1, 1), (1, 3), (1, 4), (2, 1), (3, 3)\}$$

as we found before. In cases where the number of pairs in R and S is large, the matrix method is much more reliable and systematic.

Theorem 5   Let A, B, C, and D be sets, R a relation from A to B, S a relation from B to C, and T a relation from C to D. Then

$$T \circ (S \circ R) = (T \circ S) \circ R$$

Proof. The relations R, S, and T are determined by their Boolean matrices $\mathbf{M}_R$, $\mathbf{M}_S$, and $\mathbf{M}_T$, respectively. As we showed after Example 11, the matrix of the composition is the Boolean matrix product, that is, $\mathbf{M}_{S \circ R} = \mathbf{M}_R \odot \mathbf{M}_S$. Thus

$$\mathbf{M}_{T \circ (S \circ R)} = \mathbf{M}_{S \circ R} \odot \mathbf{M}_T = (\mathbf{M}_R \odot \mathbf{M}_S) \odot \mathbf{M}_T$$

Similarly,

$$\mathbf{M}_{(T \circ S) \circ R} = \mathbf{M}_R \odot (\mathbf{M}_S \odot \mathbf{M}_T)$$

Since matrix multiplication is associative [see Theorem 2(a) of Section 3.3] we must have

$$(\mathbf{M}_R \odot \mathbf{M}_S) \odot \mathbf{M}_T = \mathbf{M}_R \odot (\mathbf{M}_S \odot M_T)$$

and therefore

$$\mathbf{M}_{T \circ (S \circ R)} = \mathbf{M}_{(T \circ S) \circ R}$$

Then

$$T \circ (S \circ R) = (T \circ S) \circ R$$

since these relations have the same matrices.

In general, $R \circ S \neq S \circ R$, as shown in the following example.

Example 13   Let $A = \{a, b\}$, $R = \{(a, a), (b, a), (b, b)\}$, and $S = \{(a, b), (b, a), (b, b)\}$. Then $S \circ R = \{(a, b), (b, a), (b, b)\}$ while $R \circ S = \{(a, a), (a, b), (b, a), (b, b)\}$.

Theorem 6   Let A, B, and C be sets, R a relation from A to B, and S a relation from B to C. Then $(S \circ R)^{-1} = R^{-1} \circ S^{-1}$.

Proof.   Let $c \in C$ and $a \in A$. Then $(c, a) \in (S \circ R)^{-1}$ if and only if $(a, c) \in S \circ R$, that is, if and only if there is a $b \in B$ with $(a, b) \in R$ and $(b, c) \in S$. Finally, this is equivalent with the statement that $(c, b) \in S^{-1}$ and $(b, a) \in R^{-1}$, that is, $(c, a) \in R^{-1} \circ S^{-1}$.

EXERCISE SET 5.3

In Exercises 1–4, let R and S be the given relations from A to B. Compute (a) $\bar{R}$; (b) $R \cup S$; (c) $R \cap S$; (d) S^{-1}.

1. $A = \{1, 2, 3, 4\}$; $B = \{1, 2, 3\}$,
    $R = \{(1, 1), (1, 2), (2, 1), (2, 3), (3, 2), (4, 3)\}$;
    $S = \{(1, 3), (2, 3), (3, 2), (3, 3)\}$
2. $A = \{a, b, c\}$; $B = \{a, b\}$;
    $R = \{(a, a), (a, b), (b, a), (c, b)\}$,
    $S = \{(b, b), (c, a)\}$
3. $A = B = \{1, 2, 3\}$;
    $R = \{(1, 1), (1, 2), (2, 3), (3, 1)\}$;
    $S = \{(2, 1), (3, 1), (3, 2), (3, 3)\}$
4. $A = \{a, b, c\}$; $B = \{1, 2, 3\}$; $R = \{(a, 1), (b, 1), (c, 2), (c, 3)\}$; $S = \{(a, 1), (a, 2), (b, 1), (b, 2)\}$

In Exercises 5 and 6, let R and S be two relations whose corresponding digraphs are shown. Compute (a) $\bar{R}$; (b) $R \cap S$; (c) $R \cup S$; (d) S^{-1}.

5.

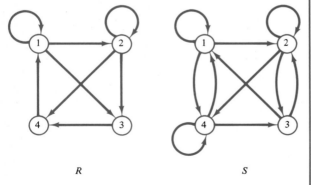

R   S

Figure 3

6.

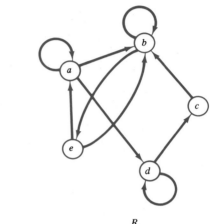

R

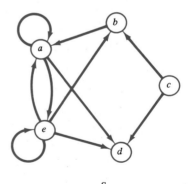

S

Figure 4

In Exercises 7 and 8, let $A = \{1, 2, 3\}$ and $B = \{1, 2, 3, 4\}$. Let R and S be the relations from A to B

whose matrices are given. Compute (a) $\bar{S}$; (b) $R \cap S$; (c) $R \cup S$; (d) R^{-1}.

7. $\mathbf{M}_R = \begin{bmatrix} 1 & 1 & 0 & 1 \\ 0 & 0 & 0 & 1 \\ 1 & 1 & 1 & 0 \end{bmatrix}$, $\mathbf{M}_S = \begin{bmatrix} 0 & 1 & 1 & 0 \\ 1 & 0 & 0 & 1 \\ 1 & 1 & 0 & 0 \end{bmatrix}$

8. $\mathbf{M}_R = \begin{bmatrix} 1 & 0 & 1 & 0 \\ 0 & 0 & 0 & 1 \\ 1 & 1 & 1 & 0 \end{bmatrix}$, $\mathbf{M}_S = \begin{bmatrix} 1 & 1 & 1 & 1 \\ 0 & 0 & 0 & 1 \\ 0 & 1 & 0 & 1 \end{bmatrix}$

9. Let $A = B = \{1, 2, 3, 4\}$, and let R be the relation on A given by $R = \{(1, 3), (2, 1), (2, 3), (3, 1), (3, 2), (3, 4), (4, 2)\}$. Compute (a) R^{-1}; (b) the digraph of R^{-1}; (c) the matrix of $\bar{R}$.

In Exercises 10 and 11, let $A = \{1, 2, 3, 4\}$ and $B = \{1, 2, 3\}$. Given the matrices $\mathbf{M}_R$ and $\mathbf{M}_S$ of the relations R and S from A to B, compute (a) $\mathbf{M}_{R \cap S}$; (b) $\mathbf{M}_{R \cup S}$; (c) $\mathbf{M}_{R^{-1}}$; (d) $\mathbf{M}_{\bar{S}}$.

10. $\mathbf{M}_R = \begin{bmatrix} 1 & 0 & 1 \\ 0 & 1 & 1 \\ 0 & 1 & 0 \\ 1 & 0 & 1 \end{bmatrix}$, $\mathbf{M}_S = \begin{bmatrix} 0 & 1 & 0 \\ 1 & 0 & 1 \\ 1 & 0 & 1 \\ 1 & 1 & 1 \end{bmatrix}$

11. $\mathbf{M}_R = \begin{bmatrix} 0 & 1 & 0 \\ 0 & 1 & 1 \\ 0 & 0 & 1 \\ 1 & 1 & 1 \end{bmatrix}$, $\mathbf{M}_S = \begin{bmatrix} 1 & 0 & 1 \\ 1 & 0 & 1 \\ 0 & 1 & 0 \\ 0 & 1 & 0 \end{bmatrix}$

12. Let $A = B = \{1, 2, 3, 4\}$, $R = \{(1, 1), (1, 3), (2, 3), (3, 1), (4, 2), (4, 4)\}$, and $S = \{(1, 2), (2, 3), (3, 1), (3, 2), (4, 3)\}$. Compute (a) $\mathbf{M}_{R \cap S}$; (b) $\mathbf{M}_{R \cup S}$; (c) $\mathbf{M}_{R^{-1}}$; (d) $\mathbf{M}_{\bar{S}}$.

13. Let $A = B = \{1, 2, 3\}$, $R = \{(1, 2), (2, 3), (3, 2)\}$, $S = \{(1, 2), (2, 3), (3, 1), (3, 2), (3, 3)\}$, and $T = \{(2, 1), (2, 3), (3, 2), (3, 3)\}$.
    (a) Verify Theorem 1(a) with R and S.
    (b) Verify Theorem 1(b) with R and S.
    (c) Verify Theorem 1(c) with S and T.
    (d) Verify Theorem 1(d) with S and T.

14. Let $A = \{1, 2, 3\}$, $R = \{(1, 1), (1, 2), (2, 1), (2, 2), (2, 3), (3, 3)\}$, and $S = \{(1, 1), (1, 2), (2, 2), (3, 1), (3, 2), (3, 3)\}$. Verify Theorem 2.

15. Let $A = \{a, b, c\}$ and $R = \{(a, a), (a, b), (b, a), (a, c), (c, a)\}$. Verify Theorem 3(a).

16. Let $A = \{1, 2, 3\}$, and $R = \{(1, 1), (1, 2), (2, 1), (2, 2), (2, 3), (3, 1), (1, 3), (3, 2)\}$. Verify Theorem 4(a).

17. Let $A = \{1, 2, 3, 4, 5, 6\}$ and

$R = \{(1, 2), (1, 1), (2, 1), (2, 2), (3, 3), (4, 4), (5, 5), (5, 6), (6, 5), (6, 6)\}$

$S = \{(1, 1), (1, 2), (1, 3), (2, 1), (2, 2), (2, 3), (3, 1), (3, 2), (3, 3), (4, 6), (4, 4), (6, 4), (6, 6), (5, 5)\}$

be equivalence relations on A. Compute the partition corresponding to $R \cap S$.

18. Let $A = \{1, 2, 3, 4, 5, 6\}$ and let R and S be the equivalence relations of Exercise 17. Compute the matrix $\mathbf{M}_{R \cup S}$.

19. Let $A = \{a, b, c, d\}$, $B = \{1, 2, 3\}$, and $C = \{\Box, \triangle, \diamondsuit\}$. Let R and S be the following relations from A to B and from B to C, respectively.

$R = \{(a, 1), (a, 2), (b, 2), (b, 3), (c, 1), (d, 3), (d, 2)\}$

$S = \{(1, \Box), (2, \triangle), (3, \triangle), (1, \diamondsuit)\}$

    (a) Is $(b, \triangle) \in S \circ R$?
    (b) Is $(c, \triangle) \in S \circ R$?
    (c) Compute $S \circ R$.

20. Let $A = B =$ the set of all real numbers, R be the relation "$<$," and S be the relation "$>$." Describe (a) $R \cap S$; (b) $R \cup S$; (c) S^{-1}.

21. Let $A =$ a set of people. Let $a \, R \, b$ if and only if a and b are brothers; let $a \, S \, b$ if and only if a and b are sisters. Describe $R \cup S$.

22. Let $A =$ a set of people. Let $a \, R \, b$ if and only if a is older than b; let $a \, S \, b$ if and only if a is a brother of b. Describe $R \cap S$.

23. Let $A =$ the set of all people in the Social Security data base. Let $a \, R \, b$ if and only if a and b receive the same benefits; let $a \, S \, b$ if and only if a and b have the same last name. Describe $R \cap S$.

24. Let $A =$ a set of people. Let $a \, R \, b$ if and only if a is the father of b; let $a \, S \, b$ if and only if a is the mother of b. Describe $R \cup S$.

25. Let $A = \{2, 3, 6, 12\}$ and let R and S be the following relations on A: $x \, R \, y$ if and only if $2 \mid (x - y)$; $x \, S \, y$ if and only if $3 \mid (x - y)$. Compute (a) $\bar{R}$; (b) $R \cap S$; (c) $R \cup S$; (d) S^{-1}.

26. Let $A = B = C =$ the set of all real numbers. Let R and S be the following relations from A to B and

from B to C, respectively:

$$R = \{(a, b) \mid a \le 2b\}$$
$$S = \{(b, c) \mid b \le 3c\}$$

(a)   Is $(1, 5) \in S \circ R$?

(b)   Is $(2, 3) \in S \circ R$?

(c)   Describe $S \circ R$.

27.   Let $A = \{1, 2, 3, 4\}$. Let

$$R = \{(1, 1), (1, 2), (2, 3), (2, 4), (3, 4), (4, 1),$$
$$(4, 2)\}$$

$$S = \{(3, 1), (4, 4), (2, 3), (2, 4), (1, 1), (1, 4)\}$$

(a)   Is $(1, 3) \in R \circ R$?

(b)   Is $(4, 3) \in S \circ R$?

(c)   Is $(1, 1) \in R \circ S$?

(d)   Compute $R \circ R$.

(e)   Compute $S \circ R$.

(f)   Compute $R \circ S$.

(g)   Compute $S \circ S$.

In Exercises 28 and 29, let $A = \{1, 2, 3, 4, 5\}$ and let $\mathbf{M}_R$ and $\mathbf{M}_S$ be the matrices of the relations R and S on A. Compute (a) $\mathbf{M}_{R \circ R}$; (b) $\mathbf{M}_{S \circ R}$; (c) $\mathbf{M}_{R \circ S}$; (d) $\mathbf{M}_{S \circ S}$.

28. $\mathbf{M}_R = \begin{bmatrix} 1 & 0 & 1 & 1 & 1 \\ 0 & 1 & 1 & 0 & 0 \\ 1 & 0 & 0 & 1 & 0 \\ 1 & 0 & 1 & 0 & 0 \\ 0 & 1 & 1 & 1 & 1 \end{bmatrix}$, $\mathbf{M}_S = \begin{bmatrix} 1 & 0 & 0 & 1 & 0 \\ 1 & 0 & 1 & 0 & 0 \\ 1 & 0 & 1 & 0 & 0 \\ 0 & 1 & 1 & 1 & 1 \\ 1 & 0 & 0 & 0 & 1 \end{bmatrix}$

29. $\mathbf{M}_R = \begin{bmatrix} 1 & 1 & 0 & 0 & 1 \\ 0 & 0 & 0 & 1 & 0 \\ 1 & 1 & 0 & 0 & 1 \\ 0 & 1 & 0 & 1 & 1 \\ 1 & 0 & 0 & 0 & 0 \end{bmatrix}$, $\mathbf{M}_S = \begin{bmatrix} 0 & 0 & 0 & 1 & 1 \\ 1 & 0 & 0 & 0 & 1 \\ 0 & 1 & 0 & 1 & 0 \\ 1 & 1 & 0 & 1 & 1 \\ 1 & 0 & 1 & 0 & 0 \end{bmatrix}$

THEORETICAL EXERCISES

T1.   If R and S are equivalence relations on a set A, is $S \circ R$ an equivalence relation on A? Prove or disprove your statement.

T2.   Which properties of relations on a set A are preserved by composition? Prove or disprove your statement.

T3.   Let R and S be relations on a set A. If R and S are asymmetric, either prove or disprove that $R \cap S$ and $R \cup S$ are asymmetric.

T4.   Let R and S be relations on a set A. If R and S are antisymmetric, either prove or disprove that $R \cap S$ and $R \cup S$ are antisymmetric.

T5.   Let R be a relation from A to B and let S and T be relations from B to C. Prove
(a) $(S \cup T) \circ R = (S \circ R) \cup (T \circ R)$
(b) $(S \cap T) \circ R = (S \circ R) \cap (T \circ R)$

T6.   Let R and S be relations from A to B and let T be a relation from B to C. Show that if $R \subseteq S$, then $T \circ R \subseteq T \circ S$.

T7.   Show that if R and S are relations on a set A, then
(a) $\mathbf{M}_{R \cap S} = \mathbf{M}_R \wedge \mathbf{M}_S$
(b) $\mathbf{M}_{R \cup S} = \mathbf{M}_R \vee \mathbf{M}_S$
(c) $\mathbf{M}_{R^{-1}} = (\mathbf{M}_R)^T$
(d) $\mathbf{M}_{\bar{R}} = \overline{\mathbf{M}}_R$

T8.   Let R be a relation from A to B. Prove
(a) $\text{Dom}\,(R^{-1}) = \text{Ran}\,(R)$
(b) $\text{Ran}\,(R^{-1}) = \text{Dom}\,(R)$

T9.   Prove Theorem 3.

In Exercises T10–T12, let R be a relation on a set A and let Δ be the relation of equality on A.

T10.   Prove that R is reflexive if and only if $R \supseteq \Delta$.

T11.   Prove that R is symmetric if and only if $R = R^{-1}$.

T12.   Prove that R is transitive if and only if $R \supseteq R^2$.

5.4 Transitive Closure and Warshall's Algorithm

Transitive Closure

In this section we consider a construction that has several interpretations and many important applications. Suppose that R is a relation on a set A and that R is not transitive. We will show that the transitive closure of R (see Section 5.3) is just the connectivity relation R^∞, defined in Section 4.4.

Theorem 1   Let R be a relation on a set A. Then R^∞ is the transitive closure of R.

Proof. We recall that if a and b are in the set A, then $a\, R^\infty\, b$ if and only if there is a path in R from a to b. Now R^∞ is certainly transitive, since if $a\, R^\infty\, b$ and $b\, R^\infty\, c$, the composition of the paths from a to b and from b to c form a path from a to c in R, and so $a\, R^\infty\, c$. To show that R^∞ is the smallest transitive relation containing R, we must show that if S is any transitive relation on A, and $R \subseteq S$, then $R^\infty \subseteq S$. Theorem 1 of Section 5.1 tells us that if S is transitive, then $S^n \subseteq S$ for all n; that is, if a and b are connected by a path of length n, then $a\, S\, b$. It follows that $S^\infty = \bigcup_{n=1}^{\infty} S^n \subseteq S$. It is also true that if $R \subseteq S$, then $R^\infty \subseteq S^\infty$, since any path in R is also a path in S. Putting these facts together, we see that if $R \subseteq S$ and S is transitive on A, then $R^\infty \subseteq S^\infty \subseteq S$. This means that R^∞ is the smallest of all transitive relations on A that contain R.

We see that R^∞ has several interpretations. From a geometric point of view, it is called the connectivity relation, since it specifies which vertices are connected (by paths) to other vertices.

On the other hand, from the algebraic point of view, R^∞ is the transitive closure of R, as we have shown in Theorem 1 above. In this form, it plays important roles in the theory of equivalence relations and in the theory of certain languages (see Section 10.1).

Example 1   Let $A = \{1, 2, 3, 4\}$, and let $R = \{(1, 2), (2, 3), (3, 4), (2, 1)\}$. Find the transitive closure of R.

Solution.

Method 1. The digraph of R is shown in Fig. 1. Since R^∞ is the transitive closure, we can proceed geometrically by computing all paths. We see that from vertex 1 we have paths to vertices 2, 3, 4, and 1. Note that the path from 1 to 1 proceeds from 1 to 2 to 1. Thus we see that the ordered pairs (1, 1), (1, 2), (1, 3), and (1, 4) are in R^∞. Starting from vertex 2, we have paths to vertices 2, 1, 3, and 4, so the

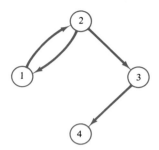

Figure 1

ordered pairs $(2, 1)$, $(2, 2)$, $(2, 3)$, and $(2, 4)$ are in R^∞. The only other path is from vertex 3 to vertex 4, so we have

$$R^\infty = \{(1, 1), (1, 2), (1, 3), (1, 4), (2, 1), (2, 2), (2, 3), (2, 4), (3, 4)\}$$

Method 2. The matrix of R is

$$\mathbf{M}_R = \begin{bmatrix} 0 & 1 & 0 & 0 \\ 1 & 0 & 1 & 0 \\ 0 & 0 & 0 & 1 \\ 0 & 0 & 0 & 0 \end{bmatrix}$$

We may proceed algebraically and compute the powers of $\mathbf{M}_R$. Thus

$$(\mathbf{M}_R)^2_\odot = \begin{bmatrix} 1 & 0 & 1 & 0 \\ 0 & 1 & 0 & 1 \\ 0 & 0 & 0 & 0 \\ 0 & 0 & 0 & 0 \end{bmatrix}, \quad (\mathbf{M}_R)^3_\odot = \begin{bmatrix} 0 & 1 & 0 & 1 \\ 1 & 0 & 1 & 0 \\ 0 & 0 & 0 & 0 \\ 0 & 0 & 0 & 0 \end{bmatrix},$$

$$(\mathbf{M}_R)^4_\odot = \begin{bmatrix} 1 & 0 & 1 & 0 \\ 0 & 1 & 0 & 1 \\ 0 & 0 & 0 & 0 \\ 0 & 0 & 0 & 0 \end{bmatrix}$$

Continuing in this way, we can see that $(\mathbf{M}_R)^n_\odot$ equals $(\mathbf{M}_R)^2_\odot$ if n is even, and equals $(\mathbf{M}_R)^3_\odot$ if n is odd. Thus

$$\mathbf{M}_{R^\infty} = \mathbf{M}_R \vee (\mathbf{M}_R)^2_\odot \vee (\mathbf{M}_R)^3_\odot = \begin{bmatrix} 1 & 1 & 1 & 1 \\ 1 & 1 & 1 & 1 \\ 0 & 0 & 0 & 1 \\ 0 & 0 & 0 & 0 \end{bmatrix}$$

and this gives the same relation as with Method 1.

In Example 1 we did not need to consider all powers R^n, to obtain R^∞. This observation is true in all cases whenever the set A is finite, as we will now prove.

Theorem 2   Let A be a set with $|A| = n$, and let R be a relation on A. Then

$$R^\infty = R \cup R^2 \cup \cdots \cup R^n$$

In other words, powers of R greater than n are not needed to compute R^∞.

Proof. Let a and b be in A, and suppose that $a, x_1, x_2, \ldots, x_m, b$ is a path from a to b in R, that is, $(a, x_1), (x_1, x_2), \ldots, (x_m, b)$ are all in R. If x_i and x_j are equal, say $i < j$, then the path can be divided into three sections. First, a path from a to x_i, then a path from x_i to x_j, and finally, a path from x_j to b. The middle path is a cycle, since $x_i = x_j$, so we simply leave it out and put the first two paths together. This gives us a shorter path from a to b (see Fig. 2).

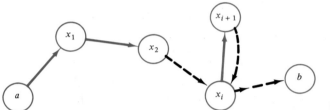

Figure 2

Now let $a, x_1, x_2, \ldots, x_k, b$ be the shortest path from a to b. If $a \neq b$, then all vertices $a, x_1, x_2, \ldots, x_k, b$ are distinct. Otherwise, the discussion above shows that we could find a shorter path. Thus the length of the path is at most $n - 1$ (since $|A| = n$). If $a = b$, then for similar reasons, the vertices $a, x_1, x_2, \ldots, x_k$ are distinct, so the length of the path is at most n. In other words, if $a \, R^\infty \, b$, then $a \, R^k \, b$ for some k from 1 to n. Thus $R^\infty = R \cup R^2 \cup \cdots \cup R^n$. This completes the proof.

The methods used to solve Example 1 each have certain difficulties. The graphical method is impractical for large sets and relations, and is not systematic. The matrix method can be used in general, and is systematic enough to program for computer, but is inefficient and, for large matrices, can be prohibitively costly. Fortunately, a more efficient algorithm for computing transitive closure is available. It is known as Warshall's algorithm, after its creator, and we describe it next.

Warshall's Algorithm

Let R be a relation on a set $A = \{a_1, a_2, \ldots, a_n\}$. If $x_1, x_2, \ldots, x_m$ is a path in R, then any vertices other than x_1 and x_m are called **interior vertices** of the path. Now if $1 \leq k \leq n$, we define a Boolean matrix $\mathbf{W}_k$ as follows. $\mathbf{W}_k$ has a 1 in position i,j

if and only if there is a path from a_i to a_j in R whose interior vertices, if any, come from the set $\{a_1, a_2, \ldots, a_k\}$.

Since any vertex must come from the set $\{a_1, a_2, \ldots, a_n\}$, it follows that the matrix $\mathbf{W}_n$ has a 1 in position i,j if and only if some path in R connects a_i with a_j. In other words, $\mathbf{W}_n = \mathbf{M}_{R^\infty}$. If we define $\mathbf{W}_0$ to be $\mathbf{M}_R$, then we will have a sequence $\mathbf{W}_0$, $\mathbf{W}_1, \ldots, \mathbf{W}_n$ whose first term is $\mathbf{M}_R$ and whose last term is $\mathbf{M}_{R^\infty}$. We will show how to compute each matrix $\mathbf{W}_k$ from the previous matrix $\mathbf{W}_{k-1}$. Then we can begin with the matrix of R and proceed one step at a time until, in n steps, we reach the matrix of R^∞. This procedure is called **Warshall's algorithm**. The matrices $\mathbf{W}_k$ are different from the powers of the matrix $\mathbf{M}_R$, and this difference results in a considerable savings of steps in the computation of the transitive closure of R.

Suppose that $\mathbf{W}_k = [t_{ij}]$ and $\mathbf{W}_{k-1} = [s_{ij}]$. If $t_{ij} = 1$, then there must be a path from a_i to a_j whose interior vertices come from the set $\{a_1, a_2, \ldots, a_k\}$. If the vertex a_k is not an interior vertex of this path, then all interior vertices must actually come from the set $\{a_1, a_2, \ldots, a_{k-1}\}$, so $s_{ij} = 1$. If a_k is an interior vertex of the path, then the situation is as shown in Fig. 3. As in the proof of Theorem 2, we may assume that all interior vertices are distinct. Thus a_k appears only once in the path, so all interior vertices of subpaths 1 and 2 must come from the set $\{a_1, a_2, \ldots, a_{k-1}\}$. This means that $s_{ik} = 1$ and $s_{kj} = 1$.

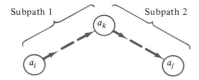

Figure 3

Thus $t_{ij} = 1$ if and only if either

$$(1)\ s_{ij} = 1 \text{ or}$$

$$(2)\ s_{ik} = 1 \text{ and } s_{kj} = 1$$

This is the basis for Warshall's algorithm. If $\mathbf{W}_{k-1}$ has a 1 in position i,j, then by (1) so will $\mathbf{W}_k$. By (2), a new 1 can be added in position i,j of $\mathbf{W}_k$ if and only if column k of $\mathbf{W}_{k-1}$ has a 1 in position i and row k of $\mathbf{W}_{k-1}$ has a 1 in position j. Thus we have the following procedure for computing $\mathbf{W}_k$ from $\mathbf{W}_{k-1}$.

To compute $\mathbf{W}_k$ from $\mathbf{W}_{k-1}$ in Warshall's algorithm

Step 1. First transfer to $\mathbf{W}_k$ all 1's in $\mathbf{W}_{k-1}$.

Step 2. List the locations $p_1, p_2, \ldots$, in column k of $\mathbf{W}_{k-1}$, where the entry is 1, and the locations $q_1, q_2, \ldots$, in row k of $\mathbf{W}_{k-1}$, where the entry is 1.

Step 3. Put 1's in all the positions p_i, q_j of $\mathbf{W}_k$ (if they are not already there.)

Example 2   Consider the relation R defined in Example 1. Then

$$\mathbf{W}_0 = \mathbf{M}_R = \begin{bmatrix} 0 & 1 & 0 & 0 \\ 1 & 0 & 1 & 0 \\ 0 & 0 & 0 & 1 \\ 0 & 0 & 0 & 0 \end{bmatrix}$$

and $n = 4$.

First we find $\mathbf{W}_1$, so that $k = 1$. $\mathbf{W}_0$ has 1's in location 2 of column 1 and location 2 of row 1. Thus $\mathbf{W}_1$ is just $\mathbf{W}_0$ with a new 1 in position $2, 2$.

$$\mathbf{W}_1 = \begin{bmatrix} 0 & 1 & 0 & 0 \\ 1 & 1 & 1 & 0 \\ 0 & 0 & 0 & 1 \\ 0 & 0 & 0 & 0 \end{bmatrix}$$

Now we compute $\mathbf{W}_2$, so that $k = 2$. We must consult column 2 and row 2 of $\mathbf{W}_1$. Matrix $\mathbf{W}_1$ has 1's in locations 1 and 2 of column 2 and locations 1, 2, and 3 of row 2.

Thus to obtain $\mathbf{W}_2$, we must put 1's in positions $1, 1$, $1, 2$, $1, 3$, $2, 1$, $2, 2$, and $2, 3$ of matrix $\mathbf{W}_1$ (if 1's are not already there). We see that

$$\mathbf{W}_2 = \begin{bmatrix} 1 & 1 & 1 & 0 \\ 1 & 1 & 1 & 0 \\ 0 & 0 & 0 & 1 \\ 0 & 0 & 0 & 0 \end{bmatrix}$$

Proceeding, we see that column 3 of $\mathbf{W}_2$ has 1's in locations 1 and 2, and row 3 of $\mathbf{W}_2$ has a 1 in location 4. To obtain $\mathbf{W}_3$ we must put 1's in positions $1, 4$ and $2, 4$ of $\mathbf{W}_2$, so

$$\mathbf{W}_3 = \begin{bmatrix} 1 & 1 & 1 & 1 \\ 1 & 1 & 1 & 1 \\ 0 & 0 & 0 & 1 \\ 0 & 0 & 0 & 0 \end{bmatrix}$$

Finally, $\mathbf{W}_3$ has 1's in locations 1, 2, 3 of column 4 and no 1's in row 4, so no new 1's are added and $\mathbf{M}_{R^\infty} = \mathbf{W}_4 = \mathbf{W}_3$. Thus we have obtained the same result as in Example 1.

Applications

At this point applications Chapters 9, 10 or both can be covered if desired.

EXERCISE SET 5.4

1. Let $A = \{1, 2, 3\}$ and let $R = \{(1, 1), (1, 2), (2, 3), (1, 3), (3, 1), (3, 2)\}$. Compute the matrix $\mathbf{M}_{R^\infty}$, of the transitive closure of R, by using the formula $\mathbf{M}_{R^\infty} = \mathbf{M}_R \vee (\mathbf{M}_R)^2_\odot \vee (\mathbf{M}_R)^3_\odot$.

2. List the relation R^∞ whose matrix was computed in Exercise 1.

3. For the relation R of Exercise 1, compute the transitive closure R^∞, by using Warshall's algorithm.

4. Let $A = \{a_1, a_2, a_3, a_4, a_5\}$ and let R be a relation on A whose matrix is

$$\mathbf{M}_R = \begin{bmatrix} 1 & 0 & 0 & 1 & 0 \\ 0 & 1 & 0 & 0 & 0 \\ 0 & 0 & 0 & 1 & 1 \\ 1 & 0 & 0 & 0 & 0 \\ 0 & 1 & 0 & 0 & 1 \end{bmatrix} = \mathbf{W}_0$$

Compute $\mathbf{W}_1$, $\mathbf{W}_2$, and $\mathbf{W}_3$ as in Warshall's algorithm.

In Exercises 5–7, let $A = \{1, 2, 3, 4\}$. For the relation R whose matrix is given, find the matrix of the transitive closure by using Warshall's algorithm.

5. $\mathbf{M}_R = \begin{bmatrix} 1 & 0 & 0 & 1 \\ 1 & 1 & 0 & 0 \\ 0 & 0 & 1 & 0 \\ 0 & 0 & 0 & 1 \end{bmatrix}$

6. $\mathbf{M}_R = \begin{bmatrix} 1 & 1 & 0 & 0 \\ 1 & 0 & 0 & 0 \\ 0 & 0 & 0 & 0 \\ 0 & 0 & 1 & 0 \end{bmatrix}$

7. $\mathbf{M}_R = \begin{bmatrix} 1 & 0 & 0 & 1 \\ 0 & 1 & 1 & 0 \\ 0 & 1 & 1 & 0 \\ 1 & 0 & 0 & 1 \end{bmatrix}$

THEORETICAL EXERCISES

T1. Prove that if R is reflexive and transitive, then $R^n = R$ for all n.

T2. Let R be a relation on a set A, and let $S = R^2$.

Prove that if $a, b \in A$, then $a\, S^\infty\, b$ if and only if there is a path in R from a to b having an even number of edges.

KEY IDEAS FOR REVIEW

☐ Properties of relations on a set A.

Reflexive:	$(a, a) \in R$ for all $a \in A$
Irreflexive:	$(a, a) \notin R$ for all $a \in A$
Symmetric:	$(a, b) \in R$ implies that $(b, a) \in R$
Asymmetric:	$(a, b) \in R$ implies that $(b, a) \notin R$
Antisymmetric:	$(a, b) \in R$ and $(b, a) \in R$ imply that $a = b$
Transitive:	$(a, b) \in R$ and $(b, c) \in R$ imply that $(a, c) \in R$

☐ Graph of symmetric relation: see page 102.

☐ Equivalence relation: reflexive, symmetric, transitive; see page 106.

☐ Equivalence relation determined by a partition: see page 108

☐ $a\, \bar{R}\, b$ (complement of R): $a\, \bar{R}\, b$ if and only if $a\, \not{R}\, b$.

☐ R^{-1}: $(x, y) \in R^{-1}$ if and only if $(y, x) \in R$.
☐ $R \cup S$, $R \cap S$: see page 112.
☐ $\mathbf{M}_{R \cap S} = \mathbf{M}_R \wedge \mathbf{M}_S$.
☐ $\mathbf{M}_{R \cup S} = \mathbf{M}_R \vee \mathbf{M}_S$.
☐ $\mathbf{M}_{R^{-1}} = (\mathbf{M}_R)^T$.
☐ $\mathbf{M}_{\bar{R}} = \overline{\mathbf{M}}_R$.
☐ If R and S are equivalence relations, so is $R \cap S$.
☐ $\mathbf{M}_{S \circ R} = \mathbf{M}_R \odot \mathbf{M}_S$.
☐ Theorem: R^{∞} is the smallest transitive relation on A that contains R.
☐ If $|A| = n$, $R^{\infty} = R \cup R^2 \cup \cdots \cup R^n$.
☐ Warshall's algorithm: computes $\mathbf{M}_{R^{\infty}}$ efficiently.

REVIEW EXERCISES

1. Determine whether the relation

    $R = \{(1, 1), (1, 3), (2, 1), (2, 2), (2, 3), (2, 4), (1, 4)\}$

    defined on the set $A = \{1, 2, 3, 4\}$ is reflexive, irreflexive, symmetric, asymmetric, antisymmetric, or transitive.

2. Determine whether the relation R defined by $a \, R \, b$ if and only if $a \leq b + 1$ on the set $A = \{1, 2, 3, 4, 5\}$ is reflexive, irreflexive, symmetric, asymmetric, antisymmetric, or transitive.

3. Draw the graph of the symmetric relation

    $R = \{(a, a), (a, b), (b, a), (a, c), (c, a), (c, c), (c, d), (d, c)\}$

    defined on the set $A = \{a, b, c, d\}$.

4. Is the relation R whose digraph is shown below an equivalence relation?

5. Is the relation

    $R = \{(a, a), (a, b), (b, a), (b, b), (c, b), (c, b), (a, c), (c, a), (c, c), (d, d)\}$

    defined on the set $A = \{a, b, c, d\}$ an equivalence relation?

6. Let $A = Z$ and let R be the relation defined on A by $a \, R \, b$ if and only if $a \mid b$. Is R an equivalence relation?

7. Let $R = \{(a, a), (a, c), (a, d), (c, a), (b, b), (c, d), (d, c), (c, c), (d, a), (d, d)\}$ be a relation on $A = \{a, b, c, d\}$.

    (a) Show that R is an equivalence relation.
    (b) Compute A/R.

8. If $\{\{1, 3\}, \{2, 4\}, \{5, 6\}\}$ is a partition of the set $A = \{1, 2, 3, 4, 5, 6\}$, determine the corresponding equivalence relation.

9. Let $A = \{a, b, c, d\}$ and $B = \{a, b, c\}$. Let R and S be the following relations from A to B.

    $R = \{(a, a), (a, b), (a, c), (b, a), (b, b), (c, a), (c, b), (d, a), (d, c)\}$

    $S = \{(a, c), (b, c), (a, d), (c, d), (d, a)\}$

    Compute (a) $\bar{R}$; (b) $\bar{S}$; (c) $R \cup S$; (d) $R \cap S$; (e) R^{-1}; (f) S^{-1}.

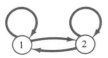 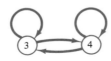 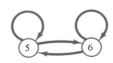

10. Let $A = \{1, 2, 3, 4\}$ and let

$R = \{(1, 1), (1, 2), (1, 3), (3, 2), (2, 4), (4, 3)\}$
$S = \{(2, 1), (2, 2), (3, 1), (3, 4), (4, 1)\}$

be relations on A. Compute
(a) $R \circ R$ (c) $R \circ S$
(b) $S \circ R$ (d) $S \circ S$

11. Let $A = \{a, b, c, d\}$ and let

$R = \{(a, a), (a, b), (b, a), (c, a), (c, b), (c, d), (d, b), (d, c)\}$

Compute $\mathbf{M}_{R^\infty}$ by using Warshall's algorithm.

CHAPTER TEST

1. Determine whether the relation R whose matrix is

$$\mathbf{M}_R = \begin{bmatrix} 1 & 0 & 0 & 1 \\ 1 & 1 & 0 & 1 \\ 0 & 1 & 1 & 0 \\ 1 & 1 & 1 & 0 \end{bmatrix}$$

defined on the set $A = \{a, b, c, d\}$ is reflexive, irreflexive, symetric, asymmetric, antisymmetric, or transitive.

2. Is the relation R whose matrix is

$$\mathbf{M}_R = \begin{bmatrix} 1 & 1 & 0 & 0 \\ 1 & 1 & 1 & 0 \\ 1 & 1 & 1 & 0 \\ 0 & 0 & 0 & 1 \end{bmatrix}$$

an equivalence relation?

3. Let R be the relation on $A = \{1, 2, 3, 4\}$ whose matrix is

$$\mathbf{M}_R = \begin{bmatrix} 1 & 1 & 1 & 0 \\ 1 & 1 & 1 & 0 \\ 1 & 1 & 1 & 0 \\ 0 & 0 & 0 & 1 \end{bmatrix}$$

(a) Show that R is an equivalence relation.
(b) Compute A/R.

4. If $\{\{a\}, \{b, c\}, \{d, e\}\}$ is a partition of the set $\{a, b, c, d, e\}$, determine the corresponding equivalence relation.

5. Let $A = \{a, b, x, y, z\}$, and let R and S be the following relations on A:

$R = \{(a, a), (a, x), (x, y), (x, b), (y, z)\}$
$S = \{(x, b), (z, a), (a, y), (b, b), (b, z)\}$

Compute (a) $\bar{S}$; (b) R^{-1}; (c) $S \circ R$; (d) $S^{-1} \circ S$.

6. Let $A = \{1, 2, 3, 4\}$ and let R be the relation on A whose matrix is given by

$$\mathbf{M}_R = \begin{bmatrix} 1 & 1 & 0 & 1 \\ 1 & 0 & 1 & 0 \\ 1 & 1 & 0 & 0 \\ 0 & 1 & 0 & 1 \end{bmatrix}$$

Compute $\mathbf{M}_{R^\infty}$ by using Warshall's algorithm.

7. Answer each of the following as true (T) or false (F).
(a) If R and S are relations from A to B and $R \subseteq S$, then $S^{-1} \subseteq R^{-1}$.
(b) If R and S are relations from A to B and $R \subseteq S$, then $\bar{S} \subseteq \bar{R}$.
(c) A relation R on a set A is antisymmetric if and only if $R \cap R^{-1} = \Delta$.
(d) A relation R on a set A is asymmetric if and only if $R \cap R^{-1} = \emptyset$.
(e) If R and S are relations on a set A, then $(R \cap S)^2 \subseteq R^2 \cap S^2$.

6

Functions

Prerequisites: Chapters 1 to 5.

In this chapter we focus our attention on a special type of relation that plays an important role in mathematics, computer science, and many applications.

6.1 Functions and Their Properties

In this section we define the notion of a function, a special type of relation. We study its basic properties and then discuss several special types of functions. A number of important applications of functions will occur in later sections of the book, so it is essential to get a good grasp of the material in this section.

Let A and B be nonempty sets. A **function** f from A to B denoted $f : A \rightarrow B$ is a relation from A to B such that for all $a \in \text{Dom } (f)$, $f(a)$ contains just one element of B. Naturally, if a is not in Dom (f), then $f(a) = \varnothing$. If $f(a) = \{b\}$, it is traditional to identify the set $\{b\}$ with the element b, and write $f(a) = b$. We will follow this custom, since no confusion results. The function f can then be described as the set of all ordered pairs $\{(a, f(a)) \,|\, a \in \text{Dom } (f)\}$. Functions are also called **mappings** or **transformations,** since they can be geometrically viewed as rules that assign to each element $a \in A$, the unique element $f(a) \in B$ (see Fig. 1). The element a is called an **argument** of the function f, and $f(a)$ is called the **value** of the function for the argument a and is also referred to as the **image** of a under f. Figure 1 is a schematic or pictorial display of the above definition of a function, and we will use several other similar diagrams. They should not be confused with the digraph of the relation f, which we will not generally display.

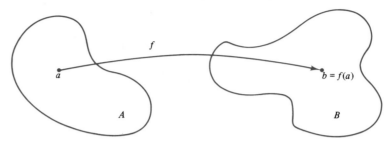

Figure 1

Example 1   Let $A = \{1, 2, 3, 4\}$ and $B = \{a, b, c, d\}$ and let

$$f = \{(1, a), (2, a), (3, d), (4, c)\}$$

Here we have

$$f(1) = a$$
$$f(2) = a$$
$$f(3) = d$$
$$f(4) = c$$

Since each set $f(n)$, $n = 1, 2, 3, 4$, is a single value, f is a function.

Note that the element $a \in B$ appears as the second element of two different ordered pairs in f. This does not conflict with the definition of a function. Thus a function may take the same value at two different elements of A.

Example 2   Let $A = \{1, 2, 3\}$ and $B = \{x, y, z\}$. Consider the relations

$$R = \{(1, x), (2, x)\} \qquad \text{and} \qquad S = \{(1, x), (1, y), (2, z), (3, y)\}$$

The relation S is not a function since $S(1) = \{x, y\}$. The relation R is a function with Dom $(R) = \{1, 2\}$ and Ran $(R) = \{x\}$.

Example 3   Let P be a computer program which accepts an integer as input and produces an integer as output. Let $A = B = Z$. Then P determines a relation f_P defined as follows:

$(m, n) \in f_P$ means that n is the output by program P when the input is m.

It is clear that f_P is a function, since any particular input corresponds to a unique output (computer results are reproducible, that is, they are the same each time the program is run).

Example 3 can be generalized to a program with any set A of possible inputs, and set B of corresponding outputs. In general, therefore, we may think of functions as **input-output** relations.

Example 4   Let $A = B =$ the set of all real numbers, and let $p(x) = a_0 + a_1 x + \cdots + a_n x^n$ be a real polynomial. Then p may be viewed as a relation. For each a in $\mathbb{R}$ we determine the relative set $p(a)$ by substituting a into the polynomial. Then since all relative sets $p(a)$ are known, the relation p is determined. Since a unique value is produced by this substitution, the relation p is actually a function.

If the formula does not make sense for all elements of A, then the domain of the resulting function is taken to be the set of elements of A for which the formula does make sense.

In elementary mathematics, the *formula* (in the case of Example 4, the polynomial) is usually confused with the *function* it produces. This may be harmful because some functions are not described by formulas, and some functions can be described by more than one formula.

Similar examples can be constructed with such elementary "functions" as e^x, $\sin x$, $\tan x$, $1/(1 + x^2)$, and so on. Suppose that, in Example 4, we used a formula that produced more than one value for $p(x)$, for example $p(x) = \pm\sqrt{x}$. Then the resulting relation would not be a function. For this reason, in older texts, relations were sometimes called "multiple-valued functions."

Example 5   A **labeled digraph** is a digraph in which the vertices or the edges (or both) are labeled with information from a set. If V is the set of vertices and L is the set of labels of a labeled digraph, then the labeling of V can be specified to be a function $f : V \to L$, where for each $v \in V, f(v)$ is the label we wish to attach to v. Similarly, we can define a labeling of the edges E, as a function $g : E \to L$, where for each $e \in E, g(e)$ is the label we wish to attach to e. An example of a labeled digraph is a map where the vertices are labeled with the names of cities and the edges are labeled with the distances or travel times between the cities. Another example is a flowchart of a program, where the vertices are labeled with the steps that are to be performed at that point in the program; the edges indicate the flow from one part of the program to another part. Figure 2 shows an example of a labeled digraph.

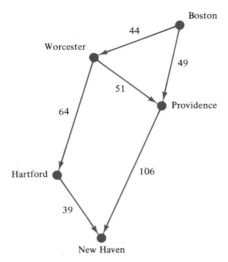

Figure 2

Example 6 Let $A = B = Z$ and let $f : A \to B$ be defined by

$$f(a) = a + 1 \qquad \text{for } a \in A$$

Here, as in Example 4, f is defined by giving a formula for the values $f(a)$.

Example 7 Let $A = Z$ and let $B = \{0, 1\}$. Let $f : A \to B$ be defined by

$$f(a) = \begin{cases} 0 & \text{if } a \text{ is even} \\ 1 & \text{if } a \text{ is odd} \end{cases}$$

Then f is a function, since each set $f(a)$ consists of a single element. Unlike the situation in Examples 4 and 6, the elements $f(a)$ are not specified through an algebraic formula. Instead, a verbal description is given.

Example 8 Let A be an arbitrary nonempty set. The **identity function** on A, denoted by 1_A, is defined by

$$1_A(a) = a$$

The reader may notice that 1_A is the relation we previously called Δ (see Section 5.1), which stands for the diagonal subset of $A \times A$. In the context of functions, the notation 1_A is preferred, since it emphasizes the input-output, or function nature of the relation. Clearly if $A_1 \subseteq A$, then $1_A(A_1) = A_1$. Suppose that f is a function from A to B and g is a function from B to C. Then the composition of f and g, $g \circ f$ (see Section 5.3) is a relation from A to C. Let $a \in \text{Dom } (g \circ f)$. Then by Theorem 4 of Section 5.3, $(g \circ f)(a) = g(f(a))$. Since f and g are functions, $f(a)$ consists of a single element $b \in B$, so $g(f(a)) = g(b)$. Since g is also a function, $g(b)$ contains just one element of C. Thus each set $(g \circ f)(a)$, for a in Dom $(g \circ f)$, contains just one element of C, so $g \circ f$ is a function. This is illustrated diagrammatically in Fig. 3.

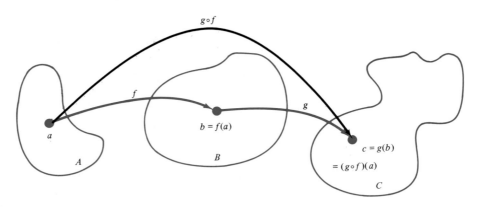

Figure 3

Example 9 Let $A = Z$, $B = Z$, and $C = $ the set of all even integers. Let $f : A \rightarrow B$ and $g : B \rightarrow C$ be defined by

$$f(a) = a + 1$$
$$g(b) = 2b$$

Find $g \circ f$.

> *Solution.* We have

$$(g \circ f)(a) = g(f(a))$$
$$= g(a + 1)$$
$$= 2(a + 1)$$

Thus if f and g are functions specified by giving formulas, then so is $g \circ f$, and the formula for $g \circ f$ is produced by substituting the formula for f into the formula for g.

Special Types of Functions

Let f be a function from A to B. Then we say that f is **everywhere defined** if Dom $(f) = A$. We say that f is **onto** if Ran $(f) = B$. Finally, we say that f is **one to one** if we cannot have $f(a) = f(a')$ for two distinct elements a and a' of A. The definition of one to one may be restated in the following equivalent form:

$$\text{If } f(a) = f(a') \qquad \text{then } a = a'$$

The latter form is often easier to verify in particular examples.

Example 10 Consider the function f defined in Example 1. Since Dom $(f) = A$, f is everywhere defined. On the other hand, Ran $(f) = \{a, c, d\} \neq B$; therefore f is not onto. Since

$$f(1) = f(2) = a$$

we can conclude that f is not o. ː to one.

Example 11 Consider the function f defined in Example 6. Which of the special properties above, if any, does f possess?

Solution. Since the formula defining f makes sense for all integers, Dom $(f) = Z = A$, so f is everywhere defined.
Suppose that

$$f(a) = f(a')$$

for a and a' in A. Then

$$a + 1 = a' + 1$$

so

$$a = a'$$

Hence f is one to one.
To see if f is onto, let b be an arbitrary element of B. Can we find an element $a \in A$ such that

$$f(a) = b \text{ ?}$$

Since

$$f(a) = a + 1$$

we need an element a in A such that

$$a + 1 = b$$

Of course,

$$a = b - 1$$

will satisfy the desired equation since $b - 1$ is in A. Hence Ran $(f) = B$, so f is onto.

Example 12   Let $A = \{a_1, a_2, a_3\}$, $B = \{b_1, b_2, b_3\}$, $C = \{c_1, c_2\}$, and $D = \{d_1, d_2, d_3, d_4\}$. Consider the following four functions, from A to B, A to D, B to C, and D to B, respectively.

(a) $f_1 = \{(a_1, b_2), (a_2, b_3), (a_3, b_1)\}$
(b) $f_2 = \{(a_1, d_2), (a_2, d_1), (a_3, d_4)\}$
(c) $f_3 = \{(b_1, c_2), (b_2, c_2), (b_3, c_1)\}$
(d) $f_4 = \{(d_1, b_2), (d_2, b_2), (d_3, b_3)\}$

Determine whether or not each function is one to one, whether each function is onto, and whether each function is everywhere defined.

Solution
(a) f_1 is everywhere defined, one to one, and onto.
(b) f_2 is everywhere defined and one to one, but not onto.
(c) f_3 is everywhere defined and onto but is not one to one.
(d) f_4 is not everywhere defined, not one to one, and not onto.

If $f : A \rightarrow B$ is a one-to-one function, then f associates to each element a of Dom (f) an element $b = f(a)$ of Ran (f). Every b in Ran (f) is matched, in this way, with one and only one element of Dom (f). For this reason, such an f is often called a **one-to-one correspondence or bijection**. If f is also everywhere defined and onto, then f is called a **one-to-one correspondence between A and B**.

Example 13   Let $\mathscr{R}$ be the set of all equivalence relations on a given set A, and let $\mathscr{P}$ be the set of all partitions on A. Then we can define a function $f : \mathscr{R} \rightarrow \mathscr{P}$ as follows. For each equivalence relation R on A, let $f(R) = A/R$, the partition of A that corresponds to R. The discussion in Section 5.2 shows that f is a one-to-one correspondence between $\mathscr{R}$ and $\mathscr{P}$.

Invertible Functions

A function $f : A \rightarrow B$ is said to be **invertible** if its inverse relation, f^{-1}, is also a function. The next example shows that a function is not necessarily invertible.

Example 14   Let f be the function of Example 1. Then

$$f^{-1} = \{(a, 1), (a, 2), (d, 3), (c, 4)\}$$

We see that f^{-1} is not a function, since $f^{-1}(a) = \{1, 2\}$.

The following theorem is frequently used.

Theorem 1   Let $f : A \rightarrow B$ be a function.
   (a) then f^{-1} is a function from B to A if and only if f is one to one.
   (b) if f^{-1} is a function, then the function f^{-1} is also one to one.
   (c) f^{-1} is everywhere defined if and only if f is onto.
   (d) f^{-1} is onto if and only if f is everywhere defined.

Proof. (a) We prove the following equivalent statement.

f^{-1} is not a function if and only if f is not one to one.

Suppose first that f^{-1} is not a function. Then for some b in B, $f^{-1}(b)$ must contain at least two distinct elements, a_1 and a_2. Then $f(a_1) = b = f(a_2)$, so f is not one to one.
   Conversely, suppose that f is not one to one. Then $f(a_1) = f(a_2) = b$ for two distinct elements a_1 and a_2 of A. Thus $f^{-1}(b)$ contains both a_1 and a_2, so f^{-1} cannot be a function.
   (b) Since $(f^{-1})^{-1}$ is the function f, part (a) shows that f^{-1} is one to one.
   (c) Recall that Dom $(f^{-1}) =$ Ran (f). Thus $B =$ Dom (f^{-1}) if and only if $B =$ Ran (f). In other words, f^{-1} is everywhere defined if and only if f is onto.
   (d) Since Ran $(f^{-1}) =$ Dom (f), $A =$ Dom (f) if and only if $A =$ Ran (f^{-1}). That is, f is everywhere defined if and only if f^{-1} is onto.

As an immediate consequence of Theorem 1 we see that if f is a one-to-one correspondence between A and B, then f^{-1} is a one-to-one correspondence between B and A. Note also that if $f : A \rightarrow B$ is one-to-one, then the equation $b = f(a)$ is equivalent with the equation $a = f^{-1}(b)$.

Example 15   Consider the function f defined in Example 6. Since f is everywhere defined, one to one, and onto, it is a one-to-one correspondence between A and B. Thus f is invertible, and f^{-1} is a one-to-one correspondence between B and A.

Example 16   Let $\mathbb{R}$ be the set of all real numbers, and let $f : \mathbb{R} \to \mathbb{R}$ be defined by $f(x) = x^2$. Is f invertible?

Solution. Since

$$f(2) = f(-2) = 4$$

we conclude that f is not one to one. Hence f is not invertible.

There are some useful results concerning the composition of functions. We summarize these in the following theorem.

Theorem 2   Let $f : A \to B$ be any function. Then
  (a) $1_B \circ f = f$
  (b) $f \circ 1_A = f$

and if f is a one-to-one correspondence, then

  (c) $f^{-1} \circ f = 1_A$
  (d) $f \circ f^{-1} = 1_B$

Proof. (a) $(1_B \circ f)(a) = 1_B(f(a)) = f(a)$, for all a in A. Thus by Theorem 1 of Section 4.2, $1_B \circ f = f$.
  (b) $(f \circ 1_A)(a) = f(1_A(a)) = f(a)$, for all a in A, so $1_A \circ f = f$.
Suppose now that f is a one-to-one correspondence between A and B. As we pointed out above, the equation $b = f(a)$ is equivalent to the equation $a = f^{-1}(b)$. Since f and f^{-1} are both everywhere defined and onto, this means that for all a in A and b in B, $f(f^{-1}(b)) = b$ and $f^{-1}(f(a)) = a$. Then
  (c) For all a in A, $1_A(a) = a = f^{-1}(f(a)) = (f^{-1} \circ f)(a)$. Thus $1_A = f^{-1} \circ f$.
  (d) For all b in B, $1_B(b) = b = f(f^{-1}(b)) = (f \circ f^{-1})(b)$. Thus $1_B = f \circ f^{-1}$.

Theorem 3   Let $f : A \to B$ and $g : B \to C$ be invertible. Then $g \circ f$ is invertible, and $(g \circ f)^{-1} = f^{-1} \circ g^{-1}$

Proof. We know that $(g \circ f)^{-1} = f^{-1} \circ g^{-1}$, since this is true for any relations. Since g^{-1} and f^{-1} are functions by assumption, so is their composition, and therefore $(g \circ f)^{-1}$ is a function. Thus $g \circ f$ is invertible.

Finally, we discuss briefly some special results that hold when A and B are finite sets. Let $A = \{a_1, \ldots, a_n\}$ and $B = \{b_1, \ldots, b_n\}$, and let f be a function from A

to B that is everywhere defined. If f is one to one, then $f(a_1), f(a_2), \cdots, f(a_n)$ are n distinct elements of B, and therefore must constitute all of B. Thus f is also onto. On the other hand, if f is onto, then $f(a_1), \ldots, f(a_n)$ form the entire set B, so they must all be different. Hence f is also one to one. We have therefore shown:

Theorem 4 Let A and B be two finite sets with the same number of elements, and let $f : A \to B$ be an everywhere defined function.
  (a) If f is one to one, then f is onto.
  (b) If f is onto, then f is one to one.

Thus for finite sets, A and B, with the same number of elements, and particularly if $A = B$, we need only prove that a function is one to one *or* onto to show that it is a bijection.

EXERCISE SET 6.1

In Exercises 1–4, let $A = \{a, b, c, d\}$ and $B = \{1, 2, 3\}$. Determine whether the relation R from A to B is a function. If it is a function, give its range.

1. $R = \{(a, 1), (b, 2), (c, 1), (d, 2)\}$
2. $R = \{(a, 1), (b, 2), (a, 2), (c, 1), (d, 2)\}$
3. $R = \{(a, 3), (b, 2), (c, 1)\}$
4. $R = \{(a, 1), (b, 1), (c, 1), (d, 1)\}$

In Exercises 5 and 6, determine whether the relation R from A to B is a function.

5. $A =$ the set of all recipients of Medicare in the United States, $B = \{x \mid x$ is a nine-digit number$\}$, $a \, R \, b$ if b is a's Social Security number.
6. $A =$ a set of people in the United States, $B = \{x \mid x$ is a nine-digit number$\}$, $a \, R \, b$ if b is a's passport number.

In Exercises 7–10, verify that the formula yields a function from A to B.

7. $A = B = Z; f(a) = a^2$
8. $A = B = \mathbb{R}; f(a) = e^a$
9. $A = \mathbb{R}, B = \{0, 1\}$; let Z be the set of integers and note that $Z \subseteq \mathbb{R}$. Then for any real number a, let

$$f(a) = \begin{cases} 0 & \text{if } a \notin Z \\ 1 & \text{if } a \in Z \end{cases}$$

10. $A = \mathbb{R}, B = Z; f(a) =$ the greatest integer less than or equal to a.

11. Let $A = B = C = \mathbb{R}$, and let $f : A \to B$, $g : B \to C$ be defined by $f(a) = a - 1$ and $g(b) = b^2$. Find
  (a) $(f \circ g)(2)$ (b) $(g \circ f)(2)$
  (c) $(g \circ f)(x)$ (d) $(f \circ g)(x)$
  (e) $(f \circ f)(y)$ (f) $(g \circ g)(y)$

12. Let $A = B = C = \mathbb{R}$ and let $f : A \to B$, $g : B \to C$ be defined by $f(a) = a + 1$ and $g(b) = b^2 + 2$. Find
  (a) $(g \circ f)(-2)$ (b) $(f \circ g)(-2)$
  (c) $(g \circ f)(x)$ (d) $(f \circ g)(x)$
  (e) $(f \circ f)(y)$ (f) $(g \circ g)(y)$

13. Let $A = B = \{x \mid x$ is a real number and $x \neq 0, 1\}$. Consider the following six functions from A to B, each defined by its formula.

$$f_1(x) = x \qquad\qquad f_2(x) = 1 - x$$

$$f_3(x) = \frac{1}{x} \qquad\qquad f_4(x) = \frac{1}{1 - x}$$

$$f_5(x) = \frac{x}{x - 1} \qquad f_6(x) = \frac{x - 1}{x}$$

Show by substituting one formula into another that the composition of any two of these six functions is another one of the six.

14. In each part, sets A and B, and a function from A to B are given. Determine whether the function is one to one or onto (or both or neither).

(a) $A = \{1, 2, 3, 4\} = B$;
    $f = \{(1, 1), (2, 3), (3, 4), (4, 2)\}$

(b) $A = \{1, 2, 3\}$; $B = \{a, b, c, d\}$;
    $f = \{(1, a), (2, a), (3, c)\}$

(c) $A = \{\frac{1}{2}, \frac{1}{3}, \frac{1}{4}\}$; $B = \{x, y, z, w\}$;
    $f = \{(\frac{1}{2}, x), (\frac{1}{4}, y), (\frac{1}{3}, w)\}$

(d) $A = \{1.1, 7, 0.06\}$; $B = \{p, q\}$;
    $f = \{(1.1, p), (7, q), (0.06, p)\}$

In Exercises 15–21, let f be a function from A to B. Determine whether each function f is one to one and whether it is onto.

15. $A = B = Z$; $f(a) = a - 1$

16. $A = B = \mathbb{R}$; $f(a) = |a|$

17. $A = \mathbb{R}$, $B = \{x \mid x \text{ is real and } x \geq 0\}$; $f(a) = |a|$

18. $A = \mathbb{R} \times \mathbb{R}$, $B = \mathbb{R}$; $f((a, b)) = a$

19. Let $S = \{1, 2, 3\}$, $T = \{a, b\}$. Let $A = B = S \times T$ and let f be defined by $f(n, a) = (n, b)$, $n = 1, 2, 3$, and $f(n, b) = (1, a)$, $n = 1, 2, 3$.

20. $A = B = \mathbb{R} \times \mathbb{R}$; $f((a, b)) = (a + b, a - b)$

21. $A = \mathbb{R}$, $B = \{x \mid x \text{ is real and } x \geq 0\}$; $f(a) = a^2$

In Exercises 22 and 23, let f be a function from $A = \{1, 2, 3, 4\}$ to $B = \{a, b, c, d\}$. Determine whether f^{-1} is a function.

22. $f = \{(1, a), (2, a), (3, c), (4, d)\}$

23. $f = \{(1, a), (2, c), (3, b), (4, d)\}$

THEORETICAL EXERCISES

T1. Let $A = B = C = \mathbb{R}$ and consider the functions $f : A \rightarrow B$ and $g : B \rightarrow C$ defined by $f(a) = 2a + 1$, $g(b) = b/3$. Verify Theorem 3: $(g \circ f)^{-1} = f^{-1} \circ g^{-1}$.

T2. If a set A has n elements, how many functions are there from A to A?

T3. If a set A has n elements, how many bijections are there from A to A?

T4. If A has m elements and B has n elements, how many functions are there from A to B?

T5. Prove that if $f : A \rightarrow B$ and $g : B \rightarrow C$ are one-to-one functions, then $g \circ f$ is one to one.

T6. Prove that if $f : A \rightarrow B$ and $g : B \rightarrow C$ are onto functions, then $g \circ f$ is onto.

T7. Let $f : A \rightarrow B$ and $g : B \rightarrow C$ be functions. Show that if $g \circ f$ is one to one, then f is one to one.

T8. Let $f : A \rightarrow B$ and $g : B \rightarrow C$ be functions. Show that if $g \circ f$ is onto, then g is onto.

6.2 Permutations

In this section we discuss bijections from a set A to itself. Of special importance is the case when A is finite. Bijections on a finite set occur in a wide variety of applications in mathematics, computer science, and physics.

A bijection from a set A to itself is called a **permutation** of A.

Example 1   Let $A = \mathbb{R}$ and let $f : A \rightarrow A$ be defined by $f(a) = 2a + 1$. Since f is one to one and onto (verify), it follows that f is a permutation of A.

If $A = \{a_1, a_2, \ldots, a_n\}$ is a finite set and p is a bijection on A, we list the elements of A, and the corresponding function values $p(a_1), p(a_2), \ldots, p(a_n)$ in the following form:

$$\begin{pmatrix} a_1 & a_2 & \cdots & a_n \\ p(a_1) & p(a_2) & \cdots & p(a_n) \end{pmatrix} \tag{1}$$

Observe that (1) completely describes p since it gives the value of p for every element of A. We often write

$$p = \begin{pmatrix} a_1 & a_2 & \cdots & a_n \\ p(a_1) & p(a_2) & \cdots & p(a_n) \end{pmatrix}$$

Thus if p is a permutation of a finite set $A = \{a_1, a_2, \ldots, a_n\}$, then the sequence $p(a_1), p(a_2), \ldots, p(a_n)$ is just a rearrangement of the elements of A, and so corresponds exactly to a permutation of A in the sense of Section 2.2.

Example 2 Let $A = \{1, 2, 3\}$. Then all the permutations of A are

$$1_A = \begin{pmatrix} 1 & 2 & 3 \\ 1 & 2 & 3 \end{pmatrix}, \qquad p_1 = \begin{pmatrix} 1 & 2 & 3 \\ 1 & 3 & 2 \end{pmatrix}, \qquad p_2 = \begin{pmatrix} 1 & 2 & 3 \\ 2 & 1 & 3 \end{pmatrix},$$

$$p_3 = \begin{pmatrix} 1 & 2 & 3 \\ 2 & 3 & 1 \end{pmatrix}, \qquad p_4 = \begin{pmatrix} 1 & 2 & 3 \\ 3 & 1 & 2 \end{pmatrix}, \qquad p_5 = \begin{pmatrix} 1 & 2 & 3 \\ 3 & 2 & 1 \end{pmatrix}$$

Example 3 Using the permutations of Example 2, compute (a) p_4^{-1}; (b) $p_3 \circ p_2$.

Solution. (a) Viewing p_4 as a function, we have

$$p_4 = \{(1, 3), (2, 1), (3, 2)\}$$

Then

$$p_4^{-1} = \{(3, 1), (1, 2), (2, 3)\}$$

or when written in increasing order of the first component of each ordered pair, we have

$$p_4^{-1} = \{(1, 2), (2, 3), (3, 1)\}$$

Thus

$$p_4^{-1} = \begin{pmatrix} 1 & 2 & 3 \\ 2 & 3 & 1 \end{pmatrix} = p_3$$

(b) The function p_2 takes 1 to 2 and p_3 takes 2 to 3, so $p_3 \circ p_2$ takes 1 to 3. Also, p_2 takes 2 to 1 and p_3 takes 1 to 2, so $p_3 \circ p_2$ takes 2 to 2. Finally, p_2 takes 3 to 3 and p_3 takes 3 to 1, so $p_3 \circ p_2$ takes 3 to 1. Thus

$$p_3 \circ p_2 = \begin{pmatrix} 1 & 2 & 3 \\ 3 & 2 & 1 \end{pmatrix}$$

We may view the process of forming $p_3 \circ p_2$ as shown in Fig. 1. Observe that $p_3 \circ p_2 = p_5$.

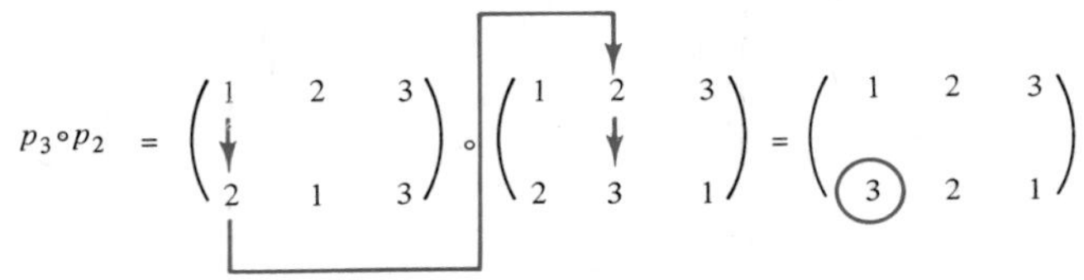

Figure 1

The composition of two permutations is another permutation usually referred to as the **product** of these permutations. In the remainder of this chapter, we will follow this convention.

Theorem 1 If $A = \{a_1, a_2, \ldots, a_n\}$ is a set containing n elements, then there are $n! = n \cdot (n - 1) \ldots 2 \cdot 1$ permutations of A.

Proof. This result follows from Theorem 3 of Section 2.2 by letting $r = n$.

Let $b_1, b_2, \ldots, b_r$ be r distinct elements of the set $A = \{a_1, a_2, \ldots, a_n\}$. The permutation $p : A \rightarrow A$ defined by

$$p(b_1) = b_2$$
$$p(b_2) = b_3$$
$$\vdots$$
$$p(b_{r-1}) = b_r$$
$$p(b_r) = b_1$$
$$p(x) = x \qquad \text{if} \qquad x \in A, \quad x \notin \{b_1, b_2, \ldots, b_r\}$$

is called a **cyclic permutation** of length r, or simply a **cycle** of length r, and will be denoted by $(b_1, b_2, \ldots, b_r)$. Do not confuse this terminology with that used for cycles in a digraph (Section 4.4). The two concepts are different and we use slightly different notations. If the elements $b_1, b_2, \ldots, b_r$ are arranged uniformly on a circle, as shown in Fig. 2, then a cycle p of length r moves these elements in a clockwise direction so that b_1 is sent to b_2, b_2 to b_3, $\ldots$, b_{r-1} to b_r, and b_r to b_1. All the other elements of A are left fixed by p.

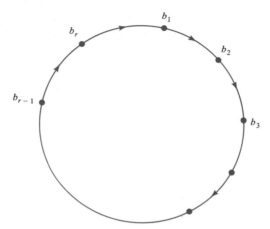

Figure 2

Example 4   Let $A = \{1, 2, 3, 4, 5\}$. The cycle $(1, 3, 5)$ denotes the permutation

$$\begin{pmatrix} 1 & 2 & 3 & 4 & 5 \\ 3 & 2 & 5 & 4 & 1 \end{pmatrix}$$

Observe that if $p = (b_1, b_2, \ldots, b_r)$ is a cycle of length r, then we can also write p by starting with any b_i, $1 \le i \le r$, and moving in a clockwise direction as shown in Fig. 2. Thus

$$(3, 5, 8, 2) = (5, 8, 2, 3) = (8, 2, 3, 5) = (2, 3, 5, 8)$$

Note also that the notation for a cycle does not indicate the number of elements in the set A. Thus the cycle $(3, 2, 1, 4)$ could be a permutation of the set $\{1, 2, 3, 4\}$ or of $\{1, 2, 3, 4, 5, 6, 7, 8\}$. We need to be told explicitly the set on which a cycle is defined. It follows from the definition that a cycle on a set A is of length 1 if and only if it is the identity permutation, 1_A.

Since cycles are permutations, we can form their product. However, as we show in the following example, the product of two cycles need not be a cycle.

Example 5   Let $A = \{1, 2, 3, 4, 5, 6\}$. Compute $(5, 6, 3) \circ (4, 1, 3, 5)$ and $(4, 1, 3, 5) \circ (5, 6, 3)$.

Solution. We have

$$(5, 6, 3) = \begin{pmatrix} 1 & 2 & 3 & 4 & 5 & 6 \\ 1 & 2 & 5 & 4 & 6 & 3 \end{pmatrix}$$

and

$$(4, 1, 3, 5) = \begin{pmatrix} 1 & 2 & 3 & 4 & 5 & 6 \\ 3 & 2 & 5 & 1 & 4 & 6 \end{pmatrix}$$

Then

$$(5, 6, 3) \circ (4, 1, 3, 5) = \begin{pmatrix} 1 & 2 & 3 & 4 & 5 & 6 \\ 1 & 2 & 5 & 4 & 6 & 3 \end{pmatrix} \circ \begin{pmatrix} 1 & 2 & 3 & 4 & 5 & 6 \\ 3 & 2 & 5 & 1 & 4 & 6 \end{pmatrix}$$

$$= \begin{pmatrix} 1 & 2 & 3 & 4 & 5 & 6 \\ 5 & 2 & 6 & 1 & 4 & 3 \end{pmatrix}$$

and

$$(4, 1, 3, 5) \circ (5, 6, 3) = \begin{pmatrix} 1 & 2 & 3 & 4 & 5 & 6 \\ 3 & 2 & 5 & 1 & 4 & 6 \end{pmatrix} \circ \begin{pmatrix} 1 & 2 & 3 & 4 & 5 & 6 \\ 1 & 2 & 5 & 4 & 6 & 3 \end{pmatrix}$$

$$= \begin{pmatrix} 1 & 2 & 3 & 4 & 5 & 6 \\ 3 & 2 & 4 & 1 & 6 & 5 \end{pmatrix}$$

Observe that

$$(5, 6, 3) \circ (4, 1, 3, 5) \neq (4, 1, 3, 5) \circ (5, 6, 3)$$

Two cycles of a set A are said to be **disjoint** if no element of A appears in both cycles.

Example 6   Let $A = \{1, 2, 3, 4, 5, 6\}$. Then the cycles $(1, 2, 5)$ and $(3, 4, 6)$ are disjoint, whereas the cycles $(1, 2, 5)$ and $(2, 4, 6)$ are not.

It is not difficult to show that if $p_1 = (a_1, a_2, \ldots, a_r)$ and $p_2 = (b_1, b_2, \ldots, b_s)$ are disjoint cycles of A, then $p_1 \circ p_2 = p_2 \circ p_1$. This can be seen by observing that p_1 affects only the a's, while p_2 affects only the b's.

We shall now present a fundamental theorem, and instead of giving its proof, we shall do an example which completely imitates the proof.

Theorem 2 A permutation of a finite set that is not the identity or a cycle can be written as a product of disjoint cycles of length ≥ 2.

Example 7 Write the permutation

$$p = \begin{pmatrix} 1 & 2 & 3 & 4 & 5 & 6 & 7 & 8 \\ 3 & 4 & 6 & 5 & 2 & 1 & 8 & 7 \end{pmatrix}$$

of the set $A = \{1, 2, 3, 4, 5, 6, 7, 8\}$ as a product of disjoint cycles.

Solution. We start with 1 and find that $p(1) = 3$, $p(3) = 6$, and $p(6) = 1$, so we have the cycle $(1, 3, 6)$. Next, we choose the first element of A that has not appeared in a previous cycle. We choose 2, and we have $p(2) = 4$, $p(4) = 5$, and $p(5) = 2$, so we obtain the cycle $(2, 4, 5)$. We now choose 7, the first element of A that has not appeared in a previous cycle. Since $p(7) = 8$ and $p(8) = 7$, we obtain the cycle $(7, 8)$. We can then write p as a product of disjoint cycles as

$$p = (1, 3, 6) \circ (2, 4, 5) \circ (7, 8)$$

It is not difficult to show that in Theorem 2, when a permutation is written as a product of disjoint cycles, the product is unique except for the order of the cycles.

Even and Odd Permutations

A cycle of length 2 is called a **transposition.** That is, a transposition is a cycle $p = (a_i, a_j)$, where $p(a_i) = a_j$ and $p(a_j) = a_i$.

Observe that if $p = (a_i, a_j)$ is a transposition of A, then $p \circ p = 1_A$, the identity permutation of A.

Every cycle can be written as a product of transpositions. In fact,

$$(b_1, b_2, \ldots, b_r) = (b_1, b_r) \circ (b_1, b_{r-1}) \circ \cdots \circ (b_1, b_3) \circ (b_1, b_2)$$

This can be verified by induction on r, as follows.

Basis Step. If $r = 2$, then the cycle is just (b_1, b_2), which already has the proper form.

Induction Step. Assume that the result is true for $r = n$, and let $(b_1, b_2, \ldots, b_n, b_{n+1})$ be a cycle of length $n + 1$. Then $(b_1, b_2, \ldots, b_n, b_{n+1}) = (b_1, b_{n+1}) \circ (b_1, b_2, \ldots, b_n)$, as may be verified by computing the composition. By the induction assumption, $(b_1, b_2, \ldots, b_n) = (b_1, b_n) \circ \cdots \circ (b_1, b_3) \circ (b_1, b_2)$. Thus, by substitution $(b_1, b_2, \ldots, b_{n+1}) = (b_1, b_{n+1}) \circ (b_1, b_n) \circ \cdots \circ (b_1, b_3) \circ (b_1, b_2)$. This completes the induction step. For example,

$$(1, 2, 3, 4, 5) = (1, 5) \circ (1, 4) \circ (1, 3) \circ (1, 2)$$

We now obtain the following corollary of Theorem 2.

Corollary 1   Every permutation of a finite set with at least two elements can be written as a product of transpositions.

Observe that the transpositions in Corollary 1 need not be disjoint.

Example 8   Write the permutation p of Example 7 as a product of transpositions.

Solution. We have

$$p = (7, 8) \circ (2, 4, 5) \circ (1, 3, 6)$$

Since we can write

$$(1, 3, 6) = (1, 6) \circ (1, 3)$$
$$(2, 4, 5) = (2, 5) \circ (2, 4)$$

we have

$$p = (7, 8) \circ (2, 5) \circ (2, 4) \circ (1, 6) \circ (1, 3)$$

We have observed above that every cycle can be written as a product of transpositions. However, this can be done in many different ways. For example,

$$\begin{aligned} (1, 2, 3) &= (1, 3) \circ (1, 2) \\ &= (2, 1) \circ (2, 3) \\ &= (1, 3) \circ (3, 1) \circ (1, 3) \circ (1, 2) \circ (3, 2) \circ (2, 3) \end{aligned}$$

It then follows that every permutation on a set of two or more elements can be written as a product of transpositions in many ways. However, the following theorem, whose proof we omit, brings some order to the situation.

Theorem 3   If a permutation of a finite set can be written as a product of an even number of transpositions, then it can never be written as a product of an odd number of transpositions, and conversely.

A permutation of a finite set is called **even** if it can be written as a product of an even number of transpositions, and it is called **odd** if it can be written as a product of an odd number of transpositions.

Example 9 Is the permutation

$$p = \begin{pmatrix} 1 & 2 & 3 & 4 & 5 & 6 & 7 \\ 2 & 4 & 5 & 7 & 6 & 3 & 1 \end{pmatrix}$$

even or odd?

Solution. We first write p as a product of disjoint cycles, obtaining

$$p = (3, 5, 6) \circ (1, 2, 4, 7)$$

Next, we write each of the cycles as a product of transpositions:

$$(1, 2, 4, 7) = (1, 7) \circ (1, 4) \circ (1, 2)$$
$$(3, 5, 6) = (3, 6) \circ (3, 5)$$

Then

$$p = (3, 6) \circ (3, 5) \circ (1, 7) \circ (1, 4) \circ (1, 2)$$

Since p is a product of an odd number of transpositions, it is an odd permutation.

From the definition of even and odd permutations, it follows (Exercise T1) that

(a) The product of two even permutations is even.
(b) The product of two odd permutations is even.
(c) The product of an even and an odd permutation is odd.

Theorem 4 Let $A = \{a_1, a_2, \ldots, a_n\}$ be a finite set with n elements, $n \geq 2$. There are $n!/2$ even permutations and $n!/2$ odd permutations.

Proof. Let A_n be the set of all even permutations of A and let B_n be the set of all odd permutations. We shall define a function $f : A_n \to B_n$ which we show is a one-to-one correspondence, and this will show that A_n and B_n have the same number of elements.

Since $n \geq 2$, we can choose a particular transposition q_0 of A. Say that $q_0 = (a_{n-1}, a_n)$. We now define the function $f : A_n \to B_n$ by

$$f(p) = q_0 \circ p, \qquad p \in A_n$$

Observe that if $p \in A_n$, then p is an even permutation, so $q_0 \circ p$ is an odd permutation and thus $f(p) \in B_n$. Suppose now that p_1 and p_2 are in A_n and

$$f(p_1) = f(p_2)$$

Then

$$q_0 \circ p_1 = q_0 \circ p_2 \qquad (2)$$

We now compose each side of equation (2) with q_0:

$$q_0 \circ (q_0 \circ p_1) = q_0 \circ (q_0 \circ p_2)$$

so by the associative property

$$(q_0 \circ q_0) \circ p_1 = (q_0 \circ q_0) \circ p_2$$

or since $q_0 \circ q_0 = 1_A$,

$$1_A \circ p_1 = 1_A \circ p_2$$
$$p_1 = p_2$$

Thus f is one to one.

Now let $q \in B$. Then $q_0 \circ q \in A_n$, and

$$f(q_0 \circ q) = q_0 \circ (q_0 \circ q)$$
$$= (q_0 \circ q_0) \circ q$$
$$= 1_A \circ q$$
$$= q$$

which means that f is an onto function. Since $f : A_n \to B_n$ is a one-to-one correspondence, we conclude that A_n and B_n have the same number of elements. Note that $A_n \cap B_n = \varnothing$, since no permutation can be both even and odd. Also, by Theorem 1, $|A_n \cup B_n| = n!$. Thus by Theorem 2 of Section 2.1.

$$n! = |A_n \cup B_n| = |A_n| + |B_n| - |A_n \cap B_n| = 2|A_n|$$

We then have

$$|A_n| = |B_n| = \frac{n!}{2}$$

EXERCISE SET 6.2

1. Which of the following functions $f : \mathbb{R} \to \mathbb{R}$ are permutations of $\mathbb{R}$?

    (a) f is defined by $f(a) = a - 1$.

    (b) f is defined by $f(a) = a^2$.
    (c) f is defined by $f(a) = a^3$.
    (d) f is defined by $f(a) = e^a$.

2. Which of the following functions $f : Z \rightarrow Z$ are permutations of Z?

(a) f is defined by $f(a) = a + 1$.

(b) f is defined by $f(a) = (a - 1)^2$.

(c) f is defined by $f(a) = a^2 + 1$.

(d) f is defined by $f(a) = a^3 - 3$.

In Exercises 3 and 4, let $A = \{1, 2, 3, 4, 5, 6\}$ and

$$p_1 = \begin{pmatrix} 1 & 2 & 3 & 4 & 5 & 6 \\ 3 & 4 & 1 & 2 & 6 & 5 \end{pmatrix},$$

$$p_2 = \begin{pmatrix} 1 & 2 & 3 & 4 & 5 & 6 \\ 2 & 3 & 1 & 5 & 4 & 6 \end{pmatrix},$$

$$p_3 = \begin{pmatrix} 1 & 2 & 3 & 4 & 5 & 6 \\ 6 & 3 & 2 & 5 & 4 & 1 \end{pmatrix}$$

3. Compute

(a) p_1^{-1} (b) $p_3 \circ p_1$

(c) $(p_2 \circ p_1) \circ p_2$ (d) $p_1 \circ (p_3 \circ p_2^{-1})$

4. Compute

(a) p_3^{-1} (b) $p_1^{-1} \circ p_2^{-1}$

(c) $(p_3 \circ p_2) \circ p_1$ (d) $p_3 \circ (p_2 \circ p_1)^{-1}$

In Exercises 5–8, let $A = \{1, 2, 3, 4, 5, 6, 7, 8\}$. Compute the products.

5. $(3, 5, 7, 8) \circ (1, 3, 2)$

6. $(2, 6) \circ (3, 5, 7, 8) \circ (2, 5, 3, 4)$

7. $(1, 4) \circ (2, 4, 5, 6) \circ (1, 4, 6, 7)$

8. $(5, 8) \circ (1, 2, 3, 4) \circ (3, 5, 6, 7)$

In Exercises 9 and 10, let $A = \{a, b, c, d, e, f, g\}$. Compute the products.

9. $(a, f, g) \circ (b, c, d, e)$

10. $(f, g) \circ (b, c, f) \circ (a, b, c)$

In Exercises 11–14, let $A = \{1, 2, 3, 4, 5, 6, 7, 8\}$. Write each permutation as the product of disjoint cycles.

11. $\begin{pmatrix} 1 & 2 & 3 & 4 & 5 & 6 & 7 & 8 \\ 4 & 3 & 2 & 5 & 1 & 8 & 7 & 6 \end{pmatrix}$

12. $\begin{pmatrix} 1 & 2 & 3 & 4 & 5 & 6 & 7 & 8 \\ 2 & 3 & 4 & 1 & 7 & 5 & 8 & 6 \end{pmatrix}$

13. $\begin{pmatrix} 1 & 2 & 3 & 4 & 5 & 6 & 7 & 8 \\ 6 & 5 & 7 & 8 & 4 & 3 & 2 & 1 \end{pmatrix}$

14. $\begin{pmatrix} 1 & 2 & 3 & 4 & 5 & 6 & 7 & 8 \\ 2 & 3 & 1 & 4 & 6 & 7 & 8 & 5 \end{pmatrix}$

In Exercises 15 and 16, let $A = \{a, b, c, d, e, f, g\}$. Write each permutation as the product of disjoint cycles.

15. $\begin{pmatrix} a & b & c & d & e & f & g \\ g & d & b & a & c & f & e \end{pmatrix}$

16. $\begin{pmatrix} a & b & c & d & e & f & g \\ d & e & a & b & g & f & c \end{pmatrix}$

In Exercises 17 and 18, let $A = \{1, 2, 3, 4, 5, 6, 7, 8\}$. Write each permutation as a product of transpositions.

17. $(2, 1, 4, 5, 8, 6)$ 18. $(3, 1, 6) \circ (4, 8, 2, 5)$

In Exercises 19–22, let $A = \{1, 2, 3, 4, 5, 6, 7, 8\}$. Determine whether the permutation is even or odd.

19. $\begin{pmatrix} 1 & 2 & 3 & 4 & 5 & 6 & 7 & 8 \\ 4 & 2 & 1 & 6 & 5 & 8 & 7 & 3 \end{pmatrix}$

20. $\begin{pmatrix} 1 & 2 & 3 & 4 & 5 & 6 & 7 & 8 \\ 7 & 3 & 4 & 2 & 1 & 8 & 6 & 5 \end{pmatrix}$

21. $(6, 4, 2, 1, 5)$

22. $(4, 8) \circ (3, 5, 2, 1) \circ (2, 4, 7, 1)$

THEORETICAL EXERCISES

T1. Prove that

(a) The product of two even permutations is even.

(b) The product of two odd permutations is even.

(c) The product of an even and an odd permutation is odd.

T2. Show that if p is a permutation of a finite set A, then $p^2 = p \circ p$ is a permutation of A.

T3. (a) Use mathematical induction to show that if p is a permutation of a finite set A, then $p^n = p \circ p \circ \cdots \circ p$ is a permutation of A for $n \in Z^+$.

(b)   If A is a finite set and p is a permutation of A, show that $p^m = 1_A$ for some $m \in Z^+$.

T4.   Let p be a permutation of a set A. Define the following relation R on A : $a R b$ if and only if $p^n(a) = b$ for some $n \in Z$. [p^0 is defined as the identity permutation and p^{-n} is defined as $(p^{-1})^n$.] Show that R is an equivalence relation, and describe the equivalence classes.

6.3 Binary Operations

We are all familiar from high school algebra with the idea of a binary operation on a set of elements, for example, addition or multiplication on the set of all integers. Earlier in this book, we defined and used binary operations on sets and relations. In this section we show how the idea of binary operation can be made precise using functions. We also briefly introduce the concepts of semigroup and group, which have numerous important uses in mathematics, physics, chemistry, and computer science.

A **binary operation** on a set A is a function $f : A \times A \to A$ that is everywhere defined. Observe the following properties that a binary operation must satisfy:

1.   Since Dom $(f) = A \times A$, f assigns an element $f(a, b)$ of A to each ordered pair (a, b) of elements of A. That is, the binary operation must be defined for each ordered pair of elements of A.

2.   Since a binary operation is a function, only one element of A is assigned to each ordered pair (a, b).

Thus we can say that a binary operation is a rule that assigns to each ordered pair of elements of A a unique element of A. We shall now turn to a number of examples.

It is customary to denote binary operations by a symbol such as $*$, instead of f, and to denote the element assigned to (a, b) by $a * b$ [instead of $*(a, b)$]. It should be emphasized that if a and b are elements in A, then $a * b \in A$; and this property is often described by saying that A is **closed** under the operation $*$.

Example 1   Let $A = Z$. Define $a * b$ as $a + b$. Then $*$ is a binary operation on Z.

Example 2   Let $A = \mathbb{R}$. Define $a * b$ as a/b. Then $*$ is not a binary operation, since it is not defined for every ordered pair of elements of A. For example, $3 * 0$ is not defined, since we cannot divide by zero.

Example 3   Let $A = Z^+$. Define $a * b$ as $a - b$. Then $*$ is not a binary operation since it does assign an element of A to every ordered pair of elements of A; for example, $2 * 5 \notin A$.

Example 4   Let $A = Z$. Define $a * b$ as a number less than both a and b. Then $*$ is not a binary operation, since it does not assign a *unique* element of A to each ordered pair of elements of A; for example, $8 * 6$ could be 5, 4, 3, 1, and so on. Thus, in this case, $*$ would be a relation from $A \times A$ to A, but not a function.

Example 5   Let $A = Z$. Define $a * b$ as max $\{a, b\}$. Then $*$ is a binary operation; for example, $2 * 4 = 4$, $-3 * (-5) = -3$.

Example 6   Let $A = P(S)$, for some set S. If V and W are subsets of S, define $V * W$ as $V \cup W$. Then $*$ is a binary operation on A. Moreover, if we define $V * W$ as $V \cap W$, then $*$ is another binary operation on A.

As Example 6 shows, it is possible to define many binary operations on the same set.

Example 7   Let M be the set of all Boolean matrices. Define $\mathbf{A} * \mathbf{B}$ as $\mathbf{A} \vee \mathbf{B}$ (see Section 3.4). Then $*$ is a binary operation. This is also true of $\mathbf{A} \wedge \mathbf{B}$.

Tables

If $A = \{a_1, a_2, \ldots, a_n\}$ is a finite set, we can define a binary operation on A by means of a table as shown in Fig. 1. The entry in position i, j denotes the element $a_i * a_j$.

$*$	a_1	a_2	$\cdots$	a_j	$\cdots$	a_n
a_1						
a_2						
$\vdots$						
a_i				$a_i * a_j$		
$\vdots$						
a_n						

Figure 1

Example 8   Let $A = \{0, 1\}$. Recall from Section 3.4 that we defined the binary operations $\vee$ and $\wedge$ by the following tables

$\vee$	0	1
0	0	1
1	1	1

$\wedge$	0	1
0	0	0
1	0	1

If $A = \{a, b\}$, we shall now determine the number of binary operations that can be defined on A. Every binary operation $*$ on A can be described by the table

$$
\begin{array}{c|cc}
* & a & b \\
\hline
a & & \\
b & &
\end{array}
$$

Since every blank can be filled in with the element a or b, we conclude that there are $2 \cdot 2 \cdot 2 \cdot 2 = 2^4 = 16$ ways to complete the table. Thus there are 16 binary operations on A.

Properties of Binary Operations

A binary operation on a set A is said to be **commutative** if

$$a * b = b * a$$

for all elements a and b in A.

Example 9   The binary operation of addition on Z (as discussed in Example 1) is commutative.

Example 10   The binary operation of subtraction on Z is not commutative, since

$$2 - 3 \neq 3 - 2$$

A binary operation that is described by a table is commutative if and only if the entries in the table are symmetric with respect to the main diagonal.

Example 11   Which of the following binary operations on $A = \{a, b, c, d\}$ are commutative?

$$
\begin{array}{c|cccc}
* & a & b & c & d \\
\hline
a & a & c & b & d \\
b & b & c & b & a \\
c & c & d & b & c \\
d & a & a & b & b
\end{array}
\qquad
\begin{array}{c|cccc}
* & a & b & c & d \\
\hline
a & a & c & b & d \\
b & c & d & b & a \\
c & b & b & a & c \\
d & d & a & c & d
\end{array}
$$

(a)   (b)

Solution. (a) Is not commutative since $a * b$ is c while $b * a$ is b.

(b) Is commutative since the entries in the table are symmetric with respect to the main diagonal.

A binary operation $*$ on a set A is said to be **associative** if

$$a * (b * c) = (a * b) * c$$

for all elements a, b, and c in A.

Example 12 The binary operation of addition on Z is associative.

Example 13 The binary operation of subtraction on Z is not associative since

$$2 - (3 - 5) \neq (2 - 3) - 5$$

Semigroups

A **semigroup** is a nonempty set S together with an associative binary operation $*$ defined on S. We shall denote the semigroup by $(S, *)$ or, when it is clear what the operation $*$ is, simply by S. We also refer to $a * b$ as the **product** of a and b. The semigroup $(S, *)$ is said to be commutative if $*$ is a commutative operation.

Example 14 Example 12 tells us that $(Z, +)$ is a semigroup. It is commutative, since $n + m = m + n$ for all m, n in Z.

Example 15 The binary operation $\cup$ (union) on $P(S)$, S a set, given in Example 6 is associative. This simply means that $A \cup (B \cup C) = (A \cup B) \cup C$ for all subsets A and B of S, and we know that this property is true. Thus $(P(S), \cup)$ is a semigroup. This semigroup is commutative since $A \cup B = B \cup A$, for all subsets A and B of S.

Example 16 The set Z with the binary operation of subtraction is not a semigroup, since subtraction is not associative.

Example 17 Let S be a fixed nonempty set, and let S^S be the set of all functions $f : S \rightarrow S$. If f and g are elements of S^S, we define $f * g$ as $f \circ g$, the composite function. Then $*$ is a binary operation on S^S and it follows from Section 5.3 that $*$ is associative. Hence $(S^S, *)$ is a semigroup. The semigroup S^S is not commutative.

Example 18 Let $A = \{a_1, a_2, \ldots, a_n\}$ be a nonempty set. Recall from Section 1.2 that A^* is the set of all finite sequences of elements of A. That is, A^* consists of all words that can be formed from the alphabet A. Let α and β be elements of A^*. Observe that catenation is a binary operation $\cdot$ on A^*. Recall that if $\alpha = a_1 a_2 \cdots a_n$ and $\beta = b_1 b_2 \cdots b_k$, then $\alpha \cdot \beta = a_1 a_2 \cdots a_n b_1 b_2 \cdots b_k$. It is easy to see that if α, β and γ are any elements of A^*, then

$$\alpha \cdot (\beta \cdot \gamma) = (\alpha \cdot \beta) \cdot \gamma$$

so that $\cdot$ is an associative binary operation, and $(A^*, \cdot)$ is a semigroup. The semigroup $(A^*, \cdot)$ is called the free semigroup generated by A.

In a semigroup $(S, *)$ we can establish the following generalization of the associative property; we omit the proof.

Theorem 1 If $a_1, a_2, \ldots, a_n$, $(n \geq 3)$, are arbitrary elements of a semigroup, then all products of the elements $a_1, a_2, \ldots, a_n$ that can be formed by inserting arbitrary parentheses are equal.

If $a_1, a_2, \ldots, a_n$ are elements in a semigroup $(S, *)$ we shall write their product as

$$a_1 * a_2 * \cdots * a_n$$

Example 19 Theorem 1 shows that the products

$$((a_1 * a_2) * a_3) * a_4, \qquad a_1 * (a_2 * (a_3 * a_4)), \qquad (a_1 * a_2 * a_3)) * a_4$$

are all equal.

An element e in a semigroup $(S, *)$ is called an **identity** element if

$$e * a = a * e = a$$

for all $a \in S$.

Example 20 The number 0 is an identity in the semigroup $(Z, +)$.

Example 21 The semigroup $(Z^+, +)$ has no identity element.

Theorem 2 If a semigroup $(S, *)$ has an identity element, it is unique.

Proof. Suppose that e and e' are identity elements in S. Then since e is an identity,

$$e * e' = e'$$

Also, since e' is an identity,

$$e * e' = e$$

Hence

$$e = e'$$

A **monoid** is a semigroup $(S, *)$ that has an identity.

Example 22   The semigroup $P(S)$ defined in Example 15 has the identity $\varnothing$, since

$$\varnothing * A = \varnothing \cup A = A = A \cup \varnothing = A * \varnothing$$

for any element $A \in P(S)$. Hence $P(S)$ is a monoid.

Example 23   The semigroup S^S defined in Example 17 has the identity 1_S, since

$$1_S * f = 1_S \circ f = f = f \circ 1_S$$

for any element $f \in S^S$. Hence S^S is a monoid.

Groups

We now examine a special type of monoid, called a group, which has applications in every area where symmetry occurs. Applications of groups can be found in mathematics, physics, and chemistry, as well as in less obvious areas such as sociology. Very recent and exciting applications of group theory have arisen in quarks (particle physics) and in the solutions of puzzles such as Rubik's Cube.

    A **group** $(G, *)$ is a monoid, with identity e, which has the additional property that for every element $a \in G$, there exists an element $a' \in G$, such that $a * a' = a' * a = e$. Thus a group is a set G together with a binary operation $*$ on G such that

1. $(a * b) * c = a * (b * c)$ for any elements a, b, and c in G.
2. There is a unique element e in G such that

$$a * e = e * a \qquad \text{for any } a \in G$$

3. For every $a \in G$ there is an element $a' \in G$, called an **inverse** of a, such that

$$a * a' = a' * a = e$$

Observe that if $(G, *)$ is a group, then $*$ is a binary operation so G must be closed under $*$; that is,

$$a * b \in G \qquad \text{for any elements } a \text{ and } b \text{ in } G$$

To simplify our notation, from now on when only one group $(G, *)$ is under consideration and there is no possibility of confusion, we shall write the product $a * b$ of the elements a and b in the group $(G, *)$ simply as ab, and we shall also refer to $(G, *)$ simply as G.

A commutative group is traditionally called **abelian** in honor of the Norwegian mathematician Niels Abel.

Example 24   The set of all integers Z with the operation of ordinary addition is an abelian group. If $a \in Z$, then an inverse of a is its negative $-a$.

Example 25   The set Z^+ under the operation of ordinary multiplication is not a group since the element 2 in Z^+ has no inverse. However, this set together with the given operation is a monoid.

Example 26   The set of all nonzero real numbers under the operation of ordinary multiplication is a group. An inverse of $a \neq 0$ is $1/a$.

Before proceeding with additional examples of groups, we develop several important properties that are satisfied in any group G.

Theorem 3   Let G be a group. Each element a in G has only one inverse in G.

Proof. Let a' and a'' be inverses of a. Then

$$a'(aa'') = a'e = a'$$

and

$$(a'a)a'' = ea'' = a''$$

Hence, by associativity,

$$a' = a''$$

From now on we shall denote the inverse of a by a^{-1}. Thus in a group G we have

$$aa^{-1} = a^{-1}a = e$$

Theorem 4   Let G be a group and let a, b, and c be elements of G. Then

(a) $ab = ac$ implies that $b = c$ (**left cancellation property**).

(b) $ba = ca$ implies that $b = c$ (**right cancellation property**).

Proof. (a) Suppose that

$$ab = ac$$

Multiplying both sides of this equation on the left by a^{-1}, we obtain

$$a^{-1}(ab) = a^{-1}(ac)$$
$$(a^{-1}a)b = (a^{-1}a)c \qquad \text{(by associativity)}$$
$$eb = ec \qquad \text{(by the definition of an inverse)}$$
$$b = c \qquad \text{(by the definition of an identity)}$$

(b) The proof is similar to that of part (a).

Theorem 5   Let G be a group and let a and b be elements of G. Then

(a) $(a^{-1})^{-1} = a$

(b) $(ab)^{-1} = b^{-1}a^{-1}$

Proof. (a) We have

$$aa^{-1} = a^{-1}a = e$$

and since the inverse of an element is unique, we conclude that $(a^{-1})^{-1} = a$.

(b) We easily verify that

$$(ab)(b^{-1}a^{-1}) = a(b(b^{-1}a^{-1})) = a((bb^{-1})a^{-1}) = a(ea^{-1}) = aa^{-1} = e$$

and similarly,

$$(b^{-1}a^{-1})(ab) = e$$

so

$$(ab)^{-1} = b^{-1}a^{-1}$$

We next turn to an important example of a group.

Example 27   Consider the equilateral triangle shown in Fig. 2 with vertices 1, 2, and 3. A **symmetry** of the triangle (or of any geometrical figure) is a one-to-one function from the set of points forming the triangle (the geometrical figure) to itself which preserves the

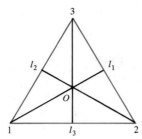

Figure 2

distance between adjacent points. Since the triangle is determined by its vertices, a symmetry of the triangle is merely a permutation of the vertices that preserves the distance between adjacent points. Let l_1, l_2, and l_3 be the angle bisectors of the corresponding angles as shown in Fig. 2, and let O be their point of intersection. We now describe the symmetries of this triangle. First, there is a counterclockwise rotation f_2 of the triangle about O through $120°$. Then f_2 can be written (see Section 6.2) as the permutation.

$$f_2 = \begin{pmatrix} 1 & 2 & 3 \\ 2 & 3 & 1 \end{pmatrix}$$

We next obtain a counterclockwise rotation f_3 about O through $240°$, which can be written as the permutation

$$f_3 = \begin{pmatrix} 1 & 2 & 3 \\ 3 & 1 & 2 \end{pmatrix}$$

Finally, there is a counterclockwise rotation f_1 about O through $360°$ which can be written as the permutation

$$f_1 = \begin{pmatrix} 1 & 2 & 3 \\ 1 & 2 & 3 \end{pmatrix}$$

Of course, f_1 can also be viewed as the result of rotating the triangle about O through $0°$.

We may also obtain three additional symmetries of the triangle, g_1, g_2, and g_3, by reflecting about the lines l_1, l_2, and l_3, respectively. Thus we may denote these reflections as the following permutations

$$g_1 = \begin{pmatrix} 1 & 2 & 3 \\ 1 & 3 & 2 \end{pmatrix}, \qquad g_2 = \begin{pmatrix} 1 & 2 & 3 \\ 3 & 2 & 1 \end{pmatrix}, \qquad g_3 = \begin{pmatrix} 1 & 2 & 3 \\ 2 & 1 & 3 \end{pmatrix}$$

Observe that the set of all symmetries of the triangle is the set of permutations of the set $\{1, 2, 3\}$, which has been considered in Section 6.2 and has been denoted by S_3. Thus

$$S_3 = \{f_1, f_2, f_3, g_1, g_2, g_3\}$$

We now introduce the operation of composition on the set S_3 and we obtain the multiplication table shown in Table 1.

Table 1

*	f_1	f_2	f_3	g_1	g_2	g_3
f_1	f_1	f_2	f_3	g_1	g_2	g_3
f_2	f_2	f_3	f_1	g_3	g_1	g_2
f_3	f_3	f_1	f_2	g_2	g_3	g_1
g_1	g_1	g_2	g_3	f_1	f_2	f_3
g_2	g_2	g_3	g_1	f_3	f_1	f_2
g_3	g_3	g_1	g_2	f_2	f_3	f_1

Each of the entries in this table can be obtained in one of two ways: algebraically or geometrically. For example, suppose that we want to compute $g_2 \circ f_2$. Proceeding algebraically, we have

$$g_2 \circ f_2 = \begin{pmatrix} 1 & 2 & 3 \\ 3 & 2 & 1 \end{pmatrix} \circ \begin{pmatrix} 1 & 2 & 3 \\ 2 & 3 & 1 \end{pmatrix} = \begin{pmatrix} 1 & 2 & 3 \\ 2 & 1 & 3 \end{pmatrix} = g_3$$

Geometrically, we proceed as in Fig. 3. Since composition of functions is always associative, we see that $\circ$ is an associative operation on S_3. Observe that f_1 is the identity in S_3 and that every element of S_3 has a unique inverse in S_3. For example, $f_2^{-1} = f_3$. Hence S_3 is a group called the **group of symmetries of the triangle.** Observe that S_3 is the first example that we have given of a group that is not abelian.

Example 28   The set of all permutations of n elements is a group of order $n!$ under the operation of composition. This group is called the **symmetric group on n letters** and is denoted by S_n. It agrees with the group of symmetries of the equilateral triangle for $n = 3$.

As in Example 27, we can also consider the group of symmetries of a square. However, it turns out that this group is of order 8, so it does not agree with the group S_4, whose order is $4! = 24$.

Applications

At this point applications Chapter 11 can be covered if desired.

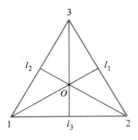

Given triangle

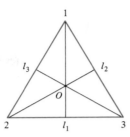

Triangle resulting after f_2

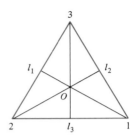

Triangle resulting after applying g_2 to the triangle at the left

Figure 3

EXERCISE SET 6.3

In Exercises 1–8, determine whether the description of $*$ is a valid definition of a binary operation on the set.

1. On $\mathbb{R}$, where $a * b$ is ab (ordinary multiplication).
2. On Z^+, where $a * b$ is a/b.
3. On Z, where $a * b$ is a^b.
4. On Z^+, where $a * b$ is a^b.
5. On Z^+, where $a * b$ is $a - b$.
6. On $\mathbb{R}$, where $a * b$ is $a\sqrt{b}$.
7. On $\mathbb{R}$, where $a * b$ is the largest rational number which is less than ab.
8. On Z, where $a * b$ is $2a + b$.

In Exercises 9–13, determine whether the binary operation $*$ is commutative and whether it is associative on the set.

9. On Z^+, where $a * b$ is $a + b + 2$.
10. On Z, where $a * b$ is ab.
11. On $\mathbb{R}$, where $a * b$ is $a|b|$.
12. On the set of nonzero real numbers, where $a * b$ is a/b.
13. On $\mathbb{R}$, where $a * b$ is the minimum of a and b.
14. Consider the binary operation $*$ defined on the set $A = \{a, b, c, d\}$ by the following table.

$*$	a	b	c	d
a	a	c	b	d
b	d	a	b	c
c	c	d	a	a
d	d	b	a	c

Compute
(a) $c * d$ and $d * c$
(b) $b * d$ and $d * b$
(c) $a * (b * c)$ and $(a * b) * c$
(d) Is $*$ commutative; associative?

15. Let A be a set with n elements.
    (a) How many binary operations can be defined on A?
    (b) How many commutative binary operations can be defined on A?

16. Let $A = \{a, b\}$. Which of the following multiplication tables define a semigroup on A? Which define a monoid on A?

$*$	a	b
a	a	b
b	a	a

(a)

$*$	a	b
a	a	b
b	b	b

(b)

$*$	a	b
a	b	a
b	a	b

(c)

$*$	a	b
a	a	b
b	b	a

(d)

$*$	a	b
a	a	a
b	b	b

(e)

$*$	a	b
a	b	b
b	a	a

(f)

In Exercises 17–23 determine whether the set together with the binary operation is a semigroup, a monoid, or neither. If it is a monoid, specify the identity. If it is a semigroup or a monoid, determine if it is commutative.

17. Z^+, where $*$ is defined as ordinary multiplication.

18. Z^+, where $a * b$ is defined as max $\{a, b\}$.

19. Z^+, where $a * b$ is defined as GCD $\{a, b\}$.

20. Z^+, where $a * b$ is defined as a.

21. The nonzero real numbers, where $*$ is ordinary multiplication.

22. $P(S)$, with S a set, where $*$ is defined as intersection.

23. Z, where $a * b = a + b - ab$.

In Exercises 24–31, determine whether the set together with the binary operation is a group. If it is a group, determine if it is abelian; specify the identity and the inverse of an element a.

24. Z, where $*$ is ordinary multiplication.

25. Z, where $*$ is subtraction.

26. Q, the set of all rational numbers under the operation of addition.

27. Q, the set of all rational numbers under the operation of multiplication.

28. $\mathbb{R}$, under the operation of multiplication.

29. $\mathbb{R}$, where $a * b = a + b + 2$.

30. Z^+, under the operation of addition.

31. The set of all $m \times n$ matrices under the operation of matrix addition.

32. Let $S = \{x \mid x$ is a real number and $x \neq 0$, $x \neq -1\}$. Consider the following functions $f_i: S \to S$, $i = 1, 2, \ldots, 6$:

$$f_1(x) = x, \qquad f_2(x) = 1 - x, \qquad f_3(x) = \frac{1}{x},$$

$$f_4 = \frac{1}{1 - x}, \qquad f_5(x) = 1 - \frac{1}{x}, \qquad f_6(x) = \frac{x}{x - 1}$$

Show that $G = \{f_1, f_2, f_3, f_4, f_5, f_6\}$ is a group under the operation of composition. Determine the multiplication table of G.

THEORETICAL EXERCISES

T.1. Let G be a group with identity e. Show that if $x^2 = x$ for some x in G, then $x = e$.

T.2. Show that a group G is abelian if and only if $(ab)^2 = a^2b^2$ for all elements a and b in G.

KEY IDEAS FOR REVIEW

☐ Function: see page 133.

☐ Function defined by a formula: see page 135.

☐ Identity function, $1_A : 1_A(a) = a$.

☐ One-to-one function f from A to B: $a \neq a'$ implies $f(a) \neq f(a')$.

☐ Onto function f from A to B: Ran $(f) = B$.

☐ Bijection: one-to-one and onto function.

☐ If f is a function from A to B, $1_B \circ f = f$; $f \circ 1_A = f$.

☐ If f is an invertible function from A to B, $f^{-1} \circ f = 1_A$; $f \circ f^{-1} = 1_B$.

☐ Theorem: A function $f: A \to B$ is invertible if and only if f is one to one and onto.

☐ $(g \circ f)^{-1} = f^{-1} \circ g^{-1}$

☐ Permutation: a bijection from a set A to itself.

☐ Theorem: If A is a set that has n elements, then there are $n!$ permutations of A.

☐ Cycle of length r: $(b_1, b_2, \ldots, b_r)$: see page 145.

☐ Theorem: A permutation of a finite set that is not the identity or a cycle can be written as a product of disjoint cycles.

☐ Transposition: a cycle of length 2.

☐ Corollary: Every permutation of a finite set with at least two elements can be written as a product of transpositions.

☐ Even (odd) permutation: one that can be written as a product of an even (odd) number of transpositions.

☐ Theorem: If a permutation of a finite set can be written as a product of an even number of transpositions, then it can never be written as a product of an odd number of transpositions, and conversely.

☐ The product of:
(a) Two even permutations is even
(b) Two odd permutations is even.
(c) An even and an odd permutation is odd.

☐ Theorem: If A is a set that has n elements, then there are $n!/2$ even permutations and $n!/2$ odd permutations of A.

☐ Binary operation on A: function $f: A \times A \rightarrow A$.

☐ Commutative binary operation: $a * b = b * a$.

☐ Associative binary operation: $a * (b * c) = (a * b) * c$.

☐ Semigroup: nonempty set S together with an associative operation $*$ defined on S.

☐ Theorem: If a semigroup $(S, *)$ has an identity element, it is unique.

☐ Monoid: semigroup that has an identity.

☐ Group $(G, *)$: monoid with identity e such that for every $a \in G$, there exists $a' \in G$ with the property that $a * a' = a' * a = e$.

☐ Theorem: let G be a group and let a, b, and c be elements of G. Then:
(a) $ab = ac$ implies that $b = c$ (left cancellation property).
(b) $ba = ca$ implies that $b = c$ (right cancellation property).

☐ Theorem: let G be a group and let a and b be elements of G. Then
(a) $(a^{-1})^{-1} = a$
(b) $(ab)^{-1} = b^{-1}a^{-1}$

☐ Order of a group G: $|G|$, the number of elements in G.

☐ S_n: the symmetric group on n letters.

REVIEW EXERCISES

1. In each part, $A = \{1, 2, 8, 9\}$, $B = \{s, t, u, v\}$. Determine whether the relation R from A to B is a function. If it is a function, give its domain and range.

(a) $R = \{(1, s), (2, u), (8, v), (9, v)\}$
(b) $R = \{(2, t), (8, v)\}$
(c) $R = \{(8, s), (9, t), (1, u), (1, v)\}$

2. In each part, sets A and B, and a function f from A to B are given. Determine whether the function f is one to one or onto (or both or neither). If f is one to one, compute the function f^{-1}.

(a) $A = \{1, 2, 3\}$, $B = \{4, 6, 8\}$
    $f = \{(1, 4), (2, 8), (3, 6)\}$

(b) $A = \{x, y, z\}$, $B = \{a, b, c, d\}$
    $f = \{(x, c), (y, a), (z, b)\}$

(c) $A = \{1, 2, 3, 4\}$, $B = \{1, 2, 3, 4\}$
    $f = \{(1, 3), (2, 4), (3, 1), (4, 3)\}$

3. Let $f\colon \mathbb{R} \to \mathbb{R}$ be defined by

$$f(x) = 3x^3$$

and let $g\colon \mathbb{R} \to \mathbb{R}$ be defined by

$$g(x) = 6x + 1$$

Describe each of the following functions by formulas.

(a) $f \circ f$
(b) $f^{-1} \circ g$
(c) $(g \circ g) \circ f$

4. Let

$$p_1 = \begin{pmatrix} 1 & 2 & 3 & 4 & 5 & 6 \\ 3 & 4 & 5 & 6 & 1 & 2 \end{pmatrix} \text{ and}$$

$$p_2 = \begin{pmatrix} 1 & 2 & 3 & 4 & 5 & 6 \\ 6 & 5 & 1 & 2 & 4 & 3 \end{pmatrix}$$

Compute

(a) p_1^{-1} (b) $p_2 \circ p_1$
(c) p_1^2 (d) $p_2^{-1} \circ p_1$

5. Let $A = \{1, 2, 3, 4, 5, 6, 7, 8, 9\}$. Compute the following products of cycles.

(a) $(1, 3, 5, 7) \circ (1, 2)$
(b) $(4, 8, 2) \circ (7, 8, 1) \circ (4, 5, 6)$
(c) $(1, 2, 9, 8) \circ (1, 2, 4, 5)$
(d) $(1, 2, 4, 5) \circ (7, 9) \circ (4, 3, 8)$

6. Write each of the following permutations as the product of disjoint cycles.

(a) $\begin{pmatrix} 1 & 2 & 3 & 4 & 5 & 6 & 7 & 8 & 9 \\ 3 & 9 & 5 & 4 & 1 & 7 & 2 & 6 & 8 \end{pmatrix}$

(b) $\begin{pmatrix} 1 & 2 & 3 & 4 & 5 & 6 & 7 & 8 \\ 1 & 3 & 5 & 2 & 8 & 7 & 6 & 4 \end{pmatrix}$

(c) $\begin{pmatrix} 1 & 2 & 3 & 4 & 5 & 6 & 7 & 8 & 9 & 10 \\ 2 & 1 & 6 & 10 & 8 & 5 & 4 & 3 & 7 & 9 \end{pmatrix}$

7. Determine whether the following permutations are even or odd.

(a) $\begin{pmatrix} 1 & 2 & 3 & 4 & 5 & 6 & 7 & 8 \\ 8 & 7 & 6 & 1 & 2 & 4 & 5 & 3 \end{pmatrix}$

(b) $(4, 5, 1) \circ (3, 5, 2) \circ (1, 2, 4)$

8. In each of the following determine whether the description of $*$ is a valid definition of a binary operation on the given set.

(a) On Z^+, where $a * b$ is a/b.
(b) On $P(S)$, where $A * B$ is $A \oplus B$.
(c) On $\mathbb{R}$, where $a * b = a + \sqrt{b}$.

9. In each of the following, determine whether the binary operation $*$ is commutative and whether it is associative on the set given.

(a) On $P(S)$, where $A * B = A - B$.
(b) On Z, where $a * b = \min \{a, b\}$.
(c) On $\mathbb{R}^+$, where $a * b = a^2 b$.

10. Let $A = \{a, b\}$. Which of the following multiplication tables define a semigroup on A? Which define a monoid? Which define a group?

(a)

$*$	a	b
a	a	a
b	a	a

(b)

$*$	a	b
a	b	a
b	a	b

(c)

$*$	a	b
a	a	b
b	b	a

11. In each of the following, determine whether the set given, together with the binary operation given is a group.

(a) $\mathbb{R}$ under the operation of subtraction.
(b) Z^+ under the operation of multiplication.
(c) $\mathbb{R}^+$ under the operation of multiplication.

CHAPTER TEST

1. Let $A = \{x, y, z, w\}$, $B = \{1, 2, 3, 4\}$. In each part determine whether the relation R from A to B is a function. If it is a function, state whether it is onto and whether it is one to one.
    (a) $R = \{(x, 2), (y, 4), (z, 1), (w, 2)\}$
    (b) $R = \{(x, 4), (y, 1), (z, 2), (w, 3)\}$
    (c) $R = \{(x, 4), (y, 2), (z, 4), (w, 3)\}$

2. Let $A = \{1, 2, 3\}$, $B = \{a, b, c\}$, and let f and g be functions from A to B given by

    $$f = \{(1, a), (2, a), (3, b)\}$$

    $$g = \{(1, c), (2, a), (3, b)\}$$

    Compute
    (a) g^{-1}
    (b) $(f \circ g^{-1}) \circ g^{-1}$
    (c) $f \circ (g^{-1} \circ f)$

3. Let $A = \{1, 2, 3, 4, 5, 6, 7, 8, 9, 10\}$. Compute the following products.
    (a) $(5, 10, 1) \circ (3, 4, 9)^{-1}$
    (b) $(9, 10, 3, 2) \circ (1, 2, 3, 6, 7)$
    (c) $(1, 2, 3, 4) \circ (8, 7, 6) \circ (1, 9, 10)$

4. Write each of the following permutations as the product of disjoint cycles.
    (a) $\begin{pmatrix} 1 & 2 & 3 & 4 & 5 & 6 \\ 1 & 3 & 5 & 6 & 2 & 4 \end{pmatrix}$
    (b) $\begin{pmatrix} 1 & 2 & 3 & 4 & 5 & 6 & 7 & 8 & 9 \\ 2 & 3 & 8 & 7 & 9 & 4 & 6 & 1 & 5 \end{pmatrix}$
    (c) $(7, 8, 4, 5, 6) \circ (1, 2, 3)$

5. In each of the following, determine whether the description of $*$ is a valid definition of a binary operation on the given set.
    (a) On the set P of all permutations of $\{1, 2, 3\}$, where $p_2 * p_1 = p_2^{-1} \circ p_1$.
    (b) On the set Z, where $a * b = (a + b)/2$.
    (c) On the set Z^+, where $a * b = a^2 - b^2$.

6. In each of the following, determine whether the given set, with the given binary operation, is a semigroup, a monoid, or a group.
    (a) $A =$ all 2×2 matrices, under the operation of matrix multiplication.
    (b) $A =$ all even integers, under the operation of multiplication.
    (c) $A =$ all 2×2 matrices under the operation of matrix addition.

7. For each of the following, answer true (T) or false (F).
    (a) If $f: A \rightarrow B$ is a function, then f must be everywhere defined.
    (b) If $f: A \rightarrow B$ is a function, and f^{-1} is also a function, then f must be one to one and onto.
    (c) The composition of two permutations of a set A is a permutation of A.
    (d) If G is a group, then the equations $ax = b$ and $ya = b$ always have solutions in G.
    (e) If G is a group with identity e, then the set of all elements x in G such that $x^2 = e$ is a semigroup with respect to the same operation as the operation in G.

7

Partially Ordered Sets

Prerequisites: Chapters 1 to 6.

> In this chapter we study partially ordered sets, including lattices and Boolean algebras. These structures are useful in set theory, algebra, sorting and searching, and, especially in the case of Boolean algebras, in the construction of logical representations for computer circuits.

7.1 Partially Ordered Sets (Posets)

> A relation R on a set A is called a **partial order** if R is reflexive, antisymmetric, and transitive. The set A together with the partial order R is called a **partially ordered set**, or simply a **poset**, and we will denote this poset by (A, R). If there is no possibility of confusion about the partial order, we may refer to the poset simply as A, rather than (A, R).

Example 1 Let A be a collection of subsets of a set S. The relation $\subseteq$ of set inclusion is a partial order on A, so $(A, \subseteq)$ is a poset.

Example 2 Let Z^+ be the set of all positive integers. The usual relation $\leq$ (less than or equal to) is a partial order on Z^+, as is $\geq$ (greater than or equal to).

Example 3 The relation of divisibility ($a\,R\,b$ if and only if $a \mid b$) is a partial order on Z^+.

Example 4   Let W be the set of all equivalence relations on a set A. Since W consists of subsets of $A \times A$, W is a partially ordered set under the partial order of set inclusion. If R and S are equivalence relations on A, the same property may be expressed in relational notation as follows.

$$R \subseteq S \text{ if and only if } x \, R \, y \quad \text{implies} \quad x \, S \, y \text{ for all } x, y \text{ in } A$$

Then $(W, \subseteq)$ is a poset.

Example 5   The relation $<$ on Z^+ is not a partial order since it is not reflexive.

Example 6   Let R be a partial order on a set A, and let R^{-1} be the inverse relation of R. Then R^{-1} is also a partial order. To see this we recall the characterizations of reflexive, anti-symmetric, and transitive given in Section 5.1. If R has these three properties, then, by Theorems 1, 2, and 3 of Section 5.3, R^{-1} also has these properties. Thus R^{-1} is also a partial order. The poset (A, R^{-1}) is called the **dual** of the poset (A, R), and the partial order R^{-1} is called the **dual** of the partial order R.

The most familiar partial orders are the relations $\leq$ and $\geq$ on Z and $\mathbb{R}$. For this reason, when speaking in general of a partial order R on a set A, we shall often use the symbols $\leq$ or $\geq$ for R. This makes the properties of R more familiar and easier to remember. Thus the reader may see the symbol $\leq$ used for many different partial orders, on different sets. Do not mistake this to mean that these relations are all the same, or that they have anything to do with the familiar relation $\leq$ on Z or $\mathbb{R}$. If it becomes absolutely necessary to distinguish partial orders from one another, we may also use symbols such as $\leq_1$, $\leq'$, $\geq_1$, $\geq'$, and so on, to denote partial orders.

We will observe the following convention. Whenever $(A, \leq)$ is a poset, we will always use the symbol $\geq$ for the partial order $\leq^{-1}$, and thus $(A, \geq)$ will be the dual poset. Similarly, the dual of poset $(A, \leq_1)$ will be denoted by $(A, \geq_1)$ and the dual of the poset $(B, \leq')$ will be denoted by $(B, \geq')$. Again, this convention is to remind us of the familiar dual posets $(Z, \leq)$ and $(Z, \geq)$, as well as the posets $(\mathbb{R}, \leq)$ and $(\mathbb{R}, \geq)$.

If $(A, \leq)$ is a poset, the elements a and b of A are said to be **comparable** if

$$a \leq b \quad \text{or} \quad b \leq a$$

Observe that in a partially ordered set every pair of elements need not be comparable. For example, consider the poset in Example 3. The elements 2 and 7 are not comparable, since $2 \not\mid 7$ and $7 \not\mid 2$. Thus the word "partial" in partially ordered set means that some elements may not be comparable. If every pair of elements in a poset A is comparable, we say that A is a **linearly ordered** set and the partial order is called a **linear order.** We also say that A is a **chain.**

Example 7 The poset of Example 2 is linearly ordered.

The following theorem is sometimes useful, since it shows how to construct a new poset from given posets.

Theorem 1 If $(A, \leq)$ and $(B, \leq)$ are posets, then $(A \times B, \leq)$ is a poset, with partial order $\leq$ defined by

$$(a, b) \leq (a', b') \qquad \text{if } a \leq a' \text{ in } A \text{ and } b \leq b' \text{ in } B$$

Note that the symbol $\leq$ is being used to denote three distinct partial orders. The reader should find it easy to determine which of the three is meant at any time.

Proof. If $(a, b) \in A \times B$, then $(a, b) \leq (a, b)$ since $a \leq a$ in A and $b \leq b$ in B, so $\leq$ satisfies the reflexive property in $A \times B$. Now, suppose that $(a, b) \leq (a', b')$ and $(a', b') \leq (a, b)$, where a and $a' \in A$, and b and $b' \in B$. Then

$$a \leq a' \qquad \text{and} \qquad a' \leq a \qquad \text{in } A$$

and

$$b \leq b' \qquad \text{and} \qquad b' \leq b \qquad \text{in } B$$

Since A and B are posets, the antisymmetry of the partial orders in A and B implies that

$$a = a' \qquad \text{and} \qquad b = b'$$

Hence $(a, b) = (a', b')$, and $\leq$ satisfies the antisymmetric property in $A \times B$. Finally, suppose that

$$(a, b) \leq (a', b') \qquad \text{and} \qquad (a', b') \leq (a'', b'')$$

where $a, a', a'' \in A$, and $b, b', b'' \in B$. Then

$$a \leq a' \qquad \text{and} \qquad a' \leq a''$$

so $a \leq a''$, by the transitive property of the partial order in A. Similarly,

$$b \leq b' \qquad \text{and} \qquad b' \leq b''$$

so $b \leq b''$, by the transitive property of the partial order in B. Hence

$$(a, b) \leq (a'', b'')$$

Consequently, the transitive property holds for the partial order in $A \times B$, and we conclude that $A \times B$ is a poset.

The partial order $\leq$ defined on the Cartesian product $A \times B$ as above is called the **product partial order**.

If $(A, \leq)$ is a poset, we say that $a < b$ if $a \leq b$ but $a \neq b$. Suppose now that $(A, \leq)$ and $(B, \leq)$ are posets. In Theorem 1 we have defined the product partial order on $A \times B$. Another useful partial order on $A \times B$, denoted by $\prec$, is defined as follows:

$$(a, b) \prec (a', b') \quad \text{if } a < a' \quad \text{or} \quad \text{if } a = a' \text{ and } b \leq b'$$

This ordering is called **lexicographic**, or "dictionary" order. The ordering of the elements in the first coordinate dominates, except in case of "ties," when attention passes on to the second coordinate. If $(A, \leq)$ and $(B, \leq)$ are linearly ordered sets, then the lexicographic order $\prec$ on $A \times B$ is also a linear order.

Example 8   Let $A = \mathbb{R}$, with the usual ordering $\leq$. Then the plane $\mathbb{R}^2 = \mathbb{R} \times \mathbb{R}$ may be given lexicographic order. This is illustrated in Fig. 1. We see that the plane is linearly ordered by lexicographic order. Each vertical line has the usual order and points on one vertical line are less than any points on a vertical line farther to the right. Thus in Fig. 1, $p_1 \prec p_2$, $p_1 \prec p_3$, and $p_2 \prec p_3$.

Lexicographic ordering is easily extended to Cartesian products $A_1 \times A_2 \times \cdots \times A_n$ of posets as follows:

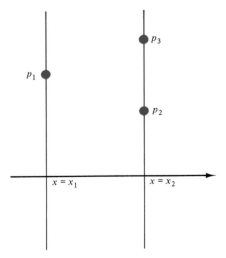

Figure 1

$$(a_1, a_2, \ldots, a_n) \prec (a'_1, a'_2, \ldots, a'_n) \qquad \text{if and only if}$$

$a_1 < a'_1$ or

$a_1 = a'_1$ and $a_2 < a'_2$ or

$a_1 = a'_1$, $a_2 = a'_2$, and $a_3 < a'_3$ or $\ldots$

$a_1 = a'_1$, $a_2 = a'_2$, $\ldots$, $a_{n-1} = a'_{n-1}$ and $a_n \leq a'_n$

Thus the first coordinate dominates except for equality, in which case we consider the second coordinate. If equality holds again, we pass to the next coordinate, and so on.

Example 9 Let $S = \{a, b, \ldots, z\}$ be the ordinary alphabet, linearly ordered in the usual way ($a \leq b$, $b \leq c$, $\ldots$, $y \leq z$). Then S^n (see Section 4.1) can be identified with the set of all words having length n. Lexicographic order on S^n has the property that if $w_1 \prec w_2$ ($w_1, w_2 \in S^n$), then w_1 would precede w_2 in a dictionary listing. This fact accounts for the name of the ordering.

Thus *park* $\prec$ *part*, *help* $\prec$ *hind*, *jump* $\prec$ *mump*. The third is true since $j < m$, the second is true since $h = h$, but $e < i$, and the first is true since $p = p$, $a = a$, $r = r$, $k \leq t$.

If S is a poset we can extend lexicographic order to S^* (see Section 1.2) in the following way.

If $x = a_1 a_2 \cdots a_n$ and $y = b_1 b_2 \cdots b_k$ are in S^* with $n \leq k$, we say that $x \prec y$ if $(a_1, \ldots, a_n) \prec (b_1, \ldots, b_n)$ in S^n under lexicographic ordering of S^n. In other words, we chop off to the length of the shorter word, and then compare.

In the previous paragraph, we use the fact that the n-tuple $(a_1, a_2, \ldots, a_n)$ $\in S^n$ and the string $a_1 a_2 \ldots a_n \in S^*$ are really the same sequence of length n, written in two different notations. The notations differ for historical reasons, and we will use them interchangeably, depending on context.

Example 10 Let $S = \{a, b, \ldots, z\}$, ordered as usual. Then S^* is the set of all possible "words" of any length, whether such words are meaningful or not.

Thus we have

$$help \prec helping$$

in S^* since

$$help \prec help$$

in S^4. Similarly, we have

$$helper \prec helping$$

since

$$helper \prec helpin$$

in S^6. As the example

$$help \prec helping$$

shows, this order includes "prefix order"; that is, any word is greater than all of its prefixes (beginning parts). This is also the way that words occur in the dictionary. Thus we have dictionary ordering again, but this time for words of any finite length.

Since a partial order is a relation, we can look at the digraph of any partial order on a finite set. We shall find that the digraphs of partial orders can be represented in a simpler manner than those of general relations. The following theorem provides the first result in this direction.

Theorem 2   The digraph of a partial order has no cycle of length greater than 1.

Proof. Suppose that the digraph of the partial order $\leq$ on the set A contains a cycle of length $n \geq 2$. Then there exist distinct elements $a_1, a_2, \ldots, a_n \in A$ such that

$$a_1 \leq a_2, a_2 \leq a_3, \ldots, a_{n-1} \leq a_n, a_n \leq a_1$$

By the transitivity of the partial order, used $n - 1$ times, $a_1 \leq a_n$. By antisymmetry, $a_n \leq a_1$ and $a_1 \leq a_n$ imply that $a_n = a_1$, a contradiction to the assumption that $a_1, a_2, \ldots, a_n$ are distinct.

Hasse Diagrams

Let A be a finite set. Theorem 2 has shown that the digraph of a partial order on A has only cycles of length 1. Indeed, since a partial order is reflexive, every vertex in the digraph of the partial order is contained in a cycle of length 1. To simplify matters, we shall delete all such cycles from the digraph. Thus the digraph shown in Fig. 2(a) would be drawn as shown in Fig. 2(b).

We shall also eliminate all edges that are implied by the transitive property. Thus, if $a \leq b$, and $b \leq c$, it follows that $a \leq c$. In this case, we omit the edge from a to c; however, we do draw the edges from a to b and from b to c. For example, the digraph shown in Fig. 2(b) would be drawn as shown in Fig. 3. We also agree to draw

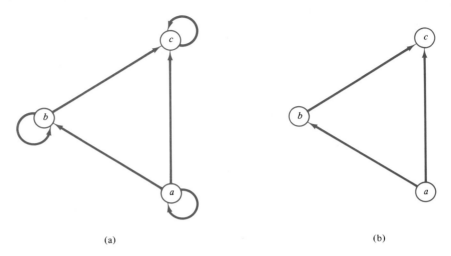

(a)   (b)

Figure 2

the digraph of a partial order with all edges pointing upward, so that arrows may be omitted from the edges. Finally, we replace the circles representing the vertices by dots. Thus the diagram shown in Fig. 4 gives the final form of the digraph shown in Fig. 2(a). The resulting diagram of a partial order, much simpler than its digraph, is called the **Hasse diagram** of the partial order or of the poset. Since the Hasse diagram completely describes the associated partial order, we shall find it to be a very useful tool.

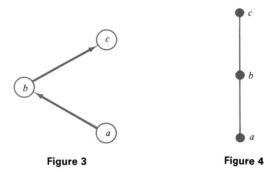

Figure 3   **Figure 4**

Example 11   Let $A = \{1, 2, 3, 4, 12\}$. Consider the partial order of divisibility on A. That is, if a and $b \in A$, $a \leq b$ if and only if $a \mid b$. Draw the Hasse diagram of the poset $(A, \leq)$.

Solution. The Hasse diagram is shown in Fig. 5.

To emphasize the simplicity of the Hasse diagram, we show in Fig. 6 the digraph of the poset in Fig. 5.

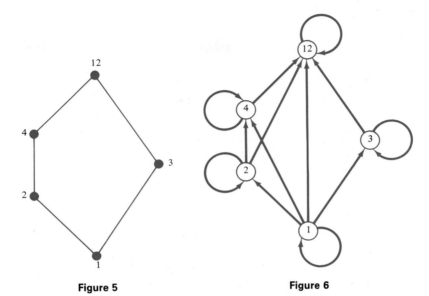

Figure 5   Figure 6

Example 12   Let $S = \{a, b, c\}$ and let $A = P(S)$. Draw the Hasse diagram of the poset A with the partial order $\subseteq$ (set inclusion).

Solution. We first determine A, obtaining

$$A = \{\varnothing, \{a\}, \{b\}, \{c\}, \{a, b\}, \{a, c\}, \{b, c\}, \{a, b, c\}\}$$

The Hasse diagram can then be drawn as shown in Fig. 7.

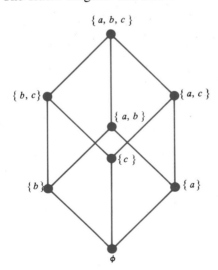

Figure 7

Observe that the Hasse diagram of a linearly ordered set is always of the form shown in Fig. 8.

Given the Hasse diagram of a partial order $\leq$, we can easily reconstruct the digraph of $\leq$. First, we recall that since every partial order is reflexive, each vertex is related to itself. Otherwise if a and b are any two vertices, then $a \leq b$ if and only if there is an *upward directed* path from a to b in the Hasse diagram.

Example 13   Figure 9(a) shows the Hasse diagram of a poset $(A, \leq)$, where $A = \{a, b, c, d, e, f\}$. We know that $a \leq a$. Since there are upward directed paths from a to d and from a to f, we must also have $a \leq d$ and $a \leq f$. A similar analysis can be made for all other vertices. The result is the digraph for $\leq$, which is shown in Figure 9(b).

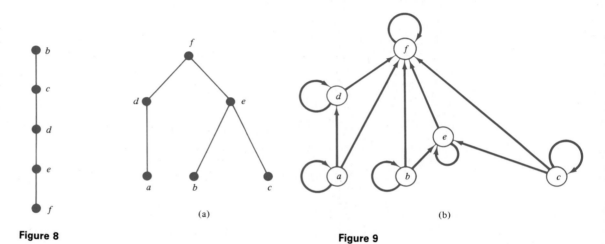

(a)   (b)

Figure 8   **Figure 9**

Example 14   Figure 10(a) shows the Hasse diagram of a poset $(A, \leq)$, where $A = \{a, b, c, d, e, f\}$. Figure 10(b) shows the Hasse diagram of the dual poset $(A, \geq)$. Notice that, as mentioned above, each of these diagrams can be constructed by turning the other upside down.

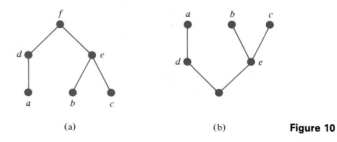

(a)   (b)   **Figure 10**

Isomorphism

Let $(A, \leq)$ and $(A', \leq)$ be posets and let $f: A \rightarrow A'$ be a one-to-one correspondence between A and A'. The function f is called an **isomorphism** from $(A \leq)$ to $(A', \leq')$ if, for any a and b in A,

$$a \leq b \quad \text{if and only if} \quad f(a) \leq' f(b).$$

If $f: A \rightarrow A'$ is an isomorphism, we say that $(A, \leq)$ and $(A', \leq')$ are **isomorphic posets**.

Example 15   Let A be the set Z^+ of all positive integers, and let $\leq$ be the usual partial order on A (see Example 2). Let A' be the set of positive even integers, and let $\leq'$ be the usual partial order on A'. Then we show that the function $f: A \rightarrow A'$ given by

$$f(a) = 2a$$

is an isomorphism from $(A, \leq)$ to $(A', \leq')$.

Solution.   First, f is one to one, since if $f(a) = f(b)$, then $2a = 2b$, so $a = b$. Next, if $c \in A'$, then $c = 2a$ for some $a \in Z^+$; therefore, $c = f(a)$. This shows that f is onto, so we see that f is a one-to-one correspondence. Finally, if a and b are elements of A, then it is clear that if $a \leq b$ then $2a \leq 2b$, so $f(a) \leq' f(b)$. Thus f is an isomorphism.

Suppose that $f: A \rightarrow A'$ is an isomorphism from a poset $(A, \leq)$ to a poset $(A', \leq')$. Suppose also that B is a subset of A and $B' = f(B)$ is the corresponding subset of A'. Then we see from the definition of isomorphism that the following general principle must hold.

Principle of Correspondence

If the elements of B have any property, relating to one another or to other elements of A, and if this property can be defined entirely in terms of the relation $\leq$, then the elements of B' must possess exactly the same property, defined in terms of $\leq'$.

Example 16   Let $(A, \leq)$ be the poset whose Hasse diagram is shown in Fig. 11 and suppose that f is an isomorphism from $(A, \leq)$ to some other poset $(A', \leq')$. Note first that $d \leq x$ for any x in A (later we will call an element such as d a "least element" of A). Then the corresponding element $f(d)$ in A' must satisfy the property $f(d) \leq' y$ for all y in

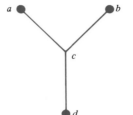

Figure 11

A'. As another example, note that $a \not\leq b$ and $b \not\leq a$. Such a pair is called **incomparable** in A. It then follows from the principle of correspondence that $f(a)$ and $f(b)$ must be incomparable in A'.

For a finite poset, one of the objects that is defined entirely in terms of the partial order is its Hasse diagram. It follows from the principle of correspondence that two finite isomorphic posets must have the same Hasse diagrams.

To be precise, let $(A, \leq)$ and $(A', \leq')$ be finite posets, let $f: A \rightarrow A'$ be a one-to-one correspondence, and let H be any Hasse diagram of $(A, \leq)$. Then

1. If f is an isomorphism and each label a of H is replaced by $f(a)$, then H will become a Hasse diagram for $(A', \leq')$.

Conversely,

2. If H becomes a Hasse diagram for $(A', \leq')$, whenever each label a is replaced by $f(a)$, then f is an isomorphism.

Example 17   Let $A = \{1, 2, 3, 6\}$ and let $\leq$ be the relation "$|$" (divides). Figure 12(a) shows a Hasse diagram for $(A, \leq)$. Let $A = P(\{a, b\}) = \{\emptyset, \{a\}, \{b\}, \{a, b\}\}$ and let $\leq'$ be set containment, $\subseteq$. If $f: A \rightarrow A'$ is defined by

$$f(1) = \emptyset, \qquad f(2) = \{a\}, \qquad f(3) = \{b\}, \qquad f(6) = \{a, b\}$$

then it is easily seen that f is a one-to-one correspondence. If each label $a \in A$ of the Hasse diagram is replaced by $f(a)$, the result is as shown in Fig. 12(b). Since this is clearly a Hasse diagram for $(A', \leq')$, the function f is an isomorphism.

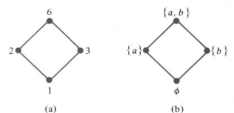

(a)   (b)   **Figure 12**

EXERCISE SET 7.1

In Exercises 1–4, determine whether the relation R is a partial order on the set A.

1. $A = Z$, and $a\,R\,b$ if and only if $a = 2b$.
2. $A = Z$, and $a\,R\,b$ if and only if $b^2 \,|\, a$.
3. $A = Z$, and $a\,R\,b$ if and only if $a = b^k$ for some $k \in Z^+$.
4. $A = \mathbb{R}$, and $a\,R\,b$ if and only if $a \le b$.

In Exercises 5–8, determine whether the relation R is a linear order on the set A.

5. $A = \mathbb{R}$, and $a\,R\,b$ if and only if $a \le b$.
6. $A = \mathbb{R}$, and $a\,R\,b$ if and only if $a \ge b$.
7. $A = P(S)$, where S is a set. The relation R is set inclusion.
8. $A = \mathbb{R} \times \mathbb{R}$, and $(a, b)\,R\,(a', b')$ if and only if $a \le a'$ and $b \le b'$, where $\le$ is the usual partial order on $\mathbb{R}$.
9. On the set $A = \{a, b, c\}$, find all partial orders $\le$ in which $a \le b$.
10. What can you say about the relation R on a set A if R is a partial order and an equivalence relation?

In Exercises 11 and 12, determine the Hasse diagram of the relation R.

11. $A = \{1, 2, 3, 4\}$, $R = \{(1, 1),\ (1, 2),\ (2, 2),\ (2, 4), (1, 3), (3, 3), (3, 4), (1, 4), (4, 4)\}$.
12. $A = \{a, b, c, d, e\}$, $R = \{(a, a),\ (b, b),\ (c, c), (a, c), (c, d), (c, e), (a, d), (d, d), (a, e), (b, c), (b, d), (b, e), (e, e)\}$.

In Exercises 13 and 14, describe the ordered pairs in the relation determined by the Hasse diagram on the set A.

13. $A = \{1, 2, 3, 4\}$

Figure 13

14. $A = \{1, 2, 3, 4\}$

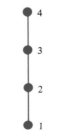

Figure 14

In Exercises 15 and 16, determine the Hasse diagram of the partial order having the given digraph.

15.

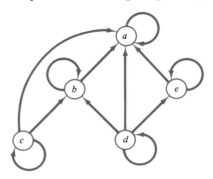

Figure 15

16.

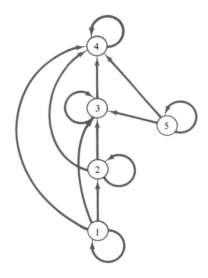

Figure 16

In Exercises 17 and 18, determine the Hasse diagram of the relations on $A = \{1, 2, 3, 4, 5\}$ whose matrix is shown.

17. $\begin{bmatrix} 1 & 1 & 1 & 1 & 1 \\ 0 & 1 & 1 & 1 & 1 \\ 0 & 0 & 1 & 1 & 1 \\ 0 & 0 & 0 & 1 & 1 \\ 0 & 0 & 0 & 0 & 1 \end{bmatrix}$

18. $\begin{bmatrix} 1 & 0 & 1 & 1 & 1 \\ 0 & 1 & 1 & 1 & 1 \\ 0 & 0 & 1 & 1 & 1 \\ 0 & 0 & 0 & 1 & 0 \\ 0 & 0 & 0 & 0 & 1 \end{bmatrix}$

In Exercises 19 and 20, determine the matrix of the partial order whose Hasse diagram is given.

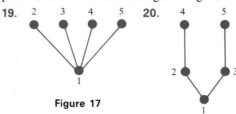

19.   2   3   4   5   **20.**   4   5

1

Figure 17

2   3

1

Figure 18

21. Let $A = Z^+ \times Z^+$ have lexicographic order. Answer each of the following as true or false.
(a) $(2, 12) \prec (5, 3)$ (b) $(3, 6) \prec (3, 24)$
(c) $(4, 8) \prec (4, 6)$ (d) $(15, 92) \prec (12, 3)$

In Exercises 22–25, consider the partial order of divisibility on the set A. Draw the Hasse diagram of the poset and determine which posets are linearly ordered.

22. $A = \{1, 2, 3, 5, 6, 10, 15, 30\}$

23. $A = \{2, 4, 8, 16, 32\}$

24. $A = \{3, 6, 12, 36, 72\}$

25. $A = \{1, 2, 3, 4, 5, 6, 10, 12, 15, 30, 60\}$

26. Let $A = \{\Box, A, B, C, E, O, M, P, S\}$ have the usual alphabetical order, where $\Box$ represents a "blank" character and $\Box \leq x$ for all $x \in A$. Arrange the following in lexicographic order (as elements of $A \times A \times A \times A$).
(a) MOP$\Box$ (b) MOPE   (c) CAP$\Box$
(d) MAP$\Box$ (e) BASE   (f) ACE$\Box$
(g) MACE   (h) CAPE

THEORETICAL EXERCISES

T1. Show that if R is a linear order on the set A, then R^{-1} is also a linear order on A.

T2. A relation R on a set A is called a **quasiorder** if it is transitive and irreflexive. Let $A = P(S)$ be the power set of a set S, and consider the following relation R on A: $U\,R\,T$ if and only if $U \subsetneq T$ (proper containment). Show that R is a quasiorder.

T3. Let $A = \{x \mid x$ is a real number and $-5 \leq x \leq 20\}$. Show that the usual relation $<$ is a quasiorder (see Exercise T2) on A.

T4. If R is a quasiorder on A (see Exercise T2), show that R^{-1} is also a quasiorder.

T5. Let $B = \{2, 3, 6, 9, 12, 18, 24\}$ and let $A = B \times B$. Define the following relation on

A: $(a, b) \prec (a', b')$ if and only if $a \mid a'$ and $b \leq b'$, where $\leq$ is the usual partial order. Show that $\prec$ is a partial order.

T6. Let $A = \{1, 2, 3, 5, 6, 10, 15, 30\}$ and consider the partial order $\leq$ of divisibility on A. That is, define $a \leq b$ to mean that $a \mid b$. Let $A' = P(S)$, where $S = \{e, f, g\}$, be the poset with partial order $\subseteq$. Show that $(A, \leq)$ and $(A', \subseteq)$ are isomorphic.

T7. Let $A = \{1, 2, 4, 8\}$ and let $\leq$ be the partial order of divisibility on A. Let $A' = \{0, 1, 2, 3\}$ and let $\leq'$ be the usual relation "less than or equal to" on integers. Show that $(A, \leq)$ and $(A', \leq')$ are isomorphic posets.

7.2 Extremal Elements of Posets

Certain elements in a poset are of special importance for many of the properties and applications of posets. In this section we discuss these elements and in later sections we shall see the important role played by them. In this section we consider a poset $(A, \leq)$ with partial order $\leq$.

An element $a \in A$ is called a **maximal element** of A if there is no element c in A such that $a < c$ (see Section 7.1). An element $b \in A$ is called a **minimal element** of A if there is no element c in A such that $c < b$.

Recall that if $(A, \leq)$ is a poset and $(A, \geq)$ is its dual poset, then the Hasse diagram of $(A, \geq)$ is the Hasse diagram of $(A, \leq)$ turned upside down. From this it follows that an element $a \in A$ is a maximal element of $(A, \leq)$ if and only if a is a minimal element of $(A, \geq)$ and a is a minimal element of $(A, \leq)$ if and only if it is a maximal element of $(A, \geq)$.

Example 1   Consider the poset A whose Hasse diagram is shown in Fig. 1. The elements a_1, a_2, and a_3 are maximal elements of A and the elements b_1, b_2, and b_3 are minimal elements. Observe that since there is no line between b_2 and b_3, we can conclude neither that $b_3 \leq b_2$ nor that $b_2 \leq b_3$.

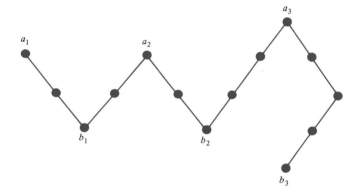

Figure 1

Example 2   Let A be the poset of all nonnegative real numbers with the usual partial order $\leq$. Then 0 is a minimal element of A. There are no maximal elements of A.

Example 3   The poset Z with the usual partial order $\leq$ has no maximal elements and has no minimal elements.

Theorem 1   Let A be a finite nonempty poset with partial order $\leq$. Then A has at least one maximal element and at least one minimal element.

Proof. Let a be any element of A. If a is not maximal, we can find an element $a_1 \in A$ such that $a < a_1$. If a_1 is not maximal, we can find an element $a_2 \in A$ such that $a_1 < a_2$. This argument cannot be continued indefinitely, since A is a finite set. Thus we eventually obtain the finite chain

$$a < a_1 < a_2 < \cdots < a_{k-1} < a_k$$

which cannot be extended. Hence we cannot have $a_k < b$ for any $b \in A$, so a_k is a maximal element of $(A, \leq)$.

The same argument says that the dual poset $(A, \geq)$ has a maximal element, so $(A, \leq)$ has a minimal element.

An element $a \in A$ is called a **greatest element** of A if $x \leq a$ for all $x \in A$. An element $a \in A$ is called a **least element** of A if $a \leq x$ for all $x \in A$.

Note that if A has a greatest element a, then a will also be a maximal element of A (the only maximal element). On the other hand, a poset may have maximal elements without having a greatest element. This is true of the poset whose Hasse diagram is shown in Fig. 11 of Section 7.1. A similar statement holds for the relationship between minimal elements of A and the least element of A.

As before, an element a of $(A, \leq)$ is a greatest (or least) element if and only if it is a least (or greatest) element of $(A, \geq)$.

Example 4 Consider the poset defined in Example 2. Then 0 is a least element; there is no greatest element.

Example 5 Let $S = \{a, b, c\}$ and consider the poset $A = P(S)$ defined in Example 12 of Section 7.1. The empty set is a least element of A and the set S is a greatest element of A.

Example 6 The poset Z with the usual partial order has neither a least nor a greatest element.

Theorem 2 A poset has at most one greatest element and at most one least element.

Proof. Suppose that a and b are greatest elements of a poset A. Then since b is a greatest element, we have $a \leq b$. Similarly, since a is a greatest element, we have $b \leq a$. Hence $a = b$ by the antisymmetry property. Thus if the poset has a greatest element, it only has one such element. A similar argument shows that if the poset has a least element, it has only one such element.

The greatest element of a poset, if it exists, is denoted by I, and is often called the **unit element**. Similarly, the least element of a poset, if it exists, is denoted by 0, and is often called the **zero element**.

Consider a poset A and a subset B of A. An element $a \in A$ is called an **upper bound** of B if $b \leq a$ for all $b \in B$. An element $a \in A$ is called a **lower bound** of B if $a \leq b$ for all $b \in B$.

Example 7 Consider the poset $A = \{a, b, c, d, e, f, g, h\}$, whose Hasse diagram is shown in Fig. 2. Find all upper and lower bounds of the following subsets of A: (a) $B_1 = \{a, b\}$; (b) $B_2 = \{c, d, e\}$.

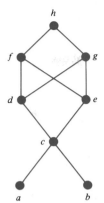

Figure 2

Solution. (a) B_1 has no lower bounds since no element of A is less than both a and b; its upper bounds are c, d, e, f, g, and h.

(b) The upper bounds of B_2 are f, g, and h; its lower bounds are c, a, and b.

As Example 7 shows, a subset B of a poset may or may not have upper or lower bounds (in A). Moreover, an upper or lower bound of B may or may not belong to B itself.

Let A be a poset and B a subset of A. An element $a \in A$ is called a **least upper bound** (LUB) of B if a is an upper bound of B and $a \leq a'$, whenever a' is an upper bound of B. Thus $a = \mathrm{LUB}\ (B)$ if $b \leq a$ for all $b \in B$, and if whenever $a' \in A$ is also an upper bound of $B (b \leq a'$ for all $b \in B)$, then $a \leq a'$.

Similarly, an element $a \in A$ is called a **greatest lower bound** (GLB) of B if a is a lower bound of B and $a' \leq a$, whenever a' is a lower bound of B. Thus $a = \mathrm{GLB}\ (B)$ if $a \leq b$ for all $b \in B$, and if whenever $a' \in A$ is also a lower bound of $B (a' \leq b$ for all $b \in B)$, then $a' \leq a$.

As usual, upper bounds in $(A, \leq)$ correspond to lower bounds in $(A, \geq)$ (for the same set of elements), and lower bounds in $(A, \leq)$ correspond to upper bounds in $(A, \geq)$. Similar statements hold for greatest lower bounds and least upper bounds.

Example 8   Let A be the poset considered in Example 7 with subsets B_1 and B_2 as defined in that example. Find all least upper bounds and all greatest lower bounds of (a) B_1; (b) B_2.

Solution. (a) Since B_1 has no lower bounds, it has no greatest lower bounds. However,

$$\mathrm{LUB}\ (B_1) = c$$

(b) Since the lower bounds of B_2 are c, a, and b, we find that

$$\mathrm{GLB}\ (B_2) = c$$

The upper bounds of B_2 are f, g, and h. Since f and g are not comparable, we conclude that B_2 has no least upper bound.

Theorem 3 Let $(A, \leq)$ be a poset. Then a subset B of A has at most one LUB and at most one GLB.

    Proof. Similar to the proof of Theorem 2.

    We conclude this section with some remarks about LUB and GLB in a finite poset A, as viewed from the Hasse diagram of A. Let $B = \{b_1, b_2, \ldots, b_r\}$. If $a = \text{LUB }(B)$, then a is the first vertex that can be reached from $b_1, b_2, \ldots, b_r$ by upward paths. Similarly, if $a = \text{GLB }(B)$, then a is the first vertex that can be reached from $b_1, b_2, \ldots, b_r$ by downward paths.

Example 9 Let $A = \{1, 2, 3, 4, 5, \ldots, 11\}$ be the poset whose Hasse diagram is shown in Fig. 3. Find the LUB and the GLB of $B = \{6, 7, 10\}$, if they exist.

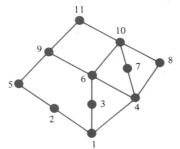

Figure 3

    Solution. Exploring all upward paths from vertices 6, 7, and 10, we find that LUB $(B) = 10$. Similarly, by examining all downward paths from 6, 7, and 10, we find that GLB $(B) = 4$.

    The next result follows immediately from the principle of correspondence (see Section 7.1.)

Theorem 4 Suppose that $(A, \leq)$ and $(A', \leq')$ are isomorphic posets under the isomorphism $f: A \to A'$.

1. If a is a maximal (minimal) element of $(A, \leq)$, then $f(a)$ is a maximal (minimal) element of $(A', \leq')$.
2. If a is the greatest (least) element of $(A, \leq)$, then $f(a)$ is the greatest (least) element of $(A', \leq')$.
3. If a is an upper bound (lower bound, least upper bound, greatest lower bound) of a subset B of A, then $f(a)$ is an upper bound (lower bound, least upper bound, greatest lower bound) of the subset $f(B)$ of A'.

4. If every subset of $(A, \leq)$ has a LUB (GLB), then every subset of $(A', \leq')$ has a LUB (GLB).

Example 10 Show that the posets $(A, \leq)$ and $(A', \leq')$, whose Hasse diagrams are shown in Fig. 4(a) and (b), respectively, are not isomorphic.

(a)   (b)   **Figure 4**

Solution. The two posets are not isomorphic because $(A, \leq)$ has a greatest element a, while $(A', \leq')$ does not have a greatest element. One could also argue that they are not isomorphic because $(A, \leq)$ does not have a least element, while $(A', \leq')$ does have a least element.

EXERCISE SET 7.2

In Exercises 1–4, determine all maximal and minimal elements of the poset whose Hasse diagram is shown.

1.

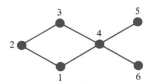

Figure 5

2.

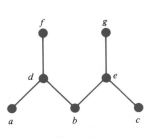

Figure 6

3.

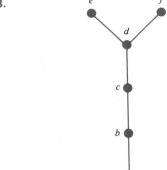

Figure 7

4.

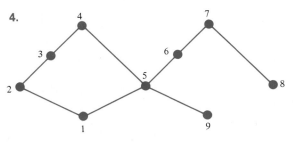

Figure 8

In Exercises 5-8, determine all maximal and minimal elements of the given poset.

5. $A = \mathbb{R}$ with the usual partial order $\leq$.

6. $A = \{x \mid x$ is a real number and $0 \leq x < 1\}$ with the usual partial order $\leq$.

7. $A = \{x \mid x$ is a real number and $0 < x \leq 1\}$ with the usual partial order $\leq$.

8. $A = \{2, 3, 4, 6, 8, 24, 48\}$ with the partial order of divisibility.

In Exercises 9–16, determine the greatest and least elements, if they exist, of the poset.

9. 10.

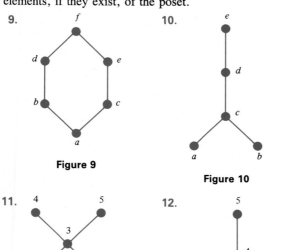

Figure 9

Figure 10

11. 12.

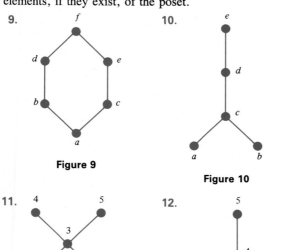

Figure 11

Figure 12

13. $A = \{x \mid x$ is a real number and $0 < x < 1\}$ with the usual partial order $\leq$.

14. $A = \{x \mid x$ is a real number and $0 \leq x \leq 1\}$ with the usual partial order $\leq$.

15. $A = \{2, 4, 6, 8, 12, 18, 24, 36, 72\}$ with the partial order of divisibility.

16. $A = \{2, 3, 4, 6, 12, 18, 24, 36\}$ with the partial order of divisibility.

In Exercises 17–27, find, if they exist, (a) all upper bounds of B; (b) all lower bounds of B; (c) the least upper bound of B; (d) the greatest lower bound of B.

17. 18.

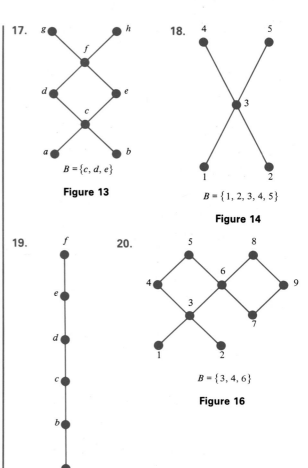

$B = \{c, d, e\}$

Figure 13

$B = \{1, 2, 3, 4, 5\}$

Figure 14

19. 20.

$B = \{3, 4, 6\}$

Figure 16

$B = \{b, c, d\}$

Figure 15

21. $(A, \leq)$ is the poset in Exercise 17; $B = \{b, g, h\}$.

22. $(A, \leq)$ is the poset in Exercise 20; $B = \{4, 6, 9\}$.

23. $(A, \leq)$ is the poset in Exercise 20; $B = \{3, 4, 8\}$.

24. $A = \mathbb{R}$ and $\leq$ denotes the usual partial order; $B = \{x \mid x$ is a real number and $1 < x < 2\}$.

25. $A = \mathbb{R}$ and $\leq$ denotes the usual partial order; $B = \{x \mid x$ is a real number and $1 \leq x < 2\}$.

26. $A = P(\{a, b, c\})$ and $\leq$ denotes the partial order of containment $B = P(\{a, b\})$.

27. $A = \{2, 3, 4, 6, 8, 12, 24, 48\}$ and $\leq$ denotes the partial order of divisibility; $B = \{4, 6, 12\}$.

7.3 Lattices

A **lattice** is a poset $(L, \leq)$ in which every subset $\{a, b\}$ consisting of two elements has a least upper bound and a greatest lower bound. We denote LUB $(\{a, b\})$ by $a \vee b$ and call it the **join** of a and b. Similarly, we denote GLB $(\{a, b\})$ by $a \wedge b$ and call it the **meet** of a and b. Lattice structures often appear in computing and mathematical applications.

Example 1   Let S be a set and let $L = P(S)$. As we have seen, $\subseteq$, containment, is a partial order on L. If A and B are two elements of L, (that is, subsets of S), then the join of A and B is their union $A \cup B$, and the meet of A and B is their intersection $A \cap B$. Hence L is a lattice.

Example 2   Consider the poset $(Z^+, \leq)$, where for a and b in Z^+, $a \leq b$ if and only if $a \mid b$. Then L is a lattice in which the join and meet of a and b are their least common multiple and greatest common divisor, respectively (see Section 3.2). That is,

$$a \vee b = \text{LCM } (a, b) \qquad \text{and} \qquad a \wedge b = \text{GCD } (a, b)$$

Example 3   Let n be a positive integer and let D_n be the set of all positive divisors of n. Then D_n is a lattice under the relation of divisibility as considered in Example 2. Thus if $n = 20$, we have $D_{20} = \{1, 2, 4, 5, 10, 20\}$. The Hasse diagram of D_{20} is shown in Fig. 1(a). If $n = 30$, we have $D_{30} = \{1, 2, 3, 5, 6, 10, 15, 30\}$. The Hasse diagram of D_{30} is shown in Fig. 1(b).

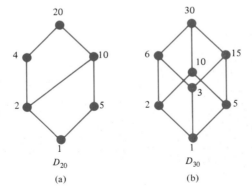

D_{20}

(a)

D_{30}

(b)   **Figure 1**

Example 4   Which of the Hasse diagrams in Fig. 2 represent lattices?

Solution. Hasse diagrams (a), (b), (d), and (e) represent lattices. Diagram (c) does not represent a lattice because $f \vee g$ does not exist. Diagram (f) does not

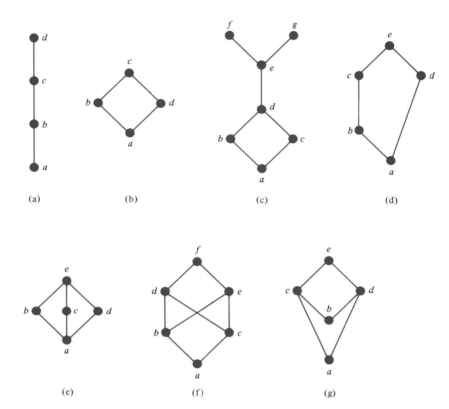

Figure 2

represent a lattice because neither $d \wedge e$ nor $b \vee c$ exist. Diagram (g) does not represent a lattice because $c \wedge d$ does not exist.

Theorem 1   If $(L_1, \leq)$ and $(L_2, \leq)$ are lattices, then $(L, \leq)$ is a lattice, where $L = L_1 \times L_2$, and the partial order $\leq$ of L is the product partial order.

    Proof. We denote the join and meet in L_1 by $\vee_1$ and $\wedge_1$, respectively, and the join and meet in L_2 by $\vee_2$ and $\wedge_2$, respectively. We already know from Theorem 1 of Section 7.1 that L is a poset. We now need to show that if (a_1, b_1) and $(a_2, b_2) \in L$, then $(a_1, b_1) \vee (a_2, b_2)$ and $(a_1, b_1) \wedge (a_2, b_2)$ exist in L. We leave it as an exercise to verify that

$$(a_1, b_1) \vee (a_2, b_2) = (a_1 \vee_1 a_2, b_1 \vee_2 b_2)$$
$$(a_1, b_1) \wedge (a_2, b_2) = (a_1 \wedge_1 a_2, b_1 \wedge_2 b_2)$$

Thus L is a lattice.

Example 5   Let L_1 and L_2 be the lattices shown in Fig. 3(a) and (b), respectively. Then $L = L_1 \times L_2$ is the lattice shown in Fig. 3(c).

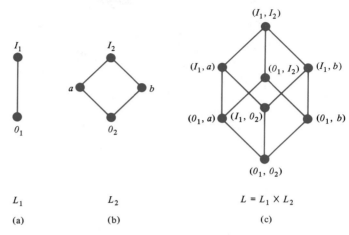

L_1 L_2 $L = L_1 \times L_2$

(a)   (b)   (c)

Figure 3

Isomorphic Lattices

If $f: L_1 \rightarrow L_2$ is an isomorphism from the poset $(L_1, \leq_1)$ to the poset $(L_2, \leq_2)$, then Theorem 4 of Section 7.2 tells us that L_1 is a lattice if and only if L_2 is a lattice. In fact, if a and b are elements of L_1, then $f(a \wedge b) = f(a) \wedge f(b)$ and $f(a \vee b) = f(a) \vee f(b)$. If two lattices are isomorphic, as posets, we say they are **isomorphic lattices**.

Example 6   Let L be the lattice D_6, and let L' be the lattice $P(S)$, under the relation of inclusion, where $S = \{a, b\}$. These posets were discussed in Example 16 of Section 7.1, where they were shown to be isomorphic. Thus, since both are lattices, they are isomorphic lattices.

Properties of Lattices

Before proving a number of properties of lattices, we recall the meaning of $a \vee b$ and $a \wedge b$.

1. $a \leq a \vee b$ and $b \leq a \vee b$ ($a \vee b$ is an upper bound of a and b).
2. If $a \leq c$ and $b \leq c$, then $a \vee b \leq c$ ($a \vee b$ is the least upper bound of a and b).
1'. $a \wedge b \leq a$ and $a \wedge b \leq b$ ($a \wedge b$ is a lower bound of a and b).

$2'$. If $c \leq a$ and $c \leq b$, then $c \leq a \wedge b$ ($a \wedge b$ is the greatest lower bound of a and b).

Theorem 2   Let L be a lattice. Then for every a and b in L:
(a) $a \vee b = b$ if and only if $a \leq b$.
(b) $a \wedge b = a$ if and only if $a \leq b$.
(c) $a \wedge b = a$ if and only if $a \vee b = b$.

Proof. (a) Suppose that $a \vee b = b$. Since $a \leq a \vee b = b$, we have $a \leq b$. Conversely, if $a \leq b$, then since $b \leq b$, b is an upper bound of a and b, so by definition of least upper bound we have $a \vee b \leq b$. Since $a \vee b$ is an upper bound, $b \leq a \vee b$, so $a \vee b = b$.

(b) The proof is analogous to the proof of part (a), and we leave it as an exercise for the student.

(c) Follows from parts (a) and (b).

Example 7   Let L be a linearly ordered set. If a and $b \in L$, then either $a \leq b$ or $b \leq a$. It follows from Theorem 2 that L is a lattice, since every pair of elements has a least upper bound and a greatest lower bound.

Theorem 3   Let L be a lattice. Then

1. (a) $a \vee a = a$ **(Idempotent Property)**
   (b) $a \wedge a = a$

2. (a) $a \vee b = b \vee a$ **(Commutative Property)**
   (b) $a \wedge b = b \wedge a$

3. (a) $a \vee (b \vee c) = (a \vee b) \vee c$ **(Associative Property)**
   (b) $a \wedge (b \wedge c) = (a \wedge b) \wedge c$

4. (a) $a \vee (a \wedge b) = a$ **(Absorption Property)**
   (b) $a \wedge (a \vee b) = a$

Proof

1. Follows from the definition of LUB and GLB.
2. The definition of LUB and GLB treat a and b symmetrically, so the results follow.

We omit the proofs of 3 and 4.

It follows from property 3 that we can write $a \vee (b \vee c)$ and $(a \vee b) \vee c$ merely as $a \vee b \vee c$, and similarly for $a \wedge b \wedge c$. Moreover, we can write

$$\text{LUB } (\{a_1, a_2, \ldots, a_n\}) \quad \text{as} \quad a_1 \vee a_2 \vee \cdots \vee a_n$$
$$\text{GLB } (\{a_1, a_2, \ldots, a_n\}) \quad \text{as} \quad a_1 \wedge a_2 \wedge \cdots \wedge a_n$$

since we can show by induction that these joins and meets are independent of the grouping of the terms.

Special Types of Lattices (Optional)

A lattice L is said to be **bounded** if it has a greatest element I and a least element 0 (see Section 7.2).

Example 8   The lattice Z^+ under the partial order of divisibility, as defined in Example 2, is not a bounded lattice since it has a least element, the number 1, but no greatest element.

Example 9   The lattice Z under the partial order $\leq$ is not bounded since it has neither a greatest nor a least element.

Example 10   The lattice $P(S)$ of all subsets of a set S, as defined in Example 1, is bounded. Its greatest element is S and its least element is $\varnothing$.

If L is a bounded lattice, then for all $a \in A$

$$0 \leq a \leq I$$
$$a \vee 0 = a \qquad a \wedge 0 = 0$$
$$a \vee I = I \qquad a \wedge I = a$$

Theorem 4   Let $L = \{a_1, a_2, \ldots, a_n\}$ be a finite lattice. Then L is bounded.

Proof. The greatest element of L is $a_1 \vee a_2 \vee \cdots \vee a_n$ and its least element is $a_1 \wedge a_2 \wedge \cdots \wedge a_n$.

A lattice L is called **distributive** if for any elements a, b, and c in L we have the following **distributive laws**:

(a) $a \wedge (b \vee c) = (a \wedge b) \vee (a \wedge c)$
(b) $a \vee (b \wedge c) = (a \vee b) \wedge (a \vee c)$

If L is not distributive, we say that L is **nondistributive**.
We leave it as an exercise to show that the distributive law holds when any two of the elements a, b, or c are equal, or when any one of the elements is 0 or I. This observation reduces the number of cases that must be checked in verifying that the distributive law holds. However, verification of the distributive law is generally a tedious task.

Example 11   The lattice $P(S)$ is distributive, since union and intersection (the join and meet, respectively) satisfy the distributive law as shown in Section 1.3.

Example 12 The lattice shown in Fig. 4 is distributive, as can be seen by verifying the distributive law for all ordered triples chosen from the elements a, b, c, and d.

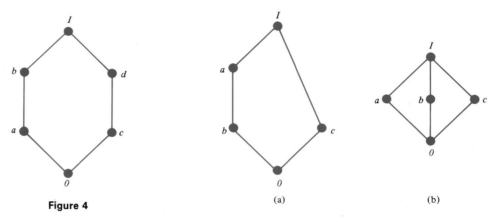

Figure 4

(a)

(b)

Figure 5

Example 13 Show that the lattices pictured in Fig. 5 are nondistributive.

Solution. For the lattice shown in Fig. 5(a), we have

$$a \wedge (b \vee c) = a \wedge I = a$$

while

$$(a \wedge b) \vee (a \wedge c) = b \vee 0 = b$$

For the lattice shown in Figure 5(b), observe that

$$a \wedge (b \vee c) = a \wedge I = a$$

while

$$(a \wedge b) \vee (a \wedge c) = 0 \vee 0 = 0$$

Let L be a bounded lattice with greatest element I and least element 0, and let $a \in L$. An element $a' \in L$ is called a **complement** of a if

$$a \vee a' = I \qquad \text{and} \qquad a \wedge a' = 0$$

Observe that

$$0' = I \qquad \text{and} \qquad I' = 0$$

Example 14   The lattice $L = P(S)$ is such that every element has a complement, since if $A \in L$, then its set complement $\bar{A}$ has the properties $A \vee \bar{A} = S$ and $A \wedge \bar{A} = \varnothing$.

Example 15   The lattices in Fig. 5 each have the property that every element has a complement. The element c in both cases has two complements, a and b.

Example 16   Consider the lattices D_{20} and D_{30} discussed in Example 3 and shown in Fig. 1. Observe that every element in D_{30} has a complement. For example, if $a = 5$, then $a' = 6$. However, the elements 2 and 10 in D_{20} have no complements.

Examples 15 and 16 show that an element a in a lattice need not have a complement and it may have more than one complement. However, for a bounded distributive lattice, the situation is more restrictive, as shown by the following theorem.

Theorem 5   Let L be a bounded distributive lattice. If a complement exists, it is unique.

Proof. Let a' and a'' be complements of the element $a \in L$. Then

$$a \vee a' = I \qquad a \vee a'' = I$$
$$a \wedge a' = 0 \qquad a \wedge a'' = 0$$

Using the distributive laws, we obtain

$$a' = a' \vee 0 = a' \vee (a \wedge a'') = (a' \vee a) \wedge (a' \vee a'')$$
$$= (a \vee a') \wedge (a' \vee a'')$$
$$= I \wedge (a' \vee a'') = a' \vee a''$$

Also,

$$a'' = a'' \vee 0 = a'' \vee (a \wedge a') = (a'' \vee a) \wedge (a'' \vee a')$$
$$= (a \vee a'') \wedge (a' \vee a'')$$
$$= I \wedge (a' \vee a'') = a' \vee a''$$

Hence

$$a' = a''$$

A lattice L is called **complemented** if it is bounded and if every element in L has a complement.

Example 17 $L = P(S)$ is complemented. Observe that in this case, each element of L has a unique complement, which can be seen directly or is implied by Theorem 5.

Example 18 The lattices discussed in Example 15 and shown in Fig. 5 are complemented. In this case, the complements are not unique.

EXERCISE SET 7.3

In Exercises 1–6, determine whether the Hasse diagram represents a lattice.

1.

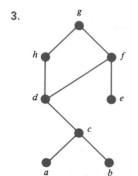

Figure 6

2.

Figure 7

3.

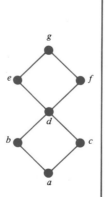

Figure 8

4.

Figure 9

5.

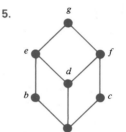

Figure 10

6.

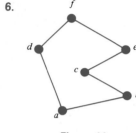

Figure 11

7. Is the poset $A = \{2, 3, 6, 12, 24, 36, 72\}$ under the relation of divisibility a lattice?

8. If L_1 and L_2 are the lattices shown in Fig. 12, draw the Hasse diagram of $L_1 \times L_2$, with the product partial order.

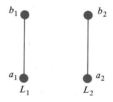

Figure 12

9. Give the Hasse diagrams of all nonisomorphic lattices that have one, two, three, four, or five elements.

10. Find the complement of each element in D_{42}.

11. Find the complement of each element in D_{105}.

In Exercises 12–16, determine whether each lattice is distributive, complemented, or both.

12.

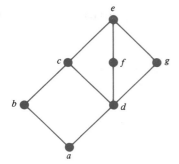

Figure 13

15.

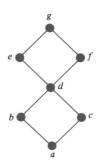

Figure 16

16. Consider the complemented lattice shown in Fig. 17. Give the complements of each element.

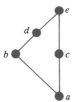

Figure 17

13.   **14.**

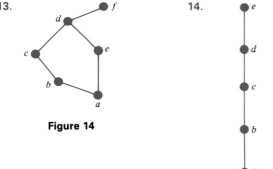

Figure 14

Figure 15

THEORETICAL EXERCISES

T1. Show that if a bounded lattice has two or more elements, then $0 \neq I$.

T2. Prove Theorem 2(b).

T3. Show that the lattice Z^+ under the usual partial order $\leq$ is distributive.

T4. Show that the lattice D_n is distributive.

T5. Show that a linearly ordered poset is a distributive lattice.

T6. Show that if L_1 and L_2 are distributive lattices, then $L = L_1 \times L_2$ is also distributive, where the order of L is the product of the orders in L_1 and L_2.

T7. Is the dual of a distributive lattice also distributive?

T8. Show that if $a \leq (b \wedge c)$ for some a, b, and c in a poset L, then the distributive properties of a lattice are satisfied by a, b, and c.

T9. Prove that if a and b are elements in a bounded, distributive lattice, and if a has a complement a', then

$$a \vee (a' \wedge b) = a \vee b$$
$$a \wedge (a' \vee b) = a \wedge b$$

T10. Let L be a distributive lattice. Show that if $a \wedge x = a \wedge y$ or $a \vee x = a \vee y$ for all a, then $x = y$.

T11. A lattice is said to be **modular** if for all a, b, c, $a \leq c$ implies that $a \vee (b \wedge c) = (a \vee b) \wedge c$.

(a) Show that a distributive lattice is modular.

(b) Show that the lattice shown in Fig. 18 is a nondistributive lattice that is modular.

T12. Let L be a bounded lattice with at least two elements. Show that no element of L is its own complement.

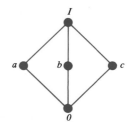

Figure 18

7.4 Finite Boolean Algebras

In this section we discuss a certain type of lattice that has a great many applications in computer science. We have seen in Example 6 of Section 7.3 that if S is a set, $L = P(S)$, and $\subseteq$ is the usual relation of inclusion, then the poset $(L, \subseteq)$ is a lattice. These lattices have many properties that are not shared by lattices in general. For this reason, they are easier to work with, and they play a more important role in various applications.

We will restrict our attention to the lattices $(P(S), \subseteq)$, where S is a finite set, and we begin by finding all essentially different examples.

Theorem 1 If $S_1 = \{x_1, x_2, \ldots, x_n\}$ and $S_2 = \{y_1, y_2, \ldots, y_n\}$ are any two finite sets with n elements, then the lattices $(P(S_1), \subseteq)$ and $(P(S_2), \subseteq)$ are isomorphic. In particular, the Hasse diagrams of these lattices may be drawn identically.

Proof. Arrange the sets as shown in Fig. 1, so that each element of S_1 is directly over the correspondingly numbered element in S_2. For each subset A of S_1, let $f(A)$ be the subset of S_2 consisting of all elements that correspond to the elements of A. Figure 2 shows a typical subset A of S_1 and the corresponding subset $f(A)$ of S_2. It is easily seen that the function f, described above, is a one-to-one correspondence from subsets of S_1 to subsets of S_2. Equally clear is the fact that if A and B are any subsets of S_1, then $A \subseteq B$ if and only if $f(A) \subseteq f(B)$. We omit the details. Thus the lattices $(P(S_1), \subseteq)$ and $(P(S_2), \subseteq)$ are isomorphic.

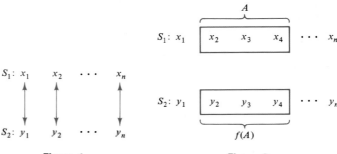

Figure 1

Figure 2

The essential point of this theorem is that the lattice $(P(S), \subseteq)$ is completely determined as a poset by the number $|S|$ and does not depend in any way on the nature of the elements in S.

Example 1   Figure 3(a) and (b) show Hasse diagrams for the lattices $(P(S), \subseteq)$ and $(P(T), \subseteq)$, respectively, where $S = \{a, b, c\}$ and $T = \{2, 3, 5\}$. It is clear from this figure that the two lattices are isomorphic. In fact, we see that one possible isomorphism $f: S \to T$ is given by

$$f(\{a\}) = \{2\} \qquad f(\{b\}) = \{3\} \qquad f(\{c\}) = \{5\}$$

$$f(\{a, b\}) = \{2, 3\} \qquad f(\{b, c\}) = \{3, 5\} \qquad f(\{a, c\}) = \{2, 5\}$$

$$f(\{a, b, c\}) = \{2, 3, 5\}$$

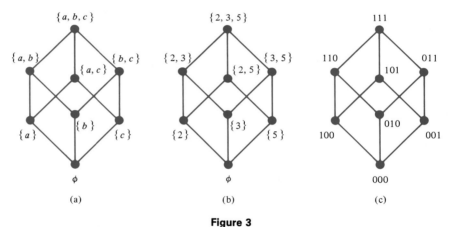

(a)   (b)   (c)

Figure 3

Thus for each $n = 0, 1, 2, \ldots$ there is only one type of lattice having the form $(P(S), \subseteq)$. This lattice depends only on n, not on S, and it has 2^n elements, as was shown in Example 2 of Section 2.2. Recall from Section 1.2 that if a set S has n elements, then all subsets of S can be represented by sequences of 0's and 1's of length n. We can therefore label the Hasse diagram of a lattice $(P(S), \subseteq)$ by such sequences. In doing so we free the diagram from dependence on a particular set S and emphasize the fact that it depends only on n.

Example 2   Figure 3(c) shows how the diagram that appears in Fig. 3(a) and (b) can be labeled by sequences of 0's and 1's. This labeling serves equally well to describe the lattice of Fig. 3(a) or (b), or for that matter the lattice $(P(S), \subseteq)$ that arises from any set S having three elements.

If the Hasse diagram of the lattice corresponding to a set with n elements is labeled by sequences of 0's and 1's of length n, as described above, then the resulting lattice is named B_n. The properties of the partial order in B_n can be described directly as follows. If $x = a_1 a_2 \cdots a_n$ and $y = b_1 b_2 \cdots b_n$ are two elements of B_n, that is two sequences of 0's and 1's then

1. $x \leq y$ if and only if $a_k \leq b_k$ (as numbers 0 or 1) for $k = 1, 2, \ldots, n$.
2. $x \wedge y = c_1 c_2 \cdots c_n$, where $c_k = \min \{a_k, b_k\}$.
3. $x \vee y = d_1 d_2 \cdots d_n$, where $d_k = \max \{a_k, b_k\}$.
4. x has a complement $x' = z_1 z_2 \cdots z_n$, where $z_k = 1$ if $x_k = 0$ and $z_k = 0$ if $x_k = 1$.

The truth of these statements can be seen by noting that $(B, \leq)$ is isomorphic with $(P(S), \subseteq)$ so that each x and y in B_n corresponds to subsets A and B of S. Then $x \leq y$, $x \wedge y$, $x \vee y$, and x', as defined above, correspond to $A \subseteq B$, $A \cap B$, $A \cup B$, and $\overline{A}$ (set complement), respectively (verify). Figure 4 shows the Hasse diagrams of the lattices B_n, for $n = 0, 1, 2, 3$.

We have seen that each lattice $(P(S), \subseteq)$ is isomorphic with B_n, where $n = |S|$. Other lattices may also be isomorphic with one of the B_n, and thus possess all the special properties that the B_n possess.

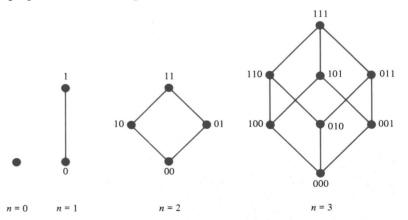

Figure 4

Example 3 In Example 1b of Section 7.1 we considered the lattice D_6 consisting of all positive integer divisors of 6, under the partial order of divisibility. The Hasse diagram of D_6 is shown in that example, and we now see that D_6 is isomorphic with B_2. In fact, $f: D_6 \rightarrow B_2$ is an isomorphism, where

$$f(1) = 00, \quad f(2) = 10, \quad f(3) = 01, \quad f(6) = 11$$

We are therefore led to make the following definition. A finite lattice is called a **Boolean algebra** if it is isomorphic with B_n, for some nonnegative integer n. Thus each B_n is a Boolean algebra, and so is each lattice $(P(S), \subseteq)$, where S is a finite set. Example 3 shows that D_6 is also a Boolean algebra.

We will work only with finite posets in this section. For the curious, however, we note that there are infinite posets that share all the relevant properties of the lattices $(P(S), \subseteq)$ (for infinite sets S of course), but which are not isomorphic with one of these lattices. This necessitates the restriction of our definition of Boolean algebra to the finite case, which is quite sufficient for the applications that we present.

Example 4   Consider the lattices D_{20} and D_{30} of all positive integer divisors of 20 and 30, respectively, under the partial order of divisibility. These posets were introduced in Example 3 of Section 7.3 and their Hasse diagrams were shown in Fig. 1 of that section. Since D_{20} has five elements and $5 \neq 2^n$ for any integer $n \geq 0$, we conclude that D_{20} is not a Boolean algebra. The poset D_{30} has eight elements, and since $8 = 2^3$, it could be a Boolean algebra. By comparing Fig. 1(b) of Section 7.3 and Fig. 4 above, we see that D_{30} is isomorphic with B_3. In fact, we see that the one-to-one correspondence $f: D_{30} \to B_3$ defined by

$$f(1) = 000 \qquad f(2) = 100 \qquad f(3) = 010$$
$$f(5) = 001 \qquad f(6) = 110 \qquad f(10) = 101$$
$$f(15) = 011 \qquad f(30) = 111$$

is an isomorphism. Thus D_{30} is a Boolean algebra.

If a finite lattice L does not contain 2^n elements for some nonnegative integer n, then we know that L cannot be a Boolean algebra. If $|L| = 2^n$, then L may or may not be a Boolean algebra. If L is relatively small, we may be able to compare its Hasse diagram with the Hasse diagram of B_n. In this way we saw in Example 4 that D_{30} is a Boolean algebra. However, this technique may not be practical if $|L|$ is large. In that case, one may be able to show that L is a Boolean algebra by directly constructing an isomorphism with some B_n or, equivalently, with $(P(S), \subseteq)$ for some finite set S. Suppose, for example, that we want to know whether a lattice D_n is a Boolean algebra, and we want a method that works no matter how large n is. The following theorem gives a partial answer.

Theorem 2   Let

$$n = p_1 p_2 \cdots p_k$$

where the p_i are distinct primes. Then D_n is a Boolean algebra.

Proof. Let $S = \{p_1, p_2, \ldots, p_k\}$. If $T \subseteq S$ and a_T is the product of the primes in T, then $a_T \mid n$. Any divisor of n must be of the form a_T for some subset T of S (where we let $a_\emptyset = 1$). The reader may verify that if V and T are subsets of S, $V \subseteq T$ if and only if $a_V \mid a_T$. Also, it follows from the proof of Theorem 7 of Section 3.2 that $a_{V \cap T} = a_V \wedge a_T$ [that is, GCD (a_V, a_T)] and $a_{V \cup T} = a_V \vee a_T$ [that is, LCM (a_V, a_T)]. Thus the function $f: P(S) \to D_n$ given by $f(T) = a_T$ is an isomorphism from $P(S)$ to D_n. Since $P(S)$ is a Boolean algebra, so is D_n.

Note that since $20 = 2 \times 2 \times 5$ is not the product of *distinct* primes, the lattice D_{20} mentioned in Example 4 is not a Boolean algebra.

Example 5 Since $210 = 2 \cdot 3 \cdot 5 \cdot 7$, $66 = 2 \cdot 3 \cdot 11$, and $646 = 2 \cdot 17 \cdot 19$, we see from Theorem 2 that D_{210}, D_{66}, and D_{646} are all Boolean algebras.

In other cases of large lattices L, we may be able to show that L is not a Boolean algebra by showing that the partial order of L does not have the necessary properties. A Boolean algebra is isomorphic with some B_n and therefore with some lattice $(P(S), \subseteq)$. Thus a Boolean algebra L must be a bounded lattice and a complemented lattice (see Section 7.3). In other words, it will have a greatest element I, corresponding to the set S, and a least element 0, corresponding to the subset $\emptyset$. Also, every element x of L will have a complement x'. The principle of correspondence (see Section 7.1) then tells us that the following rule holds.

Substitution Rule for Boolean Algebras

Any formula involving $\cup$, $\cap$, or $^-$ that holds for arbitrary subsets of a set S will continue to hold for arbitrary elements of a Boolean algebra L if $\wedge$ is substituted for $\cap$, $\vee$ for $\cup$, and $'$ for $^-$.

Example 6 If L is any Boolean algebra, and x, y, and z are in L, then the following three properties hold.

(a) $(x')' = x$ $\qquad\qquad$ **involution property**

$\left.\begin{array}{l} \text{(b) } (x \wedge y)' = x' \vee y' \\ \text{(c) } (x \vee y)' = x' \wedge y' \end{array}\right\}$ **De Morgan's laws**

This is true by the substitution rule for Boolean algebras, since we know that the corresponding formulas (a1) $\overline{(\overline{A})} = A$

(b1) $\overline{(A \cap B)} = \overline{A} \cup \overline{B}$

(c1) $\overline{(A \cup B)} = \overline{A} \cap \overline{B}$

hold for arbitrary subsets A and B of a set S.

In a similar way we can list other properties that must hold in any Boolean algebra by the substitution rule. Below we summarize all the basic properties of a Boolean algebra $(L, \leq)$, and next to each we list the corresponding property for subsets of a set S. We suppose that x, y, and z are arbitrary elements in L, and A, B, and C are arbitrary subsets of S. Also, we denote the greatest and least elements of L by I and 0, respectively.

1. $x \leq y$ if and only if $x \vee y = y$

2. $x \leq y$ if and only if $x \wedge y = x$

3. (a) $x \vee x = x$
   (b) $x \wedge x = x$

4. (a) $x \vee y = y \vee x$
   (b) $x \wedge y = y \wedge x$

5. (a) $x \vee (y \vee z) = (x \vee y) \vee z$
   (b) $x \wedge (y \wedge z) = (x \wedge y) \wedge z$

6. (a) $x \vee (x \wedge y) = x$
   (b) $x \wedge (x \vee y) = x$

7. $0 \leq x \leq I$ for all x in L

8. (a) $x \vee 0 = x$
   (b) $x \wedge 0 = 0$

9. (a) $x \vee I = I$
   (b) $x \wedge I = x$

10. (a) $x \wedge (y \vee z) = (x \wedge y) \vee (x \wedge z)$
    (b) $x \vee (y \wedge z) = (x \vee y) \wedge (x \vee z)$

11. Every element x has a unique complement x' satisfying
    (a) $x \vee x' = I$
    (b) $x \wedge x' = 0$

12. (a) $0' = I$
    (b) $I' = 0$

13. $(x')' = x$

14. (a) $(x \wedge y)' = x' \vee y'$
    (b) $(x \vee y)' = x' \wedge y'$

1'. $A \subseteq B$ if and only if $A \cup B = B$

2'. $A \subseteq B$ if and only if $A \cap B = A$

3'. (a) $A \cup A = A$
    (b) $A \cap A = A$

4'. (a) $A \cup B = B \cup A$
    (b) $A \cap B = B \cap A$

5'. (a) $A \cup (B \cup C) = (A \cup B) \cup C$
    (b) $A \cap (B \cap C) = (A \cap B) \cap C$

6'. (a) $A \cup (A \cap B) = A$
    (b) $A \cap (A \cup B) = A$

7'. $\varnothing \subseteq A \subseteq S$ for all A in $P(S)$

8'. (a) $A \cup \varnothing = A$
    (b) $A \cap \varnothing = \varnothing$

9'. (a) $A \cup S = S$
    (b) $A \cap S = A$

10'. (a) $A \cap (B \cup C) = (A \cap B) \cup (A \cap C)$
     (b) $A \cup (B \cap C) = (A \cup B) \cap (A \cup C)$

11'. Every element A has a unique complement $\overline{A}$ satisfying
     (a) $A \cup \overline{A} = S$
     (b) $A \cap \overline{A} = \varnothing$

12'. (a) $\overline{\varnothing} = S$
     (b) $\overline{S} = \varnothing$

13'. $\overline{(\overline{A})} = A$

14'. (a) $\overline{(A \cap B)} = \overline{A} \cup \overline{B}$
     (b) $\overline{(A \cup B)} = \overline{A} \cap \overline{B}$

Thus we may be able to show that a lattice L is not a Boolean algebra by showing that it does not possess one or more of the properties 1–14.

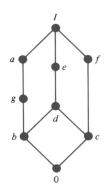

Figure 5

Example 7 Show that the lattice whose Hasse diagram is shown in Fig. 5 is not a Boolean algebra.

Solution. Elements a and g are both complements of c, that is, they both satisfy properties 11(a) and 11(b) above with respect to the element c. But property 11 says that such an element is unique in any Boolean algebra. Thus the given lattice cannot be a Boolean algebra.

Example 8 Show that if n is a positive integer and $p^2 \mid n$, where p is a prime number, then D_n is not a Boolean algebra.

Solution. Suppose that $p^2 \mid n$, so that $n = p^2 q$ for some positive integer q. Since p is also a divisor of n, p is an element of D_n. Thus, by the remarks given above, if D_n is a Boolean algebra, then p must have a complement p'. Then GCD $(p, p') = 1$ and LCM $(p, p') = n$. By Theorem 6 of Section 3.2, $pp' = n$, so $p' = n/p = pq$. This shows that GCD $(p, pq) = 1$, but this is impossible since p and pq have p as a common divisor. Hence D_n cannot be a Boolean algebra.

If we combine Example 8 and Theorem 2, we see that D_n is a Boolean algebra if and only if n is the product of distinct primes, that is, if and only if no prime divides n more than once.

Example 9 If $n = 40$, then $n = 2^3 \cdot 5$, so 2 divides n three times. If $n = 125$, then $n = 3 \cdot 5^2$, so 5 divides n twice. Thus neither D_{40} nor D_{125} are Boolean algebras.

Let us summarize what we have shown about Boolean algebras. We may attempt to show that a lattice L is a Boolean algebra by examining its Hasse diagram or constructing directly an isomorphism between L and B_n or $(P(S), \subseteq)$. We may attempt to show that L is not a Boolean algebra by checking the number of elements in L, or the properties of its partial order. If L is a Boolean algebra, then we may use any of the properties 1–14 to algebraically manipulate or simplify expressions involving elements of L. Simply proceed as if the elements were subsets and the manipulations were those that arise in set theory.

From now on we will denote the Boolean algebra B_1 simply as B. Thus B contains only the two elements 0 and 1. It is sometimes useful to know that any of the

Boolean algebras B_n can be described in terms of B. The following theorem gives this description.

Theorem 3   For any $n \geq 1$, B_n is the product $B \times B \times \cdots \times B$ of B with itself n times, where $B \times B \times \cdots \times B$ is given the product partial order.

Proof. By definition, B_n consists of all n-tuples of 0's and 1's, that is, all n-tuples of elements from B. Thus, as a set, B_n is equal to $B \times B \times \cdots \times B$ (n factors). Moreover, if $x = x_1 x_2 \cdots x_n$ and $y_1 y_2 \cdots y_n$ are two elements of B_n, then we know that

$$x \leq y \quad \text{if and only if} \quad x_k \leq y_k \qquad \text{for all } k,$$

Thus B_n, identified with $B \times B \times \cdots \times B$ (n factors), has the product partial order.

Applications

At this point applications Chapter 8 can be covered if desired.

EXERCISE SET 7.4

In Exercises 1–10, determine whether the poset is a Boolean algebra.

1.

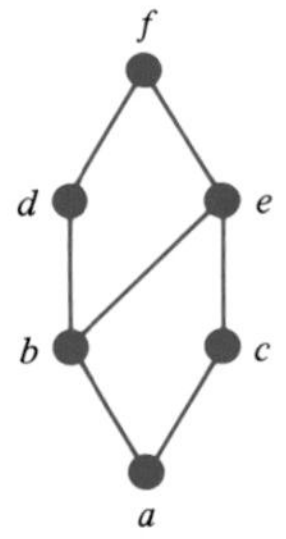

Figure 6

2.

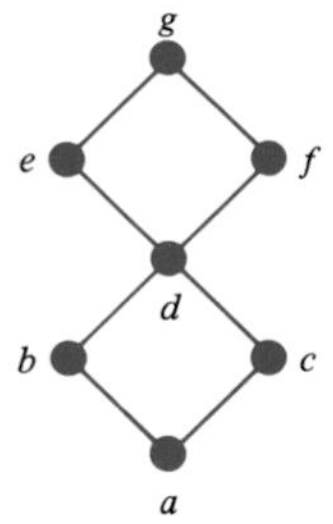

Figure 7

3.

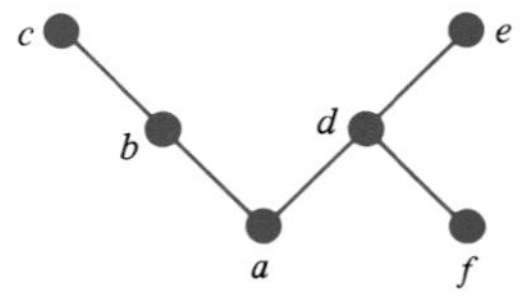

Figure 8

4.

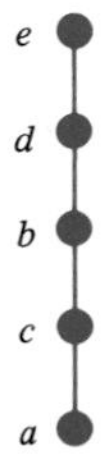

Figure 9

5.

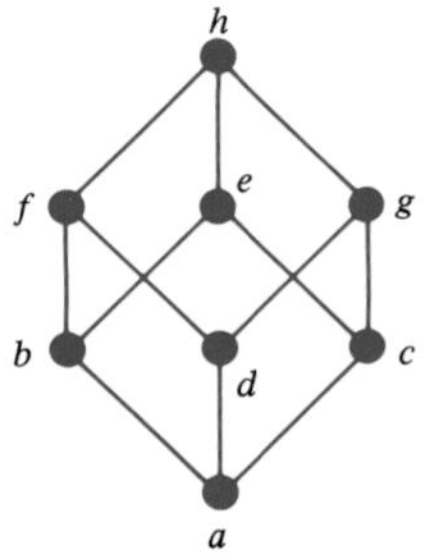

Figure 10

6.

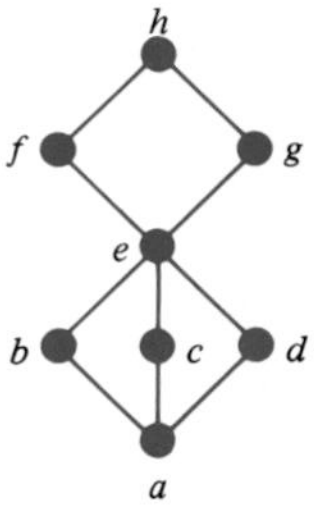

Figure 11

7.

Figure 12

8.

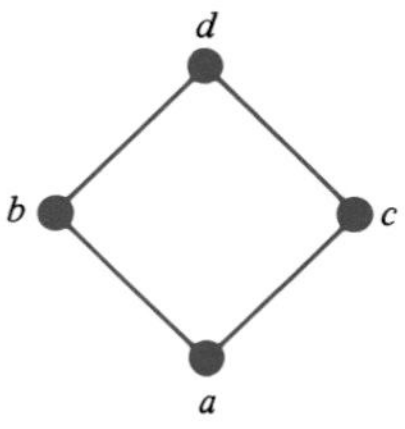

Figure 13

9. D_{385}

10. D_{60}

THEORETICAL EXERCISES

T1. Are there Boolean algebras having three elements?

T2. Show that in a Boolean algebra, for any a and b, $a \leq b$ if and only if $b' \leq a'$.

T3. Show that in a Boolean algebra, for any a and b, $a = b$ if and only if $(a \wedge b') \vee (a' \wedge b) = 0$.

T4. Show that in a Boolean algebra, for any a, b, and c:
(a) If $a \leq b$, then $a \vee c \leq b \vee c$.
(b) If $a \leq b$, then $a \wedge c \leq b \wedge c$.

T5. Show that in a Boolean algebra the following statements are equivalent for any a and b.
(a) $a \vee b = b$
(b) $a \wedge b = a$
(c) $a' \vee b = I$
(d) $a \wedge b' = 0$
(e) $a \leq b$

T6. Show that in a Boolean algebra, for any a and b:

$$(a \wedge b) \vee (a \wedge b') = a$$

T7. Show that in a Boolean algebra, for any a and b,

$$b \wedge (a \vee (a' \wedge (b \vee b'))) = b$$

T8. Show that in a Boolean algebra, for any a, b, and c,

$$(a \wedge b \wedge c) \vee (b \wedge c) = b \wedge c$$

T9. Show that in a Boolean algebra, for any a, b, and c, if $a \leq b$ then

$$((a \vee c) \wedge (b' \vee c))' = (a' \vee b) \wedge c'$$

T10. Show that in a Boolean algebra, for any a, b, and c, if $a \leq b$ then,

$$a \vee (b \wedge c) = b \wedge (a \vee c)$$

KEY IDEAS FOR REVIEW

☐ Partial order on a set: relation that is reflexive, antisymmetric, and transitive.

☐ Partially ordered set or poset: set together with a partial order.

☐ Linearly ordered set: partially ordered set in which every pair of elements are comparable.

☐ Theorem: If A and B are posets, then $A \times B$ is a poset with the product partial order.

☐ Dual of a poset $(A, \leq)$: The poset $(A, \geq)$, where $\geq$ denotes the inverse of $\leq$.

☐ Hasse diagram: see page 174.

☐ Principle of Correspondence: see page 177.

☐ Isomorphism of posets: see page 177.

☐ Maximal (minimal) element of poset: see page 180.

☐ Theorem: A finite nonempty poset has at least one maximal element and at least one minimal element.

☐ Greatest (least) element of a poset A: see page 182.

☐ Theorem: A poset has at most one greatest element and at most one least element.

☐ Upper (lower) bound of subset B of poset A: Element $a \in A$ such that $b \leq a (a \leq b)$ for all $b \in B$.

☐ Least upper bound (greatest lower bound) of subset B of poset A: Element $a \in A$ such that a is an upper (lower) bound of B and $a \leq a'(a' \leq a)$, where a' is any upper (lower) bound of B.

☐ Lattice: a poset in which every subset consisting of two elements has a LUB and a GLB.

☐ Theorem: If L_1 and L_2 are lattices, then $L = L_1 \times L_2$ is a lattice.

☐ Theorem: Let L be a lattice, and $a \in L$, $b \in L$, $c \in L$. Then
(a) $a \vee b = b$ if and only if $a \leq b$.
(b) $a \wedge b = a$ if and only if $a \leq b$.
(c) $a \wedge b = a$ if and only if $a \vee b = b$.

☐ Theorem: Let L be a lattice. Then
1. (a) $a \vee a = a$
   (b) $a \wedge a = a$ (Idempotent Property)
2. (a) $a \vee b = b \vee a$
   (b) $a \wedge b = b \wedge a$ (Commutative Property)
3. (a) $a \vee (b \vee c) = (a \vee b) \vee c$
   (b) $a \wedge (b \wedge c) = (a \wedge b) \wedge c$ (Associative Property)
4. (a) $a \vee (a \wedge b) = a$
   (b) $a \wedge (a \vee b) = a$ (Absorption Property)

☐ Theorem: Let L be a lattice, and $a \in L$, $b \in L$, $c \in L$.
1. If $a \leq b$, then
   (a) $a \vee c \leq b \vee c$
   (b) $a \wedge c \leq b \wedge c$
2. $a \leq c$ and $b \leq c$ if and only if $a \vee b \leq c$.
3. $c \leq a$ and $c \leq b$ if and only if $c \leq a \wedge b$.
4. If $a \leq b$ and $c \leq d$, then
   (a) $a \vee c \leq b \vee d$
   (b) $a \wedge c \leq b \wedge d$

☐ Isomorphic lattices: page 189.
☐ Bounded lattice: has a greatest element I and a least element 0.
☐ Theorem: A finite lattice is bounded.
☐ Distributive lattice: satisfies the Distributive Laws:

$$a \wedge (b \vee c) = (a \wedge b) \vee (a \wedge c); \qquad a \vee (b \wedge c) = (a \vee b) \wedge (a \vee c)$$

☐ Complement of a: Element $a' \in L$ (bounded lattice) such that

$$a \vee a' = I \qquad \text{and} \qquad a \wedge a' = 0$$

☐ Theorem: Let L be a bounded distributive lattice. If a complement exists, it is unique.
☐ Complemented lattice: bounded lattice in which every element has a complement.
☐ Substitution Rule for Boolean algebras: page 200.
☐ Boolean algebra: a lattice isomorphic with $(P(S), \subseteq)$, for some finite set S.
☐ Properties of a Boolean algebra: see page 201.

REVIEW EXERCISES

In Exercises 1 and 2, determine whether the relation R is a partial order on the set A.

1. $A = Z$ and $a\ R\ b$ if and only if $a < b + 2$.

2. $A = \{p,\ q,\ r,\ s,\ t\}$, $R = \{(p,\ p),\ (p,\ q),\ (p,\ r),\ (p,\ s),\ (p,\ t),\ (t,\ s),\ (t,\ r),\ (t,\ q),\ (t,\ t),\ (q,\ q),\ (s,\ s), (s,\ r,),\ (r,\ r)\}$.

3. Give the relation on the set $A = \{a,\ b,\ c,\ d,\ e\}$ determined by the Hasse diagram shown below.

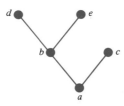

4. Determine the Hasse diagram of the partial order on the set $A = \{1,\ 2,\ 3,\ 4,\ 5\}$ whose matrix is

$$\begin{bmatrix} 1 & 1 & 1 & 1 & 1 \\ 0 & 1 & 0 & 1 & 0 \\ 0 & 0 & 1 & 0 & 1 \\ 0 & 0 & 0 & 1 & 0 \\ 0 & 0 & 0 & 0 & 1 \end{bmatrix}$$

5. Determine all maximal and minimal elements of the partial order set whose Hasse diagram is shown in the following figure

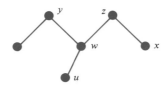

6. Let $A = \{1, 2, 3, 4, 5, 6, 7, 8, 9, 10\}, B = \{3, 4, 5\}$. For the partial order on A whose Hasse diagram is shown below list all upper bounds of B, all lower bounds of B, the greatest lower bound of B, and the least upper bound of B.

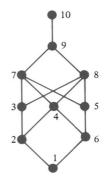

7. Is the following Hasse diagram that of a lattice? If possible, list the complements of a and of d.

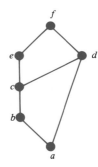

8. Is D_{220} a Boolean algebra?

9. Draw the Hasse diagram of D_{24}.

CHAPTER TEST

1. Let $A = \{2, 3, 4, 7, 8, 21, 168\}$ and let $\leq$ be the partial order "$|$", divides, on A. Draw the Hasse diagram of $(A,\ \leq)$.

2. Let $A = \{x,\ y,\ z,\ p,\ q\}$ and let $\leq$ be the partial order on A whose digraph is given below. Construct the Hasse diagram of this partial order.

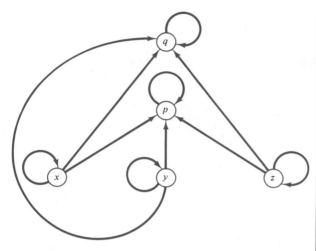

3. Let $A = \{r, s, t, u, v, w\}$. From the Hasse diagram of a partial order on A, given below, construct the matrix of the partial order.

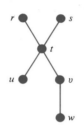

4. Let B_8 be the Boolean algebra consisting of all sequences of 0's and 1's of length 8. Let

$$A = \{(1, 1, 0, 0, 0, 1, 0, 1),$$

$$(1, 1, 1, 1, 0, 0, 0, 0), (0, 1, 1, 1, 0, 0, 0, 1)\}$$

be a subset of B_8

(a) List all upper bounds of A.

(b) Compute the LUB and GLB of A.

(c) What is the complement of
$(1, 1, 0, 0, 0, 0, 1, 1)$?

5. Is D_{595} a Boolean algebra? If it is, to which of the Boolean algebras B_n is it isomorphic?

6. For each of the following, answer true (T) or false (F).

(a) Every Boolean algebra is a lattice.

(b) Every lattice D_n is a Boolean algebra.

(c) If $(A, \leq)$ is a finite lattice and $B \subseteq A$, then LUB (B) must exist.

(d) There is a Boolean algebra with 65,535 elements.

(e) Suppose that $(A, \leq)$ is a partial order set and $B \subseteq A$. If B has only one upper bound, then this must be the least upper bound.

8

Boolean Algebras and Circuit Design

Prerequisites: Chapters 1, 4, 5, 6, and 7.

In this chapter we show that functions on Boolean algebras, and ways of representing these functions, have important applications to the specification and design of computer circuitry.

8.1 Functions on Boolean Algebras (Circuit Requirements)

Suppose that $f : B_n \to B$ is a function. Tables listing the values of the function f for all elements of B_n, such as shown in Fig. 1(a), are often called **truth tables** for f. This is because they are analogous with tables that arise in logic (see the Appendix). Suppose that the x_k represent simple declarative sentences, or **propositions**, and $f(x_1, x_2, \ldots, x_n)$ represents a compound sentence constructed from the x_k's. If we

x_1	x_2	x_3	$f(x_1, x_2, x_3)$
0	0	0	0
0	0	1	1
0	1	0	1
0	1	1	0
1	0	0	1
1	0	1	0
1	1	0	1
1	1	1	0

(a)

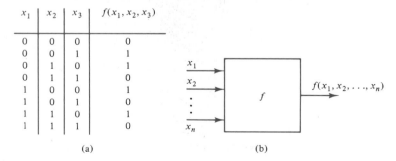

(b)

Figure 1

think of the value 0 for a sentence as meaning that the sentence is false, and 1 as meaning that the sentence is true, then tables such as Fig. 1(a) show us how the truth or falsity of $f(x_1, x_2, \ldots, x_n)$ depends on the truth or falsity of its component sentences x_k. We continue to call such tables truth tables, even when they arise in areas other than logic, such as in Boolean algebras.

The reason that such functions are important is that, as shown schematically in Fig. 1(b), they may be used to represent the output requirements of a circuit for any possible input values. Thus each x_i represents an input circuit capable of carrying two indicator voltages (one voltage for 0 and a different voltage for 1). The function f represents the desired output response in all cases. Such requirements occur at the design stage of all combinational and sequential computer circuitry.

Note carefully that the specification of a function $f : B_n \rightarrow B$ simply lists circuit requirements. It gives no indication of how these requirements can be met. One important way of producing functions from B_n to B is by using Boolean polynomials, which we now consider.

Boolean Polynomials

Let $x_1, x_2, \ldots, x_n$ be a set of n symbols or variables. A **Boolean polynomial** $p(x_1, x_2, \ldots, x_n)$, in the variables x_k, is defined recursively as follows:

Boolean Polynomials

1. $x_1, x_2, \ldots, x_n$ are all Boolean polynomials.
2. The symbols 0 and 1 are Boolean polynomials.
3. If $p(x_1, x_2, \ldots, x_n)$ and $q(x_1, x_2, \ldots, x_n)$ are two Boolean polynomials, then so are

$$(p(x_1, x_2, \ldots, x_n) \vee q(x_1, x_2, \ldots, x_n))$$

and

$$(p(x_1, x_2, \ldots, x_n) \wedge q(x_1, x_2, \ldots, x_n))$$

4. If $p(x_1, x_2, \ldots, x_n)$ is a Boolean polynomial, then so is

$$(p(x_1, x_2, \ldots, x_n))'$$

By tradition $(0)'$ is denoted $0'$, $(1)'$ is denoted $1'$, and $(x_k)'$ is denoted x_k'.
5. There are no Boolean polynomials in the variables x_k other than those that can be obtained by repeated use of rules 1, 2, 3, and 4.

Boolean polynomials are also called **Boolean expressions**.

Example 1   The following are Boolean polynomials in the variables x, y, and z.

$$p_1(x, y, z) = (x \vee y) \wedge z$$

$$p_2(x, y, z) = (x \vee y') \vee (y \wedge 1)$$

$$p_3(x, y, z) = (x \vee (y' \wedge z)) \vee (x \wedge (y \wedge 1))$$

$$p_4(x, y, z) = (x \vee (y \vee z')) \wedge ((x' \wedge z)' \wedge (y' \vee 0))$$

Ordinary polynomials in several variables, such as $x^2y + z^4$, $xy + yz + x^2y^2$, $x^3y^3 + xz^4$, and so on, are generally interpreted as expressions representing algebraic computations with unspecified numbers. As such, they are subject to the usual rules of arithmetic. Thus the polynomials $x^2 + 2x + 1$ and $(x + 1)(x + 1)$ are considered equivalent, and so are $x(xy + yz)(x + z)$ and $x^3y + 2x^2yz + xyz^2$, since in each case we can turn one into the other with algebraic manipulation.

Similarly, Boolean polynomials may be interpreted as representing Boolean computations with unspecified elements of B, that is, with 0's and 1's. As such, these polynomials are subject to the rules of Boolean arithmetic, that is, to the rules obeyed by $\wedge$, $\vee$, and $'$ in Boolean algebras. As with ordinary polynomials, two Boolean polynomials are considered equivalent if we can turn one into the other by Boolean manipulations.

In Section 6.1 we showed how ordinary polynomials could produce functions by substitution. This process works whether the polynomials involve one or several variables. Thus the polynomial $xy + yz^3$ produces a function $f : \mathbb{R}^3 \to \mathbb{R}$ by letting $f(x, y, z) = xy + yz^3$. For example, $f(3, 4, 2) = (3 \times 4) + (4 \times 2^3) = 12 + 32 = 44$. In a similar way, Boolean polynomials involving n variables produce functions from B_n to B.

Example 2   Consider the Boolean polynomial

$$p(x_1, x_2, x_3) = ((x_1 \wedge x_2) \vee (x_1 \vee (x_2' \wedge x_3)))$$

Construct the truth table for the Boolean function $f : B_3 \to B$ determined by this Boolean polynomial.

Solution. The Boolean function $f : B_3 \to B$ is described by substituting all the 2^3 ordered triples of values from B for x_1, x_2, and x_3. The truth table for the resulting function is shown in Fig. 2.

Boolean polynomials can also be written in a symbolic or schematic way. If x and y are variables, then the basic polynomials $x \vee y$, $x \wedge y$, and x' are shown schematically in Fig. 3. Each symbol has lines for the variables on the left, and a line on the right representing the polynomial as a whole. The symbol for $x \vee y$ is called an **or gate**, that for $x \wedge y$ is called an **and gate**, and the symbol for $'$ is called an

x_1	x_2	x_3	$f(x_1, x_2, x_3) = (x_1 \wedge x_2) \vee (x_1 \vee (x_2' \wedge x_3))$
0	0	0	0
0	0	1	1
0	1	0	0
0	1	1	0
1	0	0	1
1	0	1	1
1	1	0	1
1	1	1	1

Figure 2

inverter. The logical names arise because the truth tables showing the functions represented by $x \vee y$ and $x \wedge y$ are exact analogs of the truth table for the connectives "or" and "and" respectively.

Recall that functions from B_n to B can be used to describe the desired behavior of circuits with n 0 or 1 inputs, and one 0 or 1 output. In the case of the functions corresponding to the Boolean polynomials $x \vee y$, $x \wedge y$, and x', the desired circuits can be implemented, and the schematic forms of Fig. 3 are also used to represent these circuits. By repeatedly substituting these schematic forms for $\vee$, $\wedge$, and $'$, we can make a schematic form to represent any Boolean polynomial. For the reasons previously given, such diagrams are called **logic diagrams** for the polynomial.

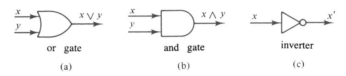

or gate	and gate	inverter
(a)	(b)	(c)

Figure 3

Example 3   Let $p(x, y, z) = ((x \wedge y) \vee (y \wedge z'))$. Figure 4(a) shows the truth table for the corresponding function $f : B_3 \to B$. Figure 4(b) shows the logic diagram for p.

Suppose that p is a Boolean polynomial in n variables, and f is the corresponding function from B_n to B. We know that f may be viewed as a description of the behavior of a circuit having n inputs and one output. In the same way, the logic diagram of p can be viewed as a description of the construction of such a circuit, at least in terms of and gates, or gates, and inverters. Thus if the function f, describing the desired behavior of a circuit, can be produced by a Boolean polynomial p, then the logic diagram for p will give one way to construct a circuit having that behavior. In general, many different polynomials will produce the same function. The logic diagrams of these polynomials will represent alternative methods for constructing the desired circuit. It is almost impossible to overestimate the importance of these facts for the study of computer circuitry.

x	y	z	$f(x, y, z) = (x \wedge y) \vee (y \wedge z')$
0	0	0	0
0	0	1	0
0	1	0	1
0	1	1	0
1	0	0	0
1	0	1	0
1	1	0	1
1	1	1	1

(a)

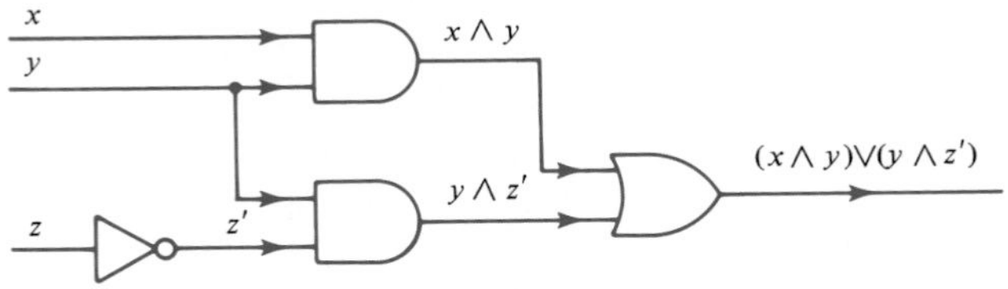

(b)

Figure 4

EXERCISE SET 8.1

1. Consider the Boolean polynomial

$$p(x, y, z) = x \wedge (y \vee z')$$

If $B = \{0, 1\}$, compute the truth table of the function $f : B_3 \to B$ defined by p.

2. Consider the Boolean polynomial

$$p(x, y, z) = (x \vee y) \wedge (z \vee x')$$

If $B = \{0, 1\}$, compute the truth table of the function $f : B_3 \to B$ defined by p.

3. Consider the Boolean polynomial

$$p(x, y, z) = (x \wedge y') \vee (y \wedge (x' \vee y))$$

If $B = \{0, 1\}$, compute the truth table of the function $f : B_3 \to B$ defined by p.

4. Construct a logic diagram implementing the function f of Exercise 1.

5. Construct a logic diagram implementing the function f of Exercise 2.

6. Construct a logic diagram implementing the function f of Exercise 3.

8.2 Expressing Boolean Functions as Boolean Polynomials (Circuit Design)

In the preceding section we considered functions from B_n to B, where B is the Boolean algebra $\{0, 1\}$. We noted that such functions can represent input-output requirements for models of many practical computer circuits. We also pointed out that if the function is given by some Boolean expression, then we can construct a logic diagram for it, and thus model the implementation of the circuit. In this section we show that all functions from B_n to B are given by Boolean expressions, and thus logic diagrams can be constructed for any such function. Our discussion illustrates a method for finding a Boolean expression that produces a given function.

If $f : B_n \to B$, we will let $S(f) = \{b \in B_n \,|\, f(b) = 1\}$. We then have the following result.

Theorem 1 Let f, f_1, and f_2 be three functions from B_n to B.
 (a) If $S(f) = S(f_1) \cup S(f_2)$, then $f(b) = f_1(b) \vee f_2(b)$ for all b in B.
 (b) If $S(f) = S(f_1) \cap S(f_2)$, then $f(b) = f_1(b) \wedge f_2(b)$ for all b in B.
($\vee$ and $\wedge$ are LUB and GLB, respectively, in B.)

Proof. (a) Let $b \in B_n$. If $b \in S(f)$, then, by definition of $S(f)$, $f(b) = 1$. Since $S(f) = S(f_1) \cup S(f_2)$, either $b \in S(f_1)$ or $b \in S(f_2)$, or both. In any case, $f_1(b) \vee f_2(b) = 1$. Now if $b \notin S(f)$, then $f(b) = 0$. Also, we must have $b \notin S(f_1)$ and $b \notin S(f_2)$, so $f_1(b) = 0$ and $f_2(b) = 0$. This means that $f_1(b) \vee f_2(b) = 0$. Thus, for all $b \in B_n$, $f(b) = f_1(b) \vee f_2(b)$.

(b) This part is proved in a manner completely analogous to that used in part (a).

Recall that a function $f : B_n \to B$ can be viewed as a function $f(x_1, x_2, \ldots, x_n)$ of n variables, each of which may assume the values 0 or 1. If $E(x_1, x_2, \ldots, x_n)$ is a Boolean expression, then the function that it produces is generated by substituting all combinations of 0's and 1's for the x_i's in the expression.

Example 1 Let $f_1 : B_2 \to B$ be produced by the expression $E(x, y) = x'$, and let $f_2 : B_2 \to B$ be produced by the expression $E(x, y) = y'$. Then the truth tables of f_1 and f_2 are shown in Fig. 1(a) and (b), respectively. Let $f : B_2 \to B$ be the function whose truth table is shown in Fig. 1(c). Clearly, $S(f) = S(f_1) \cup S(f_2)$, since f_1 is 1 at the elements $(0, 0)$,

x	y	$f_1(x, y)$
0	0	1
0	1	1
1	0	0
1	1	0

(a)

x	y	$f_2(x, y)$
0	0	1
0	1	0
1	0	1
1	1	0

(b)

x	y	$f(x, y)$
0	0	1
0	1	1
1	0	1
1	1	0

(c)

Figure 1

and $(0, 1)$ of B_2, f_2 is 1 at the elements $(0, 0)$ and $(1, 0)$ of B_2, and f is 1 at the elements $(0, 0)$, $(0, 1)$, and $(1, 0)$ of B_2. By Theorem 1, $f = f_1 \vee f_2$, so a Boolean expression that produces f is $x' \vee y'$. This is easily verified.

It is not hard to show that any function $f : B_n \rightarrow B$, for which $S(f)$ has exactly one element, is produced by a Boolean expression.

Example 2   Let $f : B_2 \rightarrow B$ be the function whose truth table is shown in Fig. 2(a). This function is equal to 1 only at the element $(0, 1)$ of B_2, that is, $S(f) = \{(0, 1)\}$. Thus $f(x, y) = 1$ only when $x = 0$ and $y = 1$. This is also true for the expression $E(x, y) = x' \wedge y$, so f is produced by this expression. The following table shows the correspondence between functions that are 1 at just one element, and the Boolean expressions that produce these functions.

$S(f)$	Expression producing f
$\{(0, 0)\}$	$x' \wedge y'$
$\{(0, 1)\}$	$x' \wedge y$
$\{(1, 0)\}$	$x \wedge y'$
$\{(1, 1)\}$	$x \wedge y$

The function $f : B_3 \rightarrow B$ whose truth table is shown in Fig. 2(b) has $S(f) = \{(0, 1, 1)\}$, that is, f equals 1 only when $x = 0$, $y = 1$, and $z = 1$. This is also true for the Boolean expression $x' \wedge y \wedge z$, which must therefore produce f.

If $b \in B_n$, then b is a sequence $(c_1, c_2, \ldots, c_n)$ of length n, where each c_k is 0 or 1. Let E_b be the Boolean expression $\bar{x}_1 \wedge \bar{x}_2 \wedge \cdots \wedge \bar{x}_n$, where $\bar{x}_k = x_k$ when $c_k = 1$ and $\bar{x}_k = x_k'$ when $c_k = 0$. Such an expression is called a **minterm**. Example 2 illustrates the fact that any function $f : B_n \rightarrow B$, for which $S(f)$ is a single element of B_n, is produced by a minterm expression. In fact, if $S(f) = \{b\}$, it is easily seen that the minterm expression E_b produces f. We then have the following result.

x	y	$f(x, y)$
0	0	0
0	1	1
1	0	0
1	1	0

x	y	z	$f(x, y, z)$
0	0	0	0
0	0	1	0
0	1	0	0
0	1	1	1
1	0	0	0
1	0	1	0
1	1	0	0
1	1	1	0

(a)   (b)   **Figure 2**

Theorem 2   Any function $f : B_n \rightarrow B$ is produced by a Boolean expression.

   Proof. Let $S(f) = \{b_1, b_2, \ldots, b_k\}$, and for each i, let $f_i : B_n \rightarrow B$ be the function defined by

$$f_i(b_i) = 1$$
$$f_i(b) = 0 \quad \text{if } b \neq b_i$$

Then $S(f_i) = \{b_i\}$, so $S(f) = S(f_1) \cup \cdots \cup S(f_n)$ and by Theorem 1

$$f = f_1 \vee f_2 \vee \cdots \vee f_n$$

By the discussion given above, each f_i is produced by the minterm E_{b_i}. Thus f is produced by the Boolean expression

$$E_{b_1} \vee E_{b_2} \vee \cdots \vee E_{b_n}$$

and this completes the proof.

Example 3   Consider the function $f : B_3 \rightarrow B$ whose truth table is shown in Fig. 3. Since $S(f) = \{(0, 1, 1), (1, 1, 1)\}$, Theorem 2 shows that f is produced by the Boolean expression $E(x, y, z) = E_{(0,1,1)} \vee E_{(1,1,1)} = (x' \wedge y \wedge z) \vee (x \wedge y \wedge z)$. This expression, however, is not the simplest Boolean expression that produces f. Using properties of Boolean algebras, we have

$$(x' \wedge y \wedge z) \vee (x \wedge y \wedge z) = (x' \vee x) \wedge (y \wedge z)$$
$$= 1 \wedge (y \wedge z) = y \wedge z$$

Thus f is also produced by the simple expression $y \wedge z$.

x	y	z	$f(x, y, z)$
0	0	0	0
0	0	1	0
0	1	0	0
0	1	1	1
1	0	0	0
1	0	1	0
1	1	0	0
1	1	1	1

Figure 3

   The process of writing a function as an "or" combination of minterms, and simplifying the resulting expression, can be systematized in various ways. We will demonstrate a graphical procedure utilizing what is known as a **Karnaugh map.** This procedure is easy for human beings to use with functions $f : B_n \rightarrow B$, if n is not too large. We will illustrate the method for $n = 2$, 3, and 4. If n is large, or if a programmable algorithm is desired, other techniques may be preferable.

We consider first the case where $n = 2$, so that f is a function of two variables, say x and y. In Fig. 4(a) we show a 2×2 matrix of "squares," with each square containing one possible input b from B_2. In Fig. 4(b) we have replaced each input b with the corresponding minterm E_b. The labeling of the squares in Fig. 4 is for reference only. In the future we will not exhibit these labels, but we will assume that the reader remembers their locations. In Fig. 4(b) we note that the x variable appears everywhere in the first row as x' and everywhere in the second row as x. We label these rows accordingly, and we perform a similar labeling of the columns.

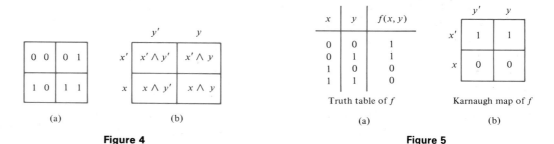

Figure 4 (a), (b)

Figure 5 (a) Truth table of f; (b) Karnaugh map of f

Example 4   Let $f : B_2 \rightarrow B$ be the function whose truth table is shown in Fig. 5(a). In Fig. 5(b) we have arranged the values of f in the appropriate squares, and we have kept the row and column labels. The resulting 2×2 array of 0's and 1's is called the **Karnaugh map of** f. Since $S(f) = \{(0, 0), (0, 1)\}$, the corresponding expression for f is $(x' \wedge y') \vee (x' \wedge y) = x' \wedge (y' \vee y) = x'$.

The outcome of Example 4 is typical. When the 1-values of a function $f : B_2 \rightarrow B$ exactly fill one row, or one column, the label of that row or column gives the Boolean expression for f. Of course, we already know that if the 1-values of f fill just one square, then f is produced by the corresponding minterm. It can be shown that the larger the rectangle of 1-values of f, the smaller the expression for f will be. Finally, if the 1-values of f do not lie in a rectangle, we can decompose these values into the union of (possibly overlapping) rectangles. Then by Theorem 1 the Boolean expression for f can be found by computing the expressions corresponding to each rectangle, and combining them with "$\vee$" symbols.

Example 5   Consider the function $f : B_2 \rightarrow B$ whose truth table is shown in Fig. 6(a). In Fig. 6(b) we show the Karnaugh map of f and decompose the 1-values into the two indicated rectangles. The expression for the function having 1's in the horizontal rectangle is x'

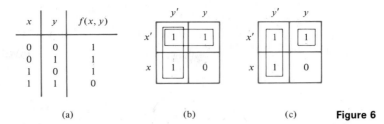

x	y	$f(x,y)$
0	0	1
0	1	1
1	0	1
1	1	0

(a)   (b)   (c)   **Figure 6**

(verify). The function having all its 1's in the vertical rectangle corresponds to the expression y' (verify). Thus f corresponds to the expression $x' \lor y'$. In Fig. 6(c) we show a different decomposition of the 1-values of f into rectangles. This decomposition is also correct, but it leads to the more complex expression $y' \lor (x' \land y)$. We see that the decomposition into rectangles is not unique, and that one should try to use the largest possible rectangles.

We now turn to the case of a function $f : B_3 \to B$, which we consider to be a function of x, y, and z. We could proceed as in the case of two variables and construct a "cube" of side 2 to contain the values of f. This would work, but three-dimensional figures are awkward to draw and use, and the idea would not generalize. Instead, we use a rectangle of size 2×4. In Fig. 7(a) and (b), respectively, we show the inputs (from B_3) and corresponding minterms for each square of such a rectangle.

Consider the rectangular areas shown in Fig. 8. If the 1-values for a function $f : B_3 \to B$ exactly fill one of the rectangles shown, then the Boolean expression for this function is one of the six expressions x, y, z, x', y', or z', as indicated in Fig. 8.

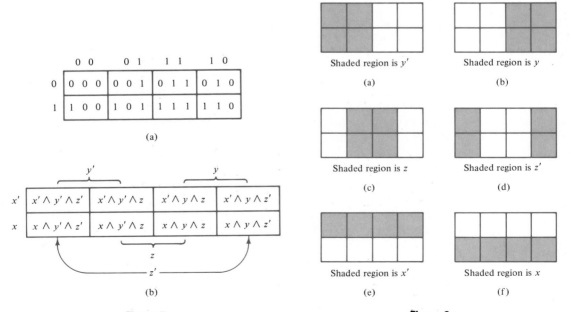

Figure 7

Figure 8

Consider the situation shown in Fig. 8(a). Theorem 1(a) shows that f can be computed by joining all the minterms corresponding to squares of the region with the symbol $\vee$. Thus f is produced by

$$(x' \wedge y' \wedge z') \vee (x' \wedge y' \wedge z) \vee (x \wedge y' \wedge z') \vee (x \wedge y' \wedge z)$$
$$= ((x' \vee x) \wedge (y' \wedge z')) \vee ((x' \vee x) \wedge (y' \wedge z))$$
$$= (1 \wedge (y' \wedge z')) \vee (1 \wedge (y' \wedge z))$$
$$= (y' \wedge z') \vee (y' \wedge z)$$
$$= y' \wedge (z' \vee z) = y' \wedge 1 = y'$$

A similar computation shows that the other five regions are correctly labeled.

If we think of the left and right edges of our basic rectangle as "glued together" to make a cylinder, as in Fig. 9, we can say that the six large regions shown in Fig. 8 consist of any two adjacent columns of the cylinder, or of the top or bottom half-cylinder.

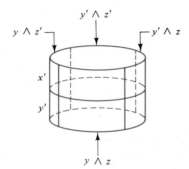

Figure 9

The six basic regions shown in Fig. 8 are the only ones whose corresponding Boolean expressions need be considered. That is why we used them to label Fig. 7(b), and we keep them as labels for all Karnaugh maps of functions from B_3 to B. Theorem 1(b) tells us that if the 1-values of a function $f : B_3 \to B$ form exactly the intersection of two or three of the basic six regions, then a Boolean expression for f can be computed by combining the expressions for these basic regions with $\wedge$ symbols.

Thus if the 1-values of the function f are as shown in Fig. 10(a), then we get them by intersecting the regions shown in Fig. 8(a) and (d). The Boolean expression for f is therefore $y' \wedge z'$. Similar derivations can be given for the other three columns. If the 1-values of f are as shown in Fig. 10(b), we get them by intersecting the regions of Fig. 8(c) and (e), so a Boolean expression for f is $z \wedge x'$. In a similar fashion we can compute the expression for any function whose 1-values fill two horizontally adjacent squares. There are eight such functions if we again consider the rectangle to be formed into a cylinder. Thus we include the case where the 1-values of f are as shown in Fig. 10(c). The resulting Boolean expression is $z' \wedge x'$.

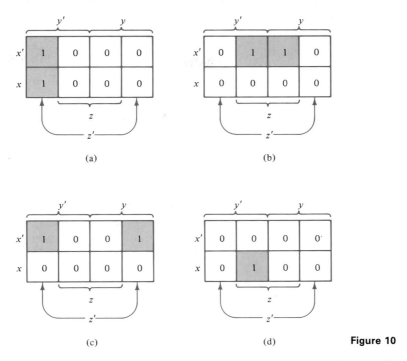

(a)   (b)

(c)   (d)   **Figure 10**

If we intersect three of the basic regions, and the intersection is not empty, the intersection must be a single square, and the resulting Boolean expression is a minterm. In Fig. 10(d), the 1-values of f form the intersection of the three regions shown in Fig. 8(a), (c), and (f). The corresponding minterm is $y' \wedge z \wedge x$. Thus one need not remember the placement of minterms in Fig. 7(b), but instead may reconstruct it.

We have seen how to compute a Boolean expression for any function $f : B_3 \rightarrow B$ whose 1-values form a rectangle of adjacent squares (in the cylinder) of length $2^n \times 2^m$, $n = 0, 1$; $m = 0, 1, 2$. In general, if the set of 1-values of f do not form such a rectangle, we may write this set as the union of such rectangles. Then a Boolean expression for f is computed by combining the expressions associated with each rectangle with $\vee$ symbols. This is true by Theorem 1(a). The discussion above shows that the larger the rectangles that are chosen, the simpler will be the resulting Boolean expression.

Example 6   Consider the function f whose truth table and corresponding Karnaugh map are shown in Fig. 11. The placement of the 1's can be derived by locating the corresponding inputs in Fig. 7(a). One decomposition of the 1-values of f is shown in Fig. 11(b). From this we see that a Boolean expression for f is $(y' \wedge z') \vee (x' \wedge y') \vee (y \wedge z)$.

x	y	z	$f(x, y, z)$
0	0	0	1
0	0	1	1
0	1	0	0
0	1	1	1
1	0	0	1
1	0	1	0
1	1	0	0
1	1	1	1

(a)

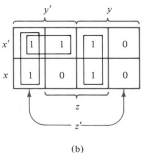

(b)   **Figure 11**

Example 7   Figure 12 shows the truth table and corresponding Karnaugh map for a function f. The decomposition into rectangles shown in Fig. 12(b) uses the idea that the first and last columns are considered adjacent (by wrapping around the cylinder). Thus the symbols are left "open ended" to signify that they join in one 2×2 rectangle, corresponding to z'. The resulting Boolean expression is $z' \lor (x \land y)$ (verify).

x	y	z	$f(x, y, z)$
0	0	0	1
0	0	1	0
0	1	0	1
0	1	1	0
1	0	0	1
1	0	1	0
1	1	0	1
1	1	1	1

(a)

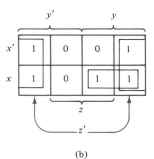

(b)   **Figure 12**

Finally, without additional comment, we present in Fig. 13 the distribution of inputs and corresponding labeling of rectangles for the case of a function $f : B_4 \rightarrow B$, considered as a function of x, y, z and w. Here again, we consider the first and last columns to be adjacent, and the first and last rows to be adjacent, both by "wrap around," and we look for rectangles with sides of length some power of 2, so the length is 1, 2, or 4. The expression corresponding to such rectangles is given by "intersecting" the large labeled rectangles of Fig. 14.

Example 8   Figure 15 shows the Karnaugh map of a function $f : B_4 \rightarrow B$. The 1-values are placed by considering the location of inputs in Fig. 13(a). Thus $f(0101) = 1, f(0001) = 0$, and so on.

The center 2×2 square represents the Boolean expression $w \land y$ (verify).

The four corners also form a square of side 2 since the right and left edges and the top and bottom edges are considered adjacent. From a geometric point of view,

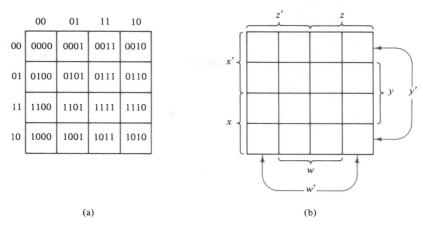

(a)   (b)

Figure 13

we can see that if we wrap the rectangle around horizontally, getting a cylinder, then when we further wrap around vertically, we will get a torus or "inner tube." On this inner tube, the four corners form a square of side 2 which represents the Boolean expression $w' \wedge y'$ (verify).

It then follows that the decomposition above leads to the Boolean expression

$$(w \wedge y) \vee (w' \wedge y')$$

for f.

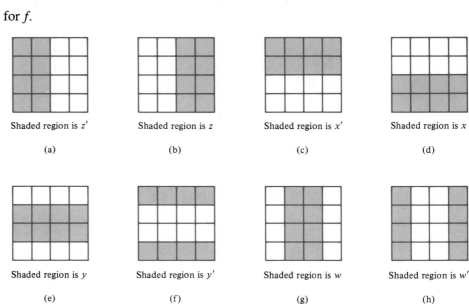

Shaded region is z'	Shaded region is z	Shaded region is x'	Shaded region is x
(a)	(b)	(c)	(d)
Shaded region is y	Shaded region is y'	Shaded region is w	Shaded region is w'
(e)	(f)	(g)	(h)

Figure 14

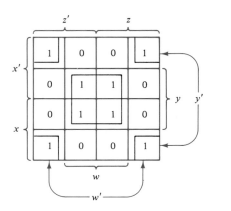

Figure 15

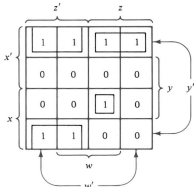

Figure 16

Example 9   We show, in Fig. 16, the Karnaugh map of a function $f : B_4 \to B$. The decomposition of 1-values into rectangles of sides 2^n, shown in this figure, again uses the "wrap around" property of top and bottom rows. The resulting expression for f is (verify)

$$(z' \wedge y') \vee (x' \wedge y' \wedge z) \vee (x \wedge y \wedge z \wedge w)$$

The first term comes from the 2×2 square formed by joining the 1×2 rectangle in the upper left-hand corner and the 1×2 rectangle in the lower left-hand corner. The second comes from the rectangle of size 1×2 in the upper right-hand corner, and the last is a minterm corresponding to the isolated square.

EXERCISE SET 8.2

In Exercises 1–6, construct Karnaugh maps for the functions whose truth tables are given.

1.

x	y	$f(x, y)$
0	0	1
0	1	0
1	0	0
1	1	1

2.

x	y	$f(x, y)$
0	0	1
0	1	0
1	0	1
1	1	0

3.

x	y	z	$f(x, y, z)$
0	0	0	1
0	0	1	1
0	1	0	0
0	1	1	0
1	0	0	1
1	0	1	0
1	1	0	1
1	1	1	0

4.

x	y	z	$f(x, y, z)$
0	0	0	0
0	0	1	1
0	1	0	1
0	1	1	1
1	0	0	0
1	0	1	0
1	1	0	0
1	1	1	1

5.

x	y	z	w	$f(x, y, z, w)$
0	0	0	0	0
0	0	0	1	0
0	0	1	0	1
0	0	1	1	0
0	1	0	0	0
0	1	0	1	0
0	1	1	0	1
0	1	1	1	0
1	0	0	0	0
1	0	0	1	0
1	0	1	0	0
1	0	1	1	1
1	1	0	0	0
1	1	0	1	0
1	1	1	0	1
1	1	1	1	1

6.

x	y	z	w	$f(x, y, z, w)$
0	0	0	0	1
0	0	0	1	0
0	0	1	0	1
0	0	1	1	0
0	1	0	0	0
0	1	0	1	1
0	1	1	0	1
0	1	1	1	0
1	0	0	0	0
1	0	0	1	0
1	0	1	0	0
1	0	1	1	0
1	1	0	0	1
1	1	0	1	0
1	1	1	0	1
1	1	1	1	0

In Exercises 7–11, Karnaugh maps of functions are given, and a decomposition of 1-values into rectangles is shown. Write the Boolean expression for these functions which arise from the maps and rectangular decompositions.

7. **8.**

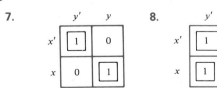

Figure 17   Figure 18

9.

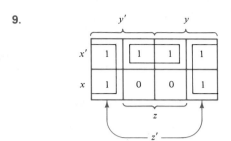

Figure 19

10.

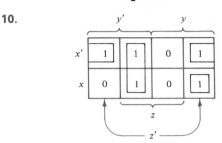

Figure 20

11.

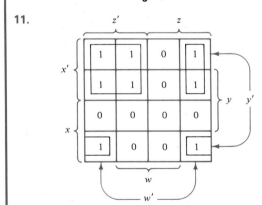

Figure 21

In Exercises 12–17, use the Karnaugh map method to find a Boolean expression for the function f.

12. Let f be the function of Exercise 1.
13. Let f be the function of Exercise 2.
14. Let f be the function of Exercise 3.
15. Let f be the function of Exercise 4.
16. Let f be the function of Exercise 5.
17. Let f be the function of Exercise 6.

KEY IDEAS FOR REVIEW

☐ Truth tables: see p. 208.

☐ Boolean expression: see p. 209.

☐ Minterm: A Boolean expression of the form $\bar{x}_1 \wedge \bar{x}_2 \wedge \cdots \wedge \bar{x}_n$, where each $\bar{x}_k$ is either x_k or x_k'.

☐ Theorem: Any function $f : B_n \to B$ is produced by a Boolean expression.

☐ Karnaugh map: see p. 216.

REVIEW EXERCISES

1. Consider the Boolean polynomial

$$p(x, y, z) = (x \wedge y) \vee (y' \wedge z)$$

If $B = \{0, 1\}$, compute the truth table of the function $f : B_3 \to B$ defined by p.

2. Consider the Boolean polynomial

$$p(x, y, z) = ((x \vee y) \wedge z') \wedge (x' \wedge y')$$

If $B = \{0, 1\}$, compute the truth table of the function $f : B_3 \to B$ defined by p.

3. Construct the logic diagram implementing the function f of Exercise 1.

4. Construct the logic diagram implementing the function f of Exercise 2.

5. Construct a Karnaugh map for the function f whose truth table is

x	y	z	$f(x, y, z)$
0	0	0	1
0	0	1	1
0	1	0	1
0	1	1	0
1	0	0	0
1	0	1	1
1	1	0	0
1	1	1	1

6. Construct a Karnaugh map for the function whose truth table is

x	y	z	$f(x, y, z)$
0	0	0	0
0	0	1	0
0	1	0	1
0	1	1	1
1	0	0	0
1	0	1	1
1	1	0	0
1	1	1	1

7. Write the Boolean expression for the function with the following Karnaugh map and decomposition of 1-values into rectangles.

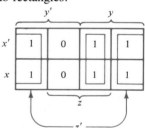

8. Write the Boolean expression for the function with the following Karnaugh map and decomposition of 1-values into rectangles.

9. Use the Karnaugh map method to find a Boolean expression for the function f given in Exercise 5.

10. Use the Karnaugh map method to find a Boolean expression for the function f given in Exercise 6.

CHAPTER TEST

1. Consider the Boolean polynomial

$$p(x, y, z) = (x' \wedge y') \vee (y \wedge z)$$

   If $B = \{0, 1\}$, compute the truth table of the function $f : B_3 \to B$ defined by p.

2. Construct the logic diagram implementing the function f given in Problem 1.

3. Construct a Karnaugh map for the function f whose truth table is

x	y	z	$f(x, y, z)$
0	0	0	1
0	0	1	0
0	1	0	0
0	1	1	0
1	0	0	1
1	0	1	1
1	1	0	0
1	1	1	0

4. Write the Boolean expression for the function with the following Karnaugh map and decomposition of 1-values into rectangles.

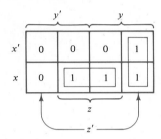

5. Use the Karnaugh map method to find a Boolean expression for the function f given in Problem 3.

6. Construct the truth table for the function whose Karnaugh map is given in Problem 4.

Prerequisites: Chapters 1, 4, and 5

In this chapter we study relations called trees, and we investigate their properties and their applications to computer algorithms.

9.1 Trees

In this section we study a special type of relation which is exceptionally useful in a variety of computer science applications, and which is usually represented by its digraph. These relations are essential for the construction of data bases and language compilers, to name just two important areas. They are called trees or sometimes rooted trees, because of the appearance of their digraphs.

Let A be a finite set, and let T be a relation on A. We say that T is a **tree** if there is a vertex v_0 in A with the property that there exists a unique path in T from v_0 to every other vertex in A, but no path from v_0 to v_0.

We show below that the vertex v_0, described in the definition above, is unique. It is often called the **root** of the tree T, and T is then referred to as a **rooted tree**. We write (T, v_0) to denote a rooted tree T with root v_0.

If (T, v_0) is a rooted tree on the set A, an element v of A will often be referred to as a **vertex in T**. This terminology simplifies the discussion, since it often happens that the underlying set A of T is of no importance.

To help us see the nature of trees, we will prove some simple properties satisfied by trees.

Theorem 1   Let (T, v_0) be a rooted tree. Then
   (a) There are no cycles in T.
   (b) v_0 is the only root of T.
   (c) Each vertex in T, other than v_0, has in-degree one, and v_0 has in-degree zero.

Proof. (a) Suppose that there is a cycle q in T, beginning and ending at vertex v. By the definition of a tree, we know that $v \neq v_0$, and there must be a path p from v_0 to v. Then $q \circ p$ (see Section 4.4) is a path from v_0 to v which is different from p, and this contradicts the definition of a tree.

(b) If v_0' is another root of T, there is a path p from v_0 to v_0' and a path q from v_0' to v_0 (since v_0' is a root). Then $q \circ p$ is a cycle from v_0 to v_0, and this is impossible by definition. Hence the vertex v_0 is the unique root.

(c) Let w_1 be a vertex in T other than v_0. Then there is a unique path $v_0, \cdots, v_k, w_1$ from v_0 to w_1 in T. This means that $(v_k, w_1) \in T$, so w_1 has in-degree at least one. If the in-degree of w_1 is more than one, there must be distinct vertices w_2 and w_3, such that (w_2, w_1) and (w_3, w_1) are both in T. If $w_2 \neq v_0$ and $w_3 \neq v_0$, there are paths p_2 from v_0 to w_2 and p_3 from v_0 to w_3, by definition. Then $(w_2, w_1) \circ p_2$ and $(w_3, w_1) \circ p_3$ are two different paths from v_0 to w_1, and this contradicts the definition of a tree with root v_0. Hence, the in-degree of w_1 is one. We leave it as an exercise to complete the proof if $w_2 = v_0$ or $w_3 = v_0$, and to show that v_0 has in-degree zero.

Theorem 1 summarizes the geometric properties of a tree. With these properties in mind, we can see how the digraph of a typical tree must look.

Let us first draw the root v_0. No edges enter v_0, but several may leave, and we draw these edges downward. The terminal vertices of the edges beginning at v_0 will be called the **level 1** vertices, while v_0 will be said to be at **level 0**. Also, v_0 is sometimes called the **parent** of these level 1 vertices, and the level 1 vertices are called the **offspring** of v_0. This is shown in Fig. 1(a). Each vertex at level 1 has no other edges entering it, by part (c) of Theorem 1, but each of these vertices may have edges leaving the vertex. The edges leaving a vertex of level 1 are drawn downward and terminate at various vertices which are said to be at **level 2**. Figure 1(b) shows the situation at this point. A parent–offspring relationship holds also for these levels (and at every consecutive pair of levels). For example, v_3 would be called the parent of the three offspring v_7, v_8, and v_9. Two vertices that are offspring of the same parent, are called **siblings**.

The above process continues for as many levels as are required to complete the digraph. If we view the digraph upside down, we will see why these relations are called trees. The number of levels of a tree is called the **height** of the tree. Thus the tree in Fig. 1(a) has two levels and the tree in Fig. 1(b) has three levels.

We should note that a tree may have infinitely many levels, and that any level other than level 0 may contain an infinite number of vertices. In fact, any vertex could have infinitely many offspring. However, in all our future discussions, trees will be assumed to have a finite number of vertices. Thus the trees will always have a bottom (highest-numbered) level. The vertices of the tree that have no offspring are, for geometric reasons, called the **leaves** of the tree.

The vertices of a tree that lie at any one level simply form a set of vertices in A. Often, however, it is useful to suppose that the offspring of each vertex of the tree are linearly ordered. Thus if a vertex v has four offspring, we may assume that they are ordered, so that we may refer to them as the first, second, third, or fourth offspring

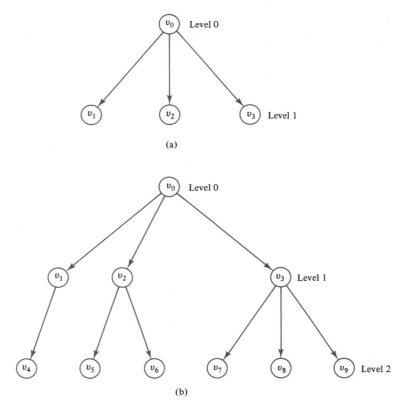

Figure 1

of v. Whenever we draw the digraph of a tree, we automatically assume some ordering at each level, by arranging offspring from left to right. Such a tree will be called an **ordered** tree. Generally, ordering of offspring in a tree is not explicitly mentioned. If ordering is needed, it is usually introduced at the time when the need arises, and it often is specified by the way the digraph of the tree is drawn.

Example 1   Let A be the set of all female descendants of a given human female v_0. We now define the following relation T on A: If v_1 and v_2 are elements of A, then $v_1 \, T \, v_2$ if and only if v_1 is the mother of v_2. The relation T on A is a rooted tree with root v_0.

Example 2   Let $A = \{v_0, v_1, v_2, v_3, v_4, v_5, v_6, v_7, v_8, v_9, v_{10}\}$ and let $T = \{(v_2, v_3), (v_2, v_1), (v_4, v_5), (v_4, v_6), (v_5, v_8), (v_6, v_7), (v_4, v_2), (v_7, v_9), (v_7, v_{10})\}$. Show that T is a rooted tree, and identify the root.

Solution. Since no paths begin at vertices v_1, v_3, v_8, v_9, and v_{10}, these vertices cannot be roots of a tree. There are no paths from vertices v_6, v_7, v_2, and v_5 to vertex v_4, so we must eliminate these vertices as possible roots. Thus if T is a rooted tree, its root must be vertex v_4. It is easy to show that there is a path from v_4 to every other vertex. For example, the path v_4, v_6, v_7, v_9 leads from v_4 to v_9, since (v_4, v_6), (v_6, v_7), and (v_7, v_9) are all in T. We draw the digraph of T, beginning with vertex v_4, and with edges shown downward. The result is shown in Fig. 2. A quick inspection of this digraph shows that paths from vertex v_4 to every other vertex are unique, and there are no paths from v_4 to v_4. Thus T is a tree with root v_4.

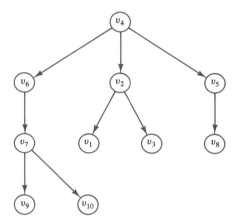

Figure 2

If n is a positive integer, we will say that a tree T is an **n-tree** if every vertex has at most n offspring. If all vertices of T, other than the leaves, have exactly n offspring, we will say that T is a **complete n-tree**. In particular, a 2-tree will often be called a **binary tree**, and a complete 2-tree will often be called a **complete binary tree**.

Binary trees are extremely important since there are quite efficient methods of implementing them and searching through them on computers.

Let (T, v_0) be a rooted tree on the set A, and let v be any vertex of T. By a **descendant** of v, we mean any vertex that can be reached by a path in T, beginning at v. Now we let $T(v)$ be the tree constructed from T in the following way. The vertices of $T(v)$ will consist of v and all of its descendants. The edges of $T(v)$ will consist of only those edges in T that begin with v or with one of the descendants of v. Then we have the following result.

Theorem 2   If (T, v_0) is a rooted tree and $v \in T$, then $T(v)$ is also a rooted tree with root v. We will say that $T(v)$ is the **subtree of T beginning at v**.

Proof. By definition of $T(v)$, we see that there is a path from v to every other vertex in $T(v)$. If there is a vertex w in $T(v)$, such that there are two distinct paths q and q' from v to w, and if p is the path in T from v_0 to v, then $q \circ p$ and $q' \circ p$ would be two distinct paths in T from v_0 to w. This is impossible, since T is a tree with root v_0. Thus each path from v to another vertex w in $T(v)$ must be unique. Also if q is a cycle at v in $T(v)$, then q is also a cycle in T. This contradicts Theorem 1(a), therefore q cannot exist. It follows that $T(v)$ is a tree with root v.

Example 3   Consider the tree T of Example 2. This tree has root v_4, and is shown in Fig. 2. In Fig. 3 we have drawn the subtrees $T(v_5)$, $T(v_2)$, and $T(v_6)$ of T.

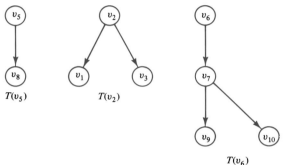

$T(v_5)$ $T(v_2)$

$T(v_6)$ **Figure 3**

EXERCISE SET 9.1

In Exercises 1–8, each relation R is defined on the set A. In each case determine if R is a tree, and if it is, find the root.

1. $A = \{a, b, c, d, e, f\}$
    $R = \{(a, d), (b, c), (c, a), (d, e)\}$

2. $A = \{a, b, c, d, e\}$
    $R = \{(a, b), (b, e), (c, d), (d, b), (c, a)\}$

3. $A = \{a, b, c, d, e, f\}$
    $R = \{(a, b), (c, e), (f, a), (f, c), (f, d)\}$

4. $A = \{1, 2, 3, 4, 5, 6\}$
    $R = \{(2, 1), (3, 4), (5, 2), (6, 5), (6, 3)\}$

5. $A = \{1, 2, 3, 4, 5, 6\}$
    $R = \{(1, 1), (2, 1), (2, 3), (3, 4), (4, 5), (4, 6)\}$

6. $A = \{1, 2, 3, 4, 5, 6\}$
    $R = \{(1, 2), (1, 3), (4, 5), (4, 6)\}$

7. $A = \{t, u, v, w, x, y, z\}$
    $R = \{(t, u), (u, w), (u, x), (u, v), (v, z), (v,y)\}$

8. $A = \{u, v, w, x, y, z\}$
    $R = \{(u, x), (u, v), (w, v), (x, z), (x, y)\}$

In Exercises 9–12, consider the rooted tree (T, v_0) shown in Fig. 4.

9. Compute the tree $T(v_2)$.

10. Compute the tree $T(v_3)$.

11. Compute the tree $T(v_8)$.

12. Compute the tree $T(v_1)$.

THEORETICAL EXERCISES

T1. Let T be a tree. Suppose that T has r vertices and s edges. Find a formula relating r to s.

T2. Show that if (T, v_0) is a rooted tree, then v_0 has in-degree zero.

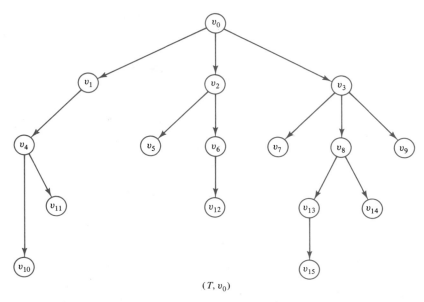

(T, v_0)

Figure 4

9.2 Labeled Trees

It is sometimes useful to label the vertices or edges of a digraph to indicate that the digraph is being used for a particular purpose. This is especially true for many uses of trees in computer science. We will now give a series of examples in which the sets of vertices of the trees are not important, but rather the utility of the tree is best emphasized by the labels on these vertices. Thus we will represent the vertices simply as dots, and show the label of each vertex next to the dot representing that vertex.

Example 1   Consider the fully parenthesized, algebraic expression

$$(3 - (2 \times x)) + ((x - 2) - (3 + x))$$

We assume, in such an expression, that no operation such as $-$, $+$, $\times$, $\div$, can be performed until both of its arguments have been evaluated, that is, until all computations inside both the left and right arguments have been performed. Thus we cannot perform the central addition until we have evaluated $(3 - (2 \times x))$ and $((x - 2) - (3 + x))$. We cannot perform the central subtraction in $((x - 2) - (3 + x))$ until we evaluate $(x - 2)$ and $(3 + x)$, and so on. It is easy to see that each such expression has a **central operator**, corresponding to the last computation which can be performed. Thus "+" is central to the main expression

above, "−" is central to $(3 - (2 \times x))$, and so on. An important graphical representation of such an expression is as a labeled binary tree. In this tree the root is labeled
with the central operator of the main expression. The two offspring of the root are
labeled with the central operator of the expressions for the left and right arguments,
respectively. If either argument is a constant or variable, instead of an expression, this
constant or variable is used to label the corresponding offspring vertex. This process
continues until the expression is exhausted. Figure 1 shows the tree for the original
expression of this example. To illustrate the technique further, we have shown in
Fig. 2 the tree corresponding to the full parenthesized expression

$$(3 \times (1 - x)) \div ((4 + (7 - (y + 2))) \times (7 + (x \div y)))$$

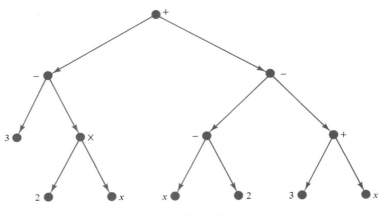

Figure 1

Our second example of a labeled tree is important for the computer implementation of a tree data structure. We start with an n-tree (T, v_0). Each vertex in T has
at most n offspring. We imagine that each vertex *potentially* has exactly n offspring,
which would be ordered from 1 to n, but that some of the offspring in the sequence
may be missing. The remaining offspring are labeled with the position they occupy
in the hypothetical sequence. Thus the offspring of any vertex are labeled with distinct
numbers from the set $\{1, 2, \ldots, n\}$.

Such a labeled digraph is sometimes called **positional**, and we will also use this
term. Note that positional trees are also ordered trees. When drawing the digraph of
a positional tree, we will imagine that the n offspring positions for each vertex are
arranged symmetrically below that vertex, and we place in its appropriate position
each offspring that actually occurs.

Figure 3 shows the digraph of a positional 3-tree, with all actually occurring
positions labeled. If offspring 1 of any vertex v actually exists, the edge from v to that
offspring is drawn sloping to the left. Offspring 2 of any vertex v is drawn vertically
downward from v, whenever it occurs. Similarly, offspring labeled 3 will be drawn
to the right. Naturally, the root is not labeled, since it is not an offspring.

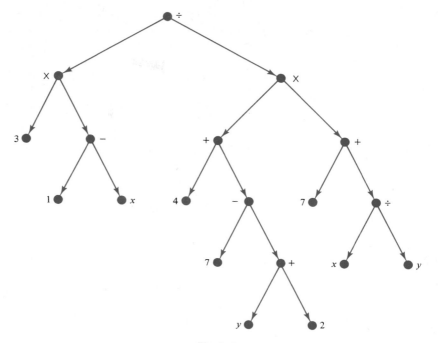

Figure 2

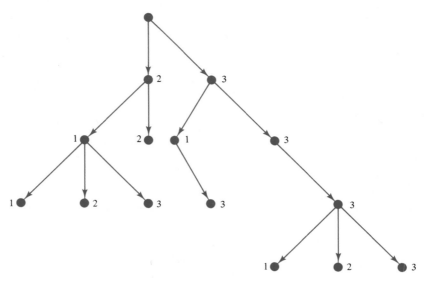

Figure 3

The positional binary tree is of special importance. In this case, for obvious reasons, the positions for potential offspring often are labeled *left* and *right*, instead of 1 and 2. Figure 4 shows the digraph of a positional binary tree with offspring labeled "*L*" for left and "*R*" for right. Labeled trees may have several sets of labels, all in force simultaneously. We will usually omit the left–right labels on a positional binary tree, in order to emphasize other useful labels. The positions of the offspring will then be indicated by the direction of the edges, as we have drawn them in Fig. 4.

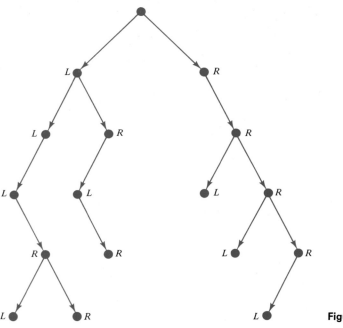

Figure 4

EXERCISE SET 9.2

In Exercises 1–10, construct the tree of the algebraic expression.

1. $(7 + (6 - 2)) - (x - (y - 4))$
2. $(x + (y - (x + y))) \times ((3 \div (2 \times 7)) \times 4)$
3. $3 - (x + (6 \times (4 \div (2 - 3))))$
4. $(((2 \times 7) + x) \div y) \div (3 - 11)$
5. $((2 + x) - (2 \times x)) - (x - 2)$
6. $(11 - (11 \times (11 + 11))) + (11 + (11 \times 11))$

7. $(3 - (2 - (11 - (9 - 4)))) \div (2 + (3 + (4 + 7)))$
8. $(x \div y) \div ((x \times 3) - (z \div 4))$
9. $((2 \times x) + (3 - (4 \times x))) - (x - (3 \times 11))$
10. $((1 + 1) + (1 - 2)) \div ((2 - x) - 1)$
11. Construct the digraphs of all distinct binary positional trees having three or fewer edges, and exactly two levels.

12. How many distinct binary positional trees are there having exactly two levels?

13. How many distinct positional 3-trees are there having exactly two levels?

14. Construct the digraphs of all distinct positional 3-trees having two or fewer edges.

9.3 Tree Searching

There are many occasions when it is useful to consider each vertex of a tree T exactly once in some specified order. As each successive vertex is encountered, we may wish to take some action or perform some computation, appropriate to the application being represented by the tree. For example, if the tree T is labeled, the label on each vertex may be displayed. If T is the tree of an algebraic expression, then at each vertex we may want to perform the computation indicated by the operator which labels that vertex. Performing appropriate tasks at a vertex will be called **visiting** the vertex. This is a convenient, nonspecific term which allows us to write algorithms without giving the details of what constitutes a "visit" in each particular case.

The process of visiting each vertex of a tree, in some specified order, will be called **searching** the tree, or performing a **tree search**. In some texts, this process is called **walking** or **traversing** the tree.

Let us consider tree searches on binary positional trees. Recall that in a binary positional tree, each vertex has two "potential" offspring. We denote these potential offspring by v_L (the left offspring) and v_R (the right offspring), and either or both may be missing. If a binary tree T is not positional, it may always be labeled (or drawn) so that it becomes positional.

Let T be a binary positional tree with root v. Then if v_L exists, the subtree $T(v_L)$ (see Section 9.1) will be called the **left subtree** of T, and if v_R exists, the subtree $T(v_R)$ will be called the **right subtree** of T.

Note that $T(v_L)$, if it exists, is a positional binary tree with root v_L, and similarly $T(v_R)$ is a positional binary tree with root v_R. This notation allows us to specify searching algorithms in a natural and powerful recursive form. Recall that recursive algorithms are those which refer to simpler versions of themselves. We first describe a method of searching called a **preorder search**. Suppose that the details of "visiting" any vertex of a tree are left unspecified, and simply referred to by the word "visit." Then the PREORDER search is defined recursively as follows.

PREORDER Search of (T, v)

1. VISIT v.
2. If v_L exists, then perform a PREORDER search of $(T(v_L), v_L)$.
3. If v_R exists, then perform a PREORDER search of $(T(v_R), v_R)$.

Example 1 Let T be the labeled, positional, binary tree whose digraph is shown in Fig. 1(a). The root of this tree is the vertex labeled "A." Suppose that for any vertex v of T, a VISIT to v simply prints out the label of v. Let us now apply the preorder search algorithm to this tree. Note first that if a tree consists only of one vertex, its root, then a search of this tree simply prints out the label of the root. In Fig. 1(b) we have placed boxes around the subtrees of T, and numbered these subtrees (in the corner of the boxes) for convenient reference.

According to the PREORDER search procedure applied to T, we will visit the root and print "A," then search subtree 1, then subtree 7. Applying the PREORDER search to subtree 1 results in visiting the root of subtree 1 and printing "B," then searching subtree 2, and finally searching subtree 4. The search of subtree 2 first prints

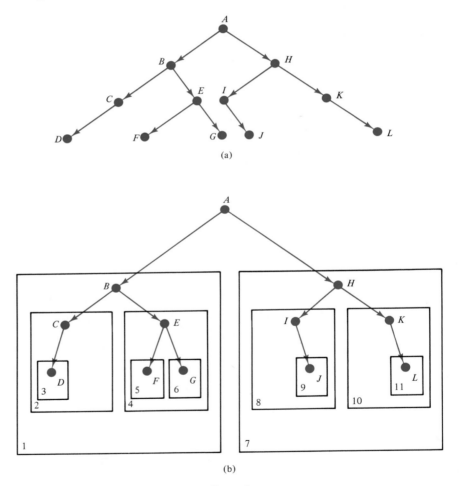

(a)

(b)

Figure 1

the symbol "*C*," then searches subtree 3. Subtree 3 has just one vertex, and so as previously mentioned, a search of this tree yields just the symbol "*D*." Up to this point, the search has yielded the string "*ABCD*." Note that we have had to interrupt the search of each tree (except subtree 3, which is a leaf of *T*) in order to apply the search procedure to a subtree. Thus we cannot finish the search of *T*, by searching subtree 7, until we apply the search procedure to subtrees 2 and 4. We could not complete the search of subtree 2 until we search subtree 3, and so on. The bookkeeping brought about by these interruptions produces the labels in the desired order, and recursion is a simple way to specify this bookkeeping.

Returning to the search, we have completed searching subtree 2, and we now must search subtree 4, since this is the right subtree of tree 1. Thus we print "*E*," and search subtrees 5 and 6 in order. These searches produce "*F*" and "*G*." The search of subtree 1 is now complete, and we go to subtree 7. Applying the same procedure, we can see that the search of subtree 7 will ultimately produce the string "*HIJKL*." The result, then, of the complete search of *T*, is to print the string "*ABCDEFGHIJKL*."

Example 2   Consider the completely parenthesized expression $(a - b) \times (c + (d \div e))$. Figure 2(a) shows the digraph of the labeled, positional binary tree representation of this expression. We apply the PREORDER search procedure to this tree, as we did to the tree in Example 1. Figure 2(b) shows the various subtrees encountered in the search. Proceeding as in Example 1, and supposing again that a VISIT to v simply prints out the label of v, we see that the string $\times - ab + c \div de$ is the result of the search. This is the **prefix** or **Polish form** of the given algebraic expression. Once again, the numbering of the boxes in Fig. 2(b) shows the order in which the subroutine PREORDER is applied to subtrees.

The Polish form of an algebraic expression is interesting because it represents the expression unambiguously, without the need for parentheses. To evaluate an expression in Polish form, proceed as follows. Move from left to right until we find a string of the form *Fxy*, where *F* is the symbol for a binary operation (say $+$, $-$, $\times$, and so on) and *x* and *y* are numbers. Evaluate *xFy*, and substitute the answer for the string *Fxy*. Continue this procedure until only one number remains.

For example, in the expression above, suppose that $a = 6$, $b = 4$, $c = 5$, $d = 2$, and $e = 2$. Then we are to evaluate $\times - 6\ 4 + 5 \div 2\ 2$. This is done in the following sequence of steps.

1. $\times - 6 + 5 \div 2\,2$
2. $\times\ 2 + 5 \div 2\,2$ (since the first string of the correct type is $-6\ 4$, and $6 - 4 = 2$)
3. $\times\ 2 + 5\ 1$ (replacing $\div 2\ 2$ by $2 \div 2 = 1$)
4. $\times\ 2\ 6$ (replacing $+5\ 1$ by $5 + 1 = 6$)
5. 12 (replacing $\times\ 2\ 6$ by 2×6)

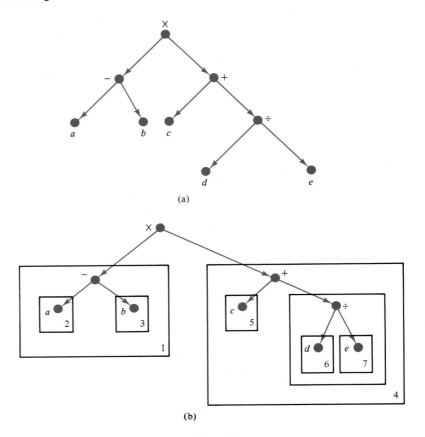

(a)

(b)

Figure 2

The example above is one of the primary reasons for calling this type of search the preorder search.

We will now consider two other recursive procedures for searching a binary positional tree T with root v.

INORDER Search of (T, v)

1. If v_L exists, then perform an INORDER search of $(T(v_L), v_L)$.
2. VISIT v.
3. If v_R exists, then perform an INORDER search of $(T(v_R), v_R)$.

POSTORDER Search of (T, v)

1. If v_L exists, then perform a POSTORDER search of $(T(v_L), v_L)$.
2. If v_R exists, then perform a POSTORDER search of $(T(v_R), v_R)$.
3. VISIT v.

Example 3   Consider the tree of Fig. 1(b) and apply the INORDER search procedure to search it. First we must search subtree 1. This requires us to first search subtree 2, and this in turn requires us to search subtree 3. As before, a search of a tree with only one vertex simply prints the label of the vertex. Thus "D" is the first symbol printed. The search of subtree 2 continues by printing "C," then stops since there is no right subtree at C. We then visit the root of subtree 1 and print "B," and proceed to the search of subtree 4, which yields "F," "E," and "G," in that order. We then visit the root of T and print "A," and proceed to search subtree 7. The reader may complete the analysis of the search of subtree 7 to show that it yields the string "$IJHKL$." Thus the complete search yields the string "$DCBFEGAIJHKL$."

Suppose now that we apply the POSTORDER search procedure to search the same tree, with the same notion of "visit." Again, the search of a tree with just one vertex will yield the label of that vertex. In general, we must search both the left and the right subtrees of a tree with root v, before we print out the label at v.

Referring again to Fig. 1(b), we see that both subtree 1 and subtree 7 must be searched before "A" is printed. Subtrees 2 and 4 must be searched before "B" is printed, and so on.

The search of subtree 2 requires us to search subtree 3, and "D" is the first symbol printed. The search of subtree 2 continues by printing "C." We now search subtree 4 yielding "F," "G," and "E." We next visit the root of subtree 1 and print "B." Proceeding with the search of subtree 7 we print the symbols "J," "I," "L," "K," and "H." Finally, we visit the root of T and print "A." Thus we print out the string "$DCFGEBJILKHA$."

Example 4   Let us now apply the inorder and postorder searches to the algebraic expression tree of Example 2 (see Fig. 2). The use of INORDER produces the string $a - b \times c + d \div e$. Notice that this is exactly the expression that we began with in Example 2, with all parentheses removed. Since the algebraic symbols lie between their arguments, this is often called the **infix notation**, and this explains the name INORDER. The expression above is ambiguous without parentheses. It could have come from the expression $a - (b \times ((c + d) \div e))$, which would have produced a different tree. Thus the tree cannot be recovered from the output of search procedure

INORDER, while it can be shown that the tree is recoverable from the Polish form produced by PREORDER. For this reason, Polish notation is often better for computer applications, although infix form is more familiar to human beings.

The use of search procedure POSTORDER on this tree produces the string $ab - cde \div + \times$. This is the **postfix** or **reverse Polish** form of the expression. It is evaluated in a manner similar to that used for Polish form, except that the arithmetic symbol is *after* its arguments rather than *before* them. If $a = 2$, $b = 1$, $c = 3$, $d = 4$, $e = 2$, the expression above is evaluated in the following sequence of steps.

1. $2\ 1 - 3\ 4\ 2 \div\ +\ \times$
2. $1\ 3\ 4\ 2 \div\ +\ \times$ (replacing $2\ 1 -$ by $2 - 1 = 1$)
3. $1\ 3\ 2 +\ \times$ (replacing $4\ 2 \div$ by $4 \div 2 = 2$)
4. $1\ 5 \times$ (replacing $3\ 2 +$ by $3 + 2 = 5$)
5. 5 (replacing $1\ 5 \times$ by $1 \times 5 = 5$)

Reverse Polish form is also parenthesis-free, and from it one can recover the tree of the expression. It is used even more frequently than the Polish form.

Searching General Trees

Until now, we have only shown how to search binary positional trees. We now show that any ordered tree T (see Section 9.1) may be represented as a binary positional tree which, although different from T, captures all the structure of T and can be used to recreate T. With the binary positional description of the tree, we may apply the computer representation and search methods previously developed. Since any tree may be ordered, we can use this technique on any (finite) tree.

Let T be any ordered tree and let A be the set of vertices of T. Define a binary positional tree $B(T)$, on the set of vertices A, as follows.

Binary Positional tree B(T) corresponding to ordered tree T

If $v \in A$, then
1. The left offspring v_L of v in $B(T)$ is the first offspring of v in T, if v has a first offspring in T.
2. The right offspring v_R of v in $B(T)$ is the next sibling of v in T (in the given order of siblings in T), if v has a next sibling in T.

Example 5   Figure 3(a) shows the digraph of a labeled tree T. We assume that each set of siblings is ordered from left to right, as they are drawn. Thus the offspring of vertex 1, namely vertices 2, 3, and 4, are ordered with vertex 2 first, 3 second, and 4 third. Similarly,

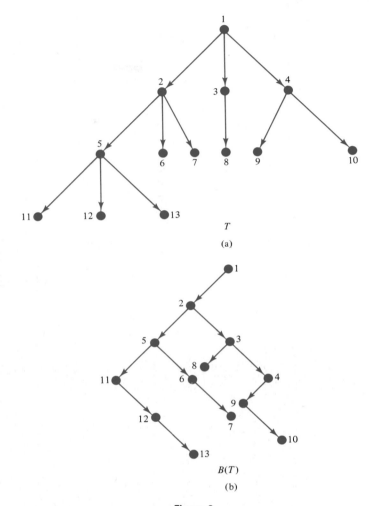

T

(a)

$B(T)$

(b)

Figure 3

the first offspring of vertex 5 is vertex 11, the second is vertex 12, and the third is vertex 13.

In Fig. 3(b) we show the digraph of the corresponding binary positional tree, $B(T)$. To obtain Fig. 3(b), we simply draw a left edge from each vertex v to its first offspring (if v has offspring). Then we draw a right edge from each vertex v to its next sibling (in the order given), if v has a next sibling. Thus the left edge from vertex 2, in Fig. 3(b), goes to vertex 5, because vertex 5 is the first offspring of vertex 2 in the tree T. Also, the right edge from vertex 2, in Fig. 3(b), goes to vertex 3, since vertex 3 is the next sibling in line (among all offspring of vertex 1).

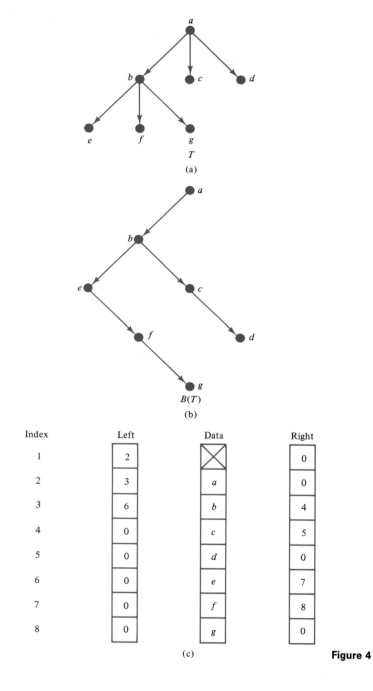

(a)

$B(T)$

(b)

Index	Left	Data	Right
1	2	✕	0
2	3	*a*	0
3	6	*b*	4
4	0	*c*	5
5	0	*d*	0
6	0	*e*	7
7	0	*f*	8
8	0	*g*	0

(c)

Figure 4

Example 6   Figure 4(a) shows the digraph of another labeled tree, with siblings ordered from left to right, as indicated. Figure 4(b) shows the digraph of the corresponding tree $B(T)$.

EXERCISE SET 9.3

In Exercises 1–4, the digraphs of labeled, binary positional trees are shown. In each case we suppose that visiting a node results in printing out the label of that node. For each exercise, show the result of performing a preorder search of the tree whose digraph is shown.

1.

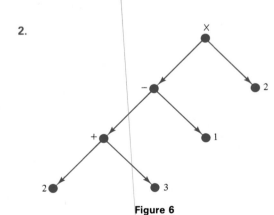

Figure 5

2.

Figure 6

3.

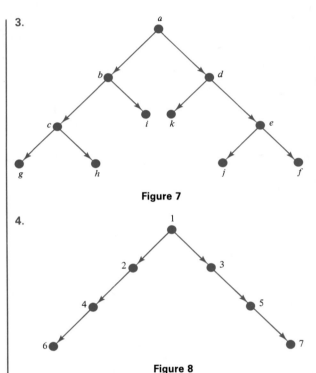

Figure 7

4.

Figure 8

In Exercises 5–12, visiting a node means printing out the label of the node.

5. Show the result of performing an inorder search of the tree shown in Fig. 5.

6. Show the result of performing an inorder search of the tree shown in Fig. 6.

7. Show the result of performing an inorder search of the tree shown in Fig. 7.

8. Show the result of performing an inorder search of the tree shown in Fig. 8.

9. Show the result of performing a postorder search of the tree shown in Fig. 5.

10. Show the result of performing a postorder search of the tree shown in Fig. 6.

11. Show the result of performing a postorder search of the tree shown in Fig. 7.

12. Show the result of performing a postorder search of the tree shown in Fig. 8.

13. Consider the tree digraph shown in Fig. 9 and the following list of words. Suppose that visiting a node of this tree means printing out the word corresponding to the number which labels that node. Print out the sequence of words that results from doing a postorder search of the tree.

1. ONE	7. I
2. COW	8. A
3. SEE	9. I
4. NEVER	10. I
5. PURPLE	11. SAW
6. NEVER	12. HOPE

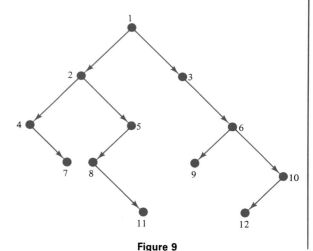

Figure 9

In Exercises 14 and 15, consider the labeled tree T whose digraph is shown. Draw the digraph of the corresponding binary positional tree $B(T)$. Label the vertices of $B(T)$ to show their correspondence to vertices of T.

14.

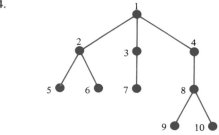

Figure 10

15.

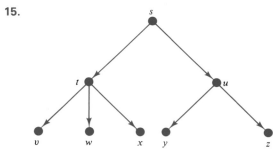

Figure 11

16. Show the result of performing a PREORDER search of $B(T)$, where T is as shown in Fig. 10.

17. Show the result of performing a POSTORDER search of $B(T)$, where T is as shown in Fig. 10.

18. Show the result of performing a POSTORDER search of $B(T)$, where T is as shown in Fig. 11.

19. Consider the digraph of a labeled binary positional tree shown in Fig. 12. If this tree is the binary form $B(T)$ of some tree T, draw the digraph of the labeled tree T.

9.4 Undirected Trees

An **undirected tree** is simply the symmetric closure of a tree (see Section 5.3); that is, it is a tree with all edges made bidirectional. As is the custom with symmetric relations, we represent an undirected tree by its graph rather than by its digraph. The

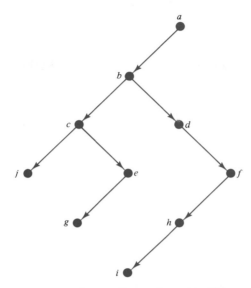

Figure 12

graph of an undirected tree T will have a single line without arrows connecting vertices a and b whenever (a, b) and (b, a) belong to T. The set $\{a, b\}$, where (a, b) and (b, a) are in T, is called an **undirected edge** of T (see Section 5.1). Thus each undirected edge $\{a, b\}$ corresponds to two ordinary edges, (a, b) and (b, a). The lines in the graph of an undirected tree T correspond to the undirected edges in T.

Example 1 Figure 1(a) shows the graph of an undirected tree T. In Fig. 1(b) and (c), we show digraphs of ordinary trees T_1 and T_2, respectively, which have T as symmetric closure. This merely shows that an undirected tree will, in general, correspond to many directed trees. Labels are included to show the correspondence of underlying vertices in the three relations. Note that the graph of T in Fig. 1(a) has six lines (undirected edges), although the relation T contains 12 pairs.

We want to present some useful alternative definitions of an undirected tree, and to do so we must make a few remarks about symmetric relations.

Let R be a symmetric relation and let $p = v_1, v_2, \ldots, v_n$ be a path in R. We will say that p is **simple** if no two edges of p correspond to the same undirected edge. If, in addition, v_1 equals v_n (so that p is a cycle), we will call p a **simple cycle**.

Example 2 Figure 2 shows the graph of a symmetric relation R. The path a, b, c, e, d, is simple but the path f, e, d, c, d, a is not simple, since d, c and c, d correspond to the same undirected edge. Also, f, e, a, d, b, a, f and d, a, b, d are simple cycles, but f, e, d, c, e, f is not a simple cycle, since f, e and e, f correspond to the same undirected edge.

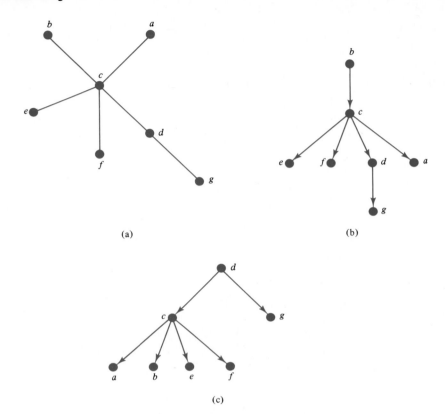

(a)

(b)

(c)

Figure 1

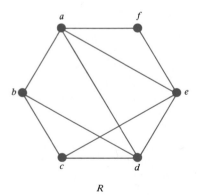

R

Figure 2

We will say that a symmetric relation R is **acyclic** if it contains no simple cycles. It can be shown that if R contains any cycles, then it contains a simple cycle. Recall (see Section 5.1) that a symmetric relation R is connected if there is a path in R from any vertex to any other vertex.

The following theorem provides a useful equivalent statement to the foregoing definition of an undirected tree.

Theorem 1   Let R be a symmetric relation on a set A. Then the following statements are equivalent.

(a) R is an undirected tree.

(b) R is connected and acyclic.

Proof. We will prove that (a) implies (b), and we will omit the proof that (b) implies (a). We suppose that R is an undirected tree, which means that R is the symmetric closure of some tree T on A. Note first that if $(a, b) \in R$, we must have either $(a, b) \in T$ or $(b, a) \in T$. In geometric terms, this means that every undirected edge in the graph of R appears in the digraph of T, directed one way or the other.

We will show by contradiction that R has no simple cycles. Suppose that R has a simple cycle $p = v_1, v_2, \ldots, v_n, v_1$. For each edge (v_i, v_j) in p, choose whichever pair (v_i, v_j) or (v_j, v_i) is in T. The result is a closed figure with edges in T, where each edge may be pointing in either direction. Now there are three possibilities. Either all arrows point "clockwise," as in Fig. 3(a), all point "counterclockwise," or some pair must be as in Fig. 3(b). Figure 3(b) is impossible, since in a tree T every vertex has in-degree ≤ 1 (see Theorem 1 of Section 9.1). But either of the other two cases would mean that T contains a cycle, which is also impossible. Thus the existence of the cycle p in R leads to a contradiction, and so is impossible.

We must also show that R is connected. Let v_0 be the root of the tree T. Then if a and b are any vertices in A, there must be paths p from v_0 to a, and q from v_0 to b, as shown in Fig. 3(c). Now all paths in T are reversible in R, so the path $p^{-1} \circ q$, shown in Fig. 3(d), connects a with b in R. Since a and b are arbitrary, R is connected, and (b) is proved. Here p^{-1} is the path p traversed in the reverse direction.

There are other useful characterizations of undirected trees. We state two of these without proof in the following theorem.

Theorem 2   Let R be a symmetric relation on a set A. Then R is an undirected tree if and only if either of the following statements is true.

(a) R is acyclic, and if any undirected edge is added to R, the new relation will not be acyclic.

(b) R is connected, and if any undirected edge is removed from R, the new relation will not be connected.

Note that in Theorem 2 adding and removing edges does not change the number of vertices.

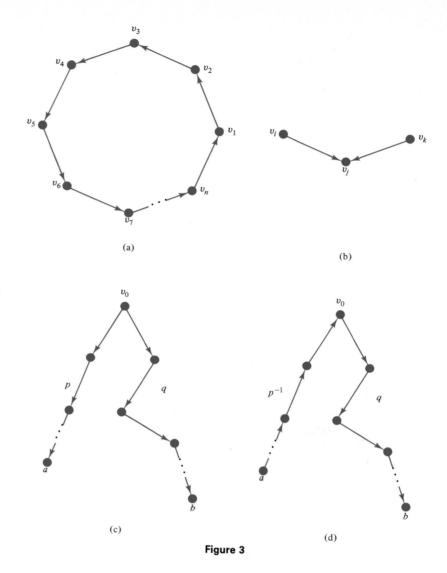

(a)

(b)

(c)

(d)

Figure 3

Spanning Trees of Connected Relations

If R is a symmetric, connected relation on a set A, we say that a tree T on A is a **spanning tree** for R if $T \subseteq R$. This simply says that T is a tree with exactly the same vertices as R and which can be obtained from R by deleting some edges of R.

Example 3   The symmetric relation R whose graph is shown in Fig. 4(a) has the tree T', whose digraph is shown in Fig. 4(b), as a spanning tree. Also, the tree T'', whose digraph is shown in Fig. 4(c), is a spanning tree for R. Since R, T', and T'' are all relations on

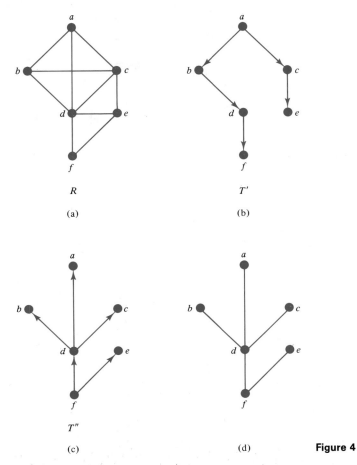

Figure 4

the same set A, we have labeled the vertices to show the correspondence of elements. As this example illustrates, spanning trees are not unique.

Sometimes there is interest in an **undirected spanning tree** for a symmetric, connected relation R. This is just the symmetric closure of a spanning tree. Figure 4(d) shows an undirected spanning tree for R that is derived from the spanning tree of Fig. 4(c). If R is a complicated relation that is symmetric and connected, it might be difficult to devise a scheme for searching R, that is, for visiting each of its vertices once in some systematic manner. If R is reduced to a spanning tree, the searching algorithms discussed in Section 9.3 can be used.

Theorem 2(b) suggests an algorithm for finding a undirected spanning tree for a relation R. Simply remove undirected edges from R until we reach a point where removal of one more undirected edge will result in a relation that is not connected. The result will be an undirected spanning tree.

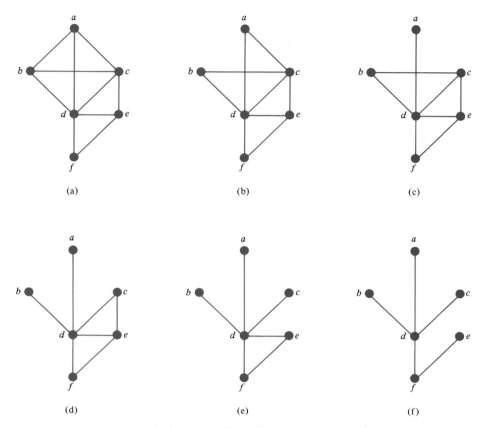

Figure 5

Example 4 In Fig. 5(a) we repeat the graph of Fig. 4(a). We then show the result of successive removal of undirected edges, culminating in Fig. 5(f), the undirected spanning tree, which agrees with Fig. 4(d).

EXERCISE SET 9.4

In Exercises 1–5, construct a spanning tree for the connected graph shown. Use the indicated vertex as the root of the tree and draw the digraph of the spanning tree produced.

1. Use e as the root.

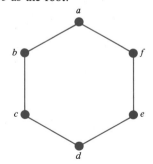

Figure 6

2. Use 5 as the root.

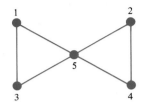

Figure 7

3. Use *c* as the root.

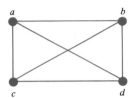

Figure 8

4. Use 4 as the root.

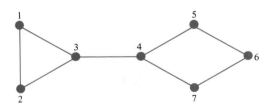

Figure 9

5. Use *e* as the root.

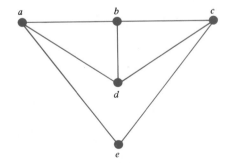

Figure 10

In Exercises 6–10, construct an undirected spanning tree for the connected graph *G* by removing edges in succession. Show the graph of the resulting undirected tree.

6. Let *G* be the graph shown in Fig. 6.
7. Let *G* be the graph shown in Fig. 7.
8. Let *G* be the graph shown in Fig. 8.
9. Let *C* be the graph shown in Fig. 9.
10. Let *G* be the graph shown in Fig. 10.
11. Consider the connected graph shown in Fig. 11. Show the graphs of three different undirected spanning trees.

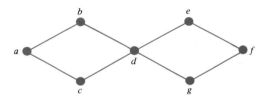

Figure 11

12. For the connected graph shown in Fig. 12, show the graphs of all undirected spanning trees.

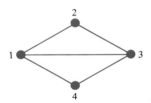

Figure 12

13. For the undirected tree shown in Fig. 13, show the digraphs of all spanning trees. How many are there?

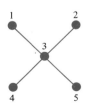

Figure 13

KEY IDEAS FOR REVIEW

☐ Tree: relation on a finite set A such that there exists a vertex $v_0 \in A$ with the property that there is a unique path from v_0 to any other vertex in A, and no path from v_0 to v_0.

☐ Root of tree: vertex v_0 in the definition of tree above.

☐ Rooted tree (T, v_0): tree T with root v_0.

☐ Theorem. Let (T, v_0) be a rooted tree. Then
(a) There are no cycles in T.
(b) v_0 is the only root of T.
(c) Each vertex in T, other than v_0, has in-degree one, and v_0 has in-degree zero.

☐ Level: see page 227.

☐ Leaves: vertices having no offspring.

☐ n-tree: tree where every vertex has at most n offspring.

☐ Binary tree: 2-tree.

☐ Theorem. If (T, v_0) is a rooted tree and $v \in T$, then $T(v)$ is also a rooted tree with root v.

☐ $T(v)$: subtree of T beginning at v.

☐ Binary positional tree: see page 234.

☐ Preorder search: see page 235.

☐ Polish form: see page 237.

☐ Inorder search: see page 238.

☐ Postorder search: see page 239.

☐ Reverse Polish form: see page 240.

☐ Searching general trees: see page 240.

☐ Undirected tree: symmetric closure of a tree.

☐ Simple path: no two edges correspond to the same undirected edge.

☐ Connected symmetric relation R: There is a path in R from any vertex to any other vertex.

☐ Spanning tree for symmetric connected relation R: tree reaching all the vertices of R, and whose edges are all edges of R.

☐ Undirected spanning tree: symmetric closure of a spanning tree.

REVIEW EXERCISES

In Exercises 1 and 2, determine whether the given relation R defined on the indicated set is a tree. If it is, find the root.

1. $A = \{a, b, c, d, e\}$;
$R = \{(a, b), (a, e)\ (c, d), (d, a)\}$

2. $A = \{a, b, c, d, e\}$;
$R = \{(a, b), (a, c), (b, d), (c, d), (d, e)\}$

In Exercises 3 and 4, consider the rooted tree (T, x) shown below. Compute the indicated tree.

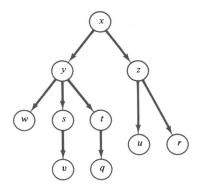

3. $T(y)$

4. $T(z)$

   In Exercises 5 and 6, construct the tree of the given algebraic expression.

5. $(x + (y - ((z + 2) - y))) + (3 - x)$

6. $(2 + 4) + ((3 + x) - (y - (z \div x)))$

7. Indicate the result of performing a PREORDER search on the labeled tree shown below.

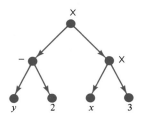

8. Indicate the result of performing a POSTORDER search on the labeled tree shown below.

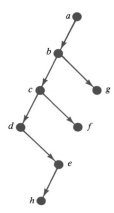

9. Construct a spanning tree for the connected graph shown below. Use a as the root.

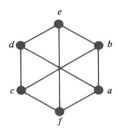

CHAPTER TEST

1. Determine whether the given relation R defined on the set $A = \{a, b, c, d, e, f\}$ is a tree. If it is, find the root.

   $R = \{(a, c), (b, c), (c, d), (c, e), (c, f), (f, e)\}$

2. Consider the rooted tree (T, a) shown below. Sketch $T(c)$.

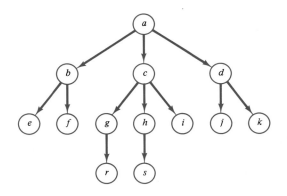

3. Construct the tree of the algebraic expression

$$((x - (y + (z - 3) - x))) + ((2 - x) \times (y \div x))$$

4. Indicate the result of performing an INORDER search on the labeled tree shown below.

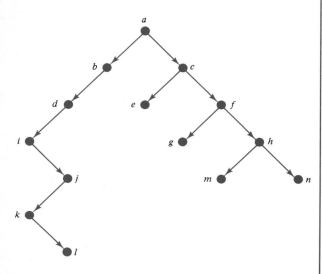

5. Construct a spanning tree for the connected graph shown below. Use 2 as the root.

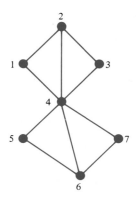

6. For each of the following, answer true (T) or false (F). For parts (a) and (b) consider the following tree.

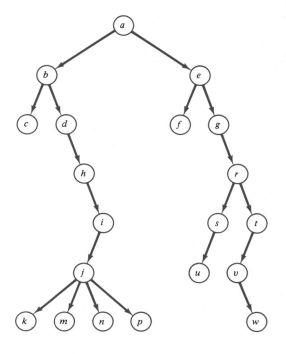

(a) The number of vertices in $T(d)$ is 6.
(b) The siblings of vertex k are vertices m, n, p, u, and v.

For parts (c) and (d) consider the following labeled tree.

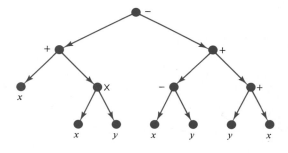

(c) The string produced by a PREORDER search is

$$- + x \times x y + - x y + y x$$

(d) The value of the string produced by a POST-ORDER search if $x = 2$ and $y = 1$ is 1.

(e) Consider the following two labeled trees. The tree on the right is the binary positional tree of the tree on the left.

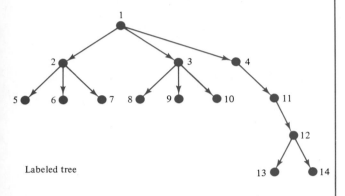

Labeled tree

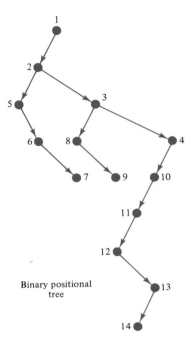

Binary positional tree

10

Languages and Grammars

Prerequisites: Chapters 1, 4, and 5.

In this chapter we study phrase structure grammars, simple devices for the construction of useful, elementary languages. We also develop several popular methods for representing these grammars.

10.1 Languages

If S is a set, then in Section 1.2 we considered the set S^* consisting of all finite strings of elements from the set S. There are many possible interpretations of the elements of S^*, depending on the nature of S. If we think of S as a set of "words," then S^* may be regarded as the collection of all possible "sentences" formed from words in S. Of course, such "sentences" do not have to be meaningful or even sensibly constructed. We may think of a language as a complete specification, at least in principle, of three things. First, there must be a set S consisting of all "words" which are to be regarded as being part of the language. Second, a subset of S^* must be designated as the set of "properly constructed sentences" in the language. The meaning of this term will depend very much on the language being constructed. Finally, it must be determined which of the properly constructed sentences have meaning, and what the meaning is.

Suppose, for example, that S consists of all English words. The specification of a properly constructed sentence involves the complete rules of English grammar; the meaning of a sentence is determined by this construction and by the meaning of the words. The sentence

"Going to the store John George to sing"

is a string in S^*, but is not a properly constructed sentence. The arrangement of nouns and verb phrases is illegal. On the other hand, the sentence

"Noiseless blue sounds sit cross-legged under the mountaintop"

is properly constructed, but completely meaningless.

For another example, S may consist of the integers, the symbols $+$, $-$, $\times$, and $\div$, and left and right parentheses. We will obtain a language if we designate as proper those strings in S^* which represent unambiguously parenthesized algebraic expressions. Thus

$$((3 - 2) + (4 \times 7)) \div 9 \qquad \text{and} \qquad (7 - (8 - (9 - 10)))$$

are properly constructed "sentences" in this language. On the other hand, $(2 - 3)) + 4$, $4 - 3 - 2$, and $)2 + (3) \times 4$ are not properly constructed. The first has too many parentheses, the second has too few (we do not know which subtraction to perform first), and the third has parentheses and numbers completely out of place. All properly constructed expressions have meaning except those involving division by zero. The meaning of an expression is the rational number it represents. Thus the meaning of $((2 - 1) \div 3) + (4 \times 6)$ is $73/3$, while $2 + (3 \div 0)$ and $(4 + 2) - (0 \div 0)$ are not meaningful.

The specification of the proper construction of sentences is called the **syntax** of a language. The specification of the meaning of sentences is called the **semantics** of a language. Among those languages that are of fundamental importance in computer science are the programming languages. These include BASIC, FORTRAN, ALGOL, PASCAL, APL, LISP, SNOBOL, ADA, FORTH, and many other general and special-purpose languages. When students are taught to program in some programming language, they are actually taught the syntax of the language. In a compiled language such as FORTRAN, most mistakes in syntax are detected by the compiler, and appropriate error messages are generated. The semantics of a programming language forms a much more difficult and advanced topic of study. The meaning of a line of programming is taken to be the entire sequence of events that take place inside the computer as a result of executing or interpreting that line.

We will not deal with semantics at all. We will study the syntax of a class of languages called phrase structure grammars. Although these are not nearly complex enough to include "real" languages such as English, they are general enough to encompass many languages of importance in computer science. This includes most aspects of programming languages, although the complete specification of some higher-level programming languages exceeds the scope of these grammars. On the other hand, phrase structure grammars are simple enough to be studied precisely, since the syntax is determined by substitution rules.

Grammars

A **phrase structure grammar** G is defined to be a 4-tuple $(V, S, v_0, \mapsto)$, where V is a finite set, S is a subset of V, $v_0 \in V - S$, and $\mapsto$ is a finite relation on V^*. The idea here is that S is, as discussed above, the set of all allowed "words" in the language, and V consists of S together with some other symbols. The element v_0 of V is a starting point for the substitutions which will shortly be discussed. Finally, the relation $\mapsto$ on V^* is a replacement relation in the sense that if $w \mapsto w'$, we may replace w by w' whenever the string w occurs either alone or as a substring of some other string. Traditionally, the statement $w \mapsto w'$ is called a **production** of G. Then w and w' are termed the **left** and **right** sides of the production, respectively. We assume that no production of G has the empty string $\wedge$ as its left side. We will call $\mapsto$ the **production relation** of G.

With these ingredients, we can introduce a substitution relation denoted by $\Rightarrow$, on V^*. We let $x \Rightarrow y$ mean that $x = l \cdot w \cdot r$, $y = l \cdot w' \cdot r$, and $w \mapsto w'$, where l and r are completely arbitrary strings in V^*. In other words, $x \Rightarrow y$ means that y results from x by using one of the allowed productions to replace part or all of x. The relation $\Rightarrow$ is usually called **direct derivability**. Finally, we let $\Rightarrow^\infty$ be the transitive closure of $\Rightarrow$ (see Section 5.4), and we decree that a string w in S^* is a syntactically correct sentence if and only if $v_0 \Rightarrow^\infty w$, that is if and only if there is a path in the relation $\Rightarrow$ from v_0 to w. In more detail, this says that a string w is a properly constructed sentence if w is in S^*, not just in V^*, and if we can get from v_0 to w by making a finite number of substitutions. This may seem complicated, but it is really quite a simple idea, as the following examples will show.

If $G = (V, S, v_0, \mapsto)$ is a phrase structure grammar, we will call S the set of **terminal symbols**, and $N = V - S$ the set of **nonterminal symbols**. Note that $V = S \cup N$.

The reader should be warned that other texts have slight variations of the definitions and notations which we have used for phrase structure grammars.

Example 1 Let $S = \{$John, Jill, drives, jogs, carelessly, rapidly, frequently$\}$, $N = \{$sentence, noun, verbphrase, verb, adverb$\}$, and let $V = S \cup N$. Let $v_0 = $ "sentence," and suppose that the relation $\mapsto$ on V^* is described by listing all productions as follows.

$$\text{sentence} \mapsto \text{noun verbphrase}$$

$$\text{noun} \mapsto \text{John}$$

$$\text{noun} \mapsto \text{Jill}$$

$$\text{verbphrase} \mapsto \text{verb adverb}$$

$$\text{verb} \mapsto \text{drives}$$

$$\text{verb} \mapsto \text{jogs}$$

$$\text{adverb} \mapsto \text{carelessly}$$

$$\text{adverb} \mapsto \text{rapidly}$$

$$\text{adverb} \mapsto \text{frequently}$$

The set S contains all the allowed words in the language; N consists of words that describe parts of sentences, but which are not actually contained in the language.

We claim that the sentence "Jill drives frequently," which we will denote by w, is an allowable or syntactically correct sentence in this language. To prove this, we consider the following sequence of strings in V^*.

sentence		
noun	verbphrase	
Jill	verbphrase	
Jill	verb	adverb
Jill	drives	adverb
Jill	drives	frequently

Now each of these strings follows from the preceding one, by using a production to make a partial or complete substitution. In other words, each string is related to the following string by the relation $\Rightarrow$, so "sentence" $\Rightarrow^\infty w$. By definition then, w is syntactically correct since, for this example, v_0 is "sentence." In phrase structure grammars, correct syntax simply refers to the process by which a sentence is formed, nothing else.

It should be noted that the sequence of substitutions which produces a valid sentence w, a sequence that will be called the **derivation** of the sentence w, is not unique. This is not surprising, since it merely means that there may be more than one path in $\Rightarrow$ from v_0 to w. The following derivation produces the sentence w of Example 1, but is not identical with the derivation given there.

sentence		
noun	verbphrase	
noun	verb	adverb
noun	verb	frequently
noun	drives	frequently
Jill	drives	frequently

The set of all properly constructed sentences that can be produced using a grammar G, is called the **language** of G, and is denoted by $L(G)$. The language of the grammar given in Example 1 is a somewhat simpleminded sublanguage of English, and it contains exactly 12 sentences. The reader can verify that "John jogs carelessly" is in the language $L(G)$ of this grammar, while "Jill frequently jogs" is not in $L(G)$.

It is also true that many different phrase structure grammars may produce the same language; that is, they may have exactly the same set of syntactically correct sentences. Thus a grammar cannot be reconstructed from its language. In Section 10.2 we will give examples in which different grammars are used to construct the same language.

Example 2 Let $V = \{v_0, w, a, b, c\}$, $S = \{a, b, c\}$, and let $\mapsto$ be the relation on V^* given by:

1. $v_0 \mapsto aw$
2. $w \mapsto bbw$
3. $w \mapsto c$

Consider the phrase structure grammar $G = (V, S, v_0, \mapsto)$. To derive a sentence of $L(G)$, it is necessary to perform successive substitutions, using (1), (2), and (3) above, until all symbols are eliminated other than the terminal symbols a, b, and c. Since we begin with the symbol v_0, we must first use production (1), or we could never eliminate v_0. This first substitution results in the string aw. We may now use (2) or (3) to substitute for w. If we use production (2), the result will contain a "w." Thus one application of (2) to aw produces the string ab^2w (a symbol b^n means n consecutive b's). If we then use (2) again, we will have the string ab^4w. We may use production (2) any number of times, but we will finally have to use production (3) to eliminate the symbol w. Once we use (3), only terminal symbols remain, so the process ends. We may summarize this analysis by saying that $L(G)$ is the subset of S^* consisting of all strings that begin with a, then have an even number $2n$, $n \geq 0$, of b's, and finally end with a c. For example, the word ab^6c is in $L(G)$. In Fig. 1, we give a derivation of this word. Note that since the "words" in this language all have exactly one letter, the "sentences" can be written as strings of letters. We omit the spaces when writing these strings.

v_0					
a	w				
a	bb	w			
a	bb	bb	w		
a	bb	bb	bb	w	
a	bb	bb	bb	c	**Figure 1**

Example 3 Let $V = \{v_0, w, a, b, c\}$, $S = \{a, b, c\}$, and let $\mapsto$ be a relation on V^* given by:

1. $v_0 \mapsto av_0b$
2. $v_0b \mapsto bw$
3. $abw \mapsto c$

Let $G = (V, S, v_0, \mapsto)$ be the corresponding phrase structure grammar. As we did in Example 2, we determine the form of allowable sentences in $L(G)$.

Since we must begin with the symbol v_0 alone, we must use production (1) first. We may continue to use (1) any number of times, but we must eventually use production (2) to eliminate v_0. Repeated use of (1) will result in a string of the form $a^n v_0 b^n$; that is, there are equal numbers of a's and b's. When (2) is used, the result is a string of the form $a^m(abw)b^m$ with $m \geq 0$. At this point the only production that can be used is (3), and we must use it to remove the nonterminal symbol w. The use of (3) finishes the substitution process and produces a string in S^*. Thus the allowable sentences $L(G)$ of the grammar G all have the form $w = a^n cb^n$, where $n \geq 0$.

Many difficulties in analyzing a grammar can arise if no restrictions are placed on the productions. For this reason, a classification of phrase structure grammars has been devised.

Let $G = (V, S, v_0, \mapsto)$ be a phrase structure grammar. Then we say that G is **Type 0**, **Type 1**, **Type 2**, or **Type 3**, according to the following descriptions.

Types of Phrase Structure Grammars

Type 0: if no restrictions are placed on the productions of G

Type 1: if for any production $w_1 \mapsto w_2$, the length of w_1 is less than or equal to the length of w_2 (where the **length** of a string is the number of words in that string)

Type 2: if the left-hand side of each production is a single, nonterminal symbol and the right-hand side consists of one or more symbols

Type 3: if the left-hand side of each production is a single, nonterminal symbol, and the right-hand side has one or more symbols including at most one nonterminal symbol, which must be at the extreme right of the string

---

In each of the preceding types, we permit the inclusion of the trivial production $v \mapsto \Lambda$, where Λ represents the empty string. This is an exception to the defining rule for types 1, 2, and 3, but it is included so that the empty string can be made part of the language. This avoids constant consideration of unimportant special cases.

It follows from the definition that each type of grammar is a special case of the type preceding it. Example 1 is a type 2 grammar, Example 2 is type 3, and Example 3 is type 0. Grammars of types 0 or 1 are quite difficult to study, and little is known about them. They include many pathological examples which are of no known practical use. We will restrict further consideration of grammars to types 2 and 3. These types are sufficiently complex to describe many aspects of actual programming languages. Type 2 grammars are sometimes called **context-free grammars**, since the symbols on the left of the productions are substituted for wherever they occur. On the other hand, a production of the type $l \cdot w \cdot r \mapsto l \cdot w' \cdot r$ (which could not occur in a type 2 grammar), is called **context sensitive**, since w' is substituted for w only in the context where it is surrounded by the strings l and r. Type 3 grammars are also called **regular grammars**.

A language will be called **type 2** or **type 3** if there is a grammar of type 2 or type 3 which produces it. This concept can cause problems. Even if a language is produced by a nontype 2 grammar, it is possible that some type 2 grammar also produces this same language. In this case, the language is type 2. The same situation may arise in the case of type 3 grammars.

The process we have considered in this section, namely deriving a sentence within a grammar, has a converse process. The converse process involves taking a sentence, and verifying that it is syntactically correct in some grammar G, by constructing a derivation that will produce it. This process is called **parsing** the sentence. Parsing is of fundamental importance for compilers, and other forms of language translation. A sentence in one language is parsed to show its structure, and corresponding sentences are generated in another language. In this way a FORTRAN program, for example, is compiled into a machine language program. The contents of this section and the next two sections are essential to the compiling process, but the complete details must be left to a more advanced course.

EXERCISE SET 10.1

In Exercises 1–7, a grammar G will be specified. In each case describe precisely the language, $L(G)$, produced by this grammar; that is, describe all syntactically correct "sentences."

1. $G = (V, S, v_0, \mapsto)$
   $V = \{v_0, v_1, x, y, z\}$, $S = \{x, y, z\}$
   $\mapsto \quad v_0 \mapsto x\, v_0$
   $\qquad v_0 \mapsto y\, v_1$
   $\qquad v_1 \mapsto y\, v_1$
   $\qquad v_1 \mapsto z$

2. $G = (V, S, v_0, \mapsto)$
   $V = \{v_0, a\}$, $S = \{a\}$
   $\mapsto \quad v_0 \mapsto a\, a\, v_0$
   $\qquad v_0 \mapsto a\, a$

3. $G = (V, S, v_0, \mapsto)$
   $V = \{v_0, a, b\}$, $S = \{a, b\}$
   $\mapsto \quad v_0 \mapsto a\, a\, v_0$
   $\qquad v_0 \mapsto a$
   $\qquad v_0 \mapsto b$

4. $G = (V, S, v_0, \mapsto)$
   $V = \{v_0, x, y, z\}$, $S = \{x, y, z,\}$
   $\mapsto : \quad v_0 \mapsto x\, v_0$
   $\qquad v_0 \mapsto y\, v_0$
   $\qquad v_0 \mapsto z$

5. $G = (V, S, v_0, \mapsto)$
   $V = \{v_0, v_1, v_2, a, + , (,)\}$
   $S = \{(,), a, + \}$
   $\mapsto : \quad v_0 \mapsto (v_0)$ (where left and right parentheses are symbols from S)
   $\qquad v_0 \mapsto a + v_1$
   $\qquad v_1 \mapsto a + v_2$
   $\qquad v_2 \mapsto a + v_2$
   $\qquad v_2 \mapsto a$

6. $G = (V, S, v_0, \mapsto)$
   $V = \{v_0, v_1, a, b\}$, $S = \{a, b\}$
   $\mapsto : \quad v_0 \mapsto a\, v_1$
   $\qquad v_1 \mapsto b\, v_0$
   $\qquad v_1 \mapsto a$

7. $G = (V, S, v_0, \mapsto)$
   $V = \{v_0, v_1, v_2, x, y, z\}$, $S = \{x , y, z\}$
   $\mapsto : \quad v_0 \mapsto v_0 v_1$
   $\qquad v_0 v_1 \mapsto v_2 v_0$
   $\qquad v_2 v_0 \mapsto x\, y$
   $\qquad v_2 \mapsto x$
   $\qquad v_1 \mapsto z$

8. For each grammar in Exercises 1–7, state whether the grammar is type 1, 2, or 3.

9. Let $G = (V, S, I, \mapsto)$, where

$$V = \{I, L, D, W, a, b, c, 0, 1, 2, 3, 4, 5, 6, 7, 8, 9\}$$

$$S = \{a, b, c, 0, 1, 2, 3, 4, 5, 6, 7, 8, 9\}$$

$\mapsto$ is given by:

1. $I \mapsto L$
2. $I \mapsto LW$
3. $W \mapsto LW$
4. $W \mapsto DW$
5. $W \mapsto L$
6. $W \mapsto D$
7. $L \mapsto a$
8. $L \mapsto b$
9. $L \mapsto c$
10. $D \mapsto 0$
11. $D \mapsto 1$

   .

   .

   .

19. $D \mapsto 9$

Which of the following statements are true for this grammar?
   (a) $ab092 \in L(G)$
   (b) $2a3b \in L(G)$
   (c) $aaaa \in L(G)$
   (d) $I \Rightarrow a$
   (e) $I \Rightarrow^\infty ab$
   (f) $DW \Rightarrow 2$
   (g) $DW \Rightarrow^\infty 2$
   (h) $W \Rightarrow^\infty 2abc$
   (i) $W \Rightarrow^\infty ba2c$

10. If G is the grammar of Exercise 9, describe $L(G)$.

11. Give two distinct derivations (sequences of substitutions starting at v_0) for the string $xyzyz \in L(G)$, where G is the grammar of Exercise 4.

12. Let G be the grammar of Exercise 5. Can you give two distinct derivations (see Exercise 11) for the string $((a + a + a))$?

13. Let G be the grammar of Exercise 9. Give two distinct derivations (see Exercise 11) of the string $a100$.

In Exercises 14–21, construct a phrase structure grammar G such that the language, $L(G)$, of G is equal to the language L.

14. $L = \{a^n b^n \mid n \geq 1\}$

15. $L = \{$strings of 0's and 1's with an equal number $n \geq 0$ of 0's and 1's$\}$

16. $L = \{a^n b^m \mid n \geq 1, m \geq 1\}$

17. $L = \{a^n b^n \mid n \geq 3\}$

18. $L = \{a^n b^m \mid n \geq 1, m \geq 3\}$

19. $L = \{x^n y^m \mid n \geq 2, m$ nonnegative and even$\}$

20. $L = \{x^n y^m \mid n$ even, m positive and odd$\}$

21. Let $G = (V, S, v_0, \mapsto)$ where
$V = \{v_0, v_1, v_2, a, b, c\}$, $S = \{a, b, c\}$
$\mapsto :$ $v_0 \mapsto aav_0$
   $v_0 \mapsto bv_1$
   $v_1 \mapsto cv_2 b$
   $v_1 \mapsto cb$
   $v_2 \mapsto bbv_2$
   $v_2 \mapsto bb$
State which of the following are in $L(G)$.
   (a) $aabcb$
   (b) $abbcb$
   (c) $aaaabcbb$
   (d) $aaaabcbbb$
   (e) $abcbbbbb$

10.2 Representations of Special Grammars and Languages

BNF Notation

For type 2 grammars (which include type 3 grammars) there are some useful, alternative methods of displaying the productions. A commonly encountered alternative is called the **BNF notation** (for Backus-Naur Form). We know that the left-hand sides of all productions in a type 2 grammar are single, nonterminal symbols. For any such symbol w, we combine all productions having w as left-hand side. The symbol w remains on the left, and all right-hand sides associated with w are listed together, separated by the symbol " | ". The relational symbol " $\mapsto$ " is replaced by the symbol " $:: =$ ". Finally, the nonterminal symbols, wherever they occur, are enclosed in pointed brackets " $\langle \ \rangle$ ". This has the additional advantage that nonterminal symbols may be permitted to have embedded spaces. Thus ⟨word1 word2⟩ shows that the string between the brackets is to be treated as one "word", not as two words. That is, we may use the space as a convenient and legitimate "letter" in a word, as long as we use pointed brackets to delimit the words.

Example 1   In BNF notation, the productions of Example 1 of Section 10.1 appear as follows.

$$\langle \text{sentence} \rangle \quad :: = \langle \text{noun} \rangle \langle \text{verbphrase} \rangle$$

$$\langle \text{noun} \rangle \quad\quad :: = \text{John} \mid \text{Jill}$$

$$\langle \text{verbphrase} \rangle :: = \langle \text{verb} \rangle \langle \text{adverb} \rangle$$

$$\langle \text{verb} \rangle \quad\quad :: = \text{drives} \mid \text{jogs}$$

$$\langle \text{adverb} \rangle \quad :: = \text{carelessly} \mid \text{rapidly} \mid \text{frequently}$$

Example 2   In BNF notation, the productions of Example 2 of Section 10.1 appear as follows.

$$\langle v_0 \rangle :: = a \langle w \rangle$$

$$\langle w \rangle :: = bb \langle w \rangle \mid c$$

Note that the left-hand side of a production may also appear in one of the strings on the right-hand side. Thus in the second line of Example 2, $\langle w \rangle$ appears on the left, and it appears in the string $bb\langle w \rangle$ on the right. When this happens, we say that the corresponding production $w \mapsto bbw$ is **recursive**. If a recursive production has w as left-hand side, we will say that the production is **normal** if w appears only once on the right-hand side and is the rightmost symbol. Other nonterminal symbols may also appear on the right side. The recursive production $w \mapsto bbw$ given in Example 2 is normal. Note that any recursive production that appears in a type 3 (regular) grammar is normal, by the definition of type 3.

Example 3   BNF notation is often used to specify actual programming languages. Both ALGOL and PASCAL had their grammars given in BNF initially. In this example, we consider a small subset common to these two grammars. This subset describes the syntax of decimal numbers, and can be viewed as a minigrammar whose corresponding language consists precisely of all properly formed decimal numbers.

Let $S = \{0, 1, 2, 3, 4, 5, 6, 7, 8, 9, \bullet\}$. Let V be the union of S with the set N, where

$$N = \{\text{decimal-number, decimal-fraction, unsigned-integer, digit}\}$$

Then let G be the grammar with symbol sets V and S, with starting symbol "decimal-number," and with productions given in BNF form as follows:

1. ⟨decimal-number⟩ :: = ⟨unsigned-integer⟩ | ⟨decimal-fraction⟩ |
   ⟨unsigned-integer⟩⟨decimal-fraction⟩
2. ⟨decimal-fraction⟩ :: = ⟨unsigned-integer⟩
3. ⟨unsigned-integer⟩ :: = ⟨digit⟩ | ⟨digit⟩⟨unsigned-integer⟩
4. ⟨digit⟩   :: = 0 | 1 | 2 | 3 | 4 | 5 | 6 | 7 | 8 | 9

Figure 1 shows a derivation, in this grammar, for the decimal number 23.14. Notice that the BNF statement numbered 3 is recursive in the second part of its right-hand side. That is, the production "unsigned-integer ↦ digit unsigned-integer" is recursive, and it is also normal. In general, we know that many different grammars may produce the same language. If the line numbered 3 above were replaced by the line

3′. ⟨unsigned-integer⟩ :: = ⟨digit⟩ | ⟨unsigned-integer⟩⟨digit⟩.

we would have a different grammar which produced exactly the same language, that is, the correctly formed decimal numbers. However, this grammar contains productions that are recursive but not normal.

```
                         decimal-number

           unsigned-integer              decimal-fraction

     digit     unsigned-integer     decimal-fraction
     digit     digit                 •    unsigned-integer
     digit     digit                 •    digit     unsigned-integer
     digit     digit                 •    digit     digit
     2         digit                 •    digit     digit
     2         3                     •    digit     digit
     2         3                     •    1         digit
     2         3                     •    1         4          Figure 1
```

Example 4   As in Example 3, we give a grammar that specifies a piece of several actual programming languages. In these languages, an identifier (a name for a variable, function, subroutine, and so on) must be composed of letters and digits, and must begin with a letter. The following grammar, with productions given in BNF, has precisely these identifiers as its language.

$$G = (V, S, \text{identifier}, \mapsto)$$

$$N = \{\text{identifier, remaining, digit, letter}\}$$

$$S = \{a, b, c, \ldots , z, 0, 1, 2, 3, \ldots , 9\}, \qquad V = N \cup S$$

1. $\langle$identifier$\rangle$ $::= \langle$letter$\rangle \mid \langle$letter$\rangle\langle$remaining$\rangle$
2. $\langle$remaining$\rangle ::= \langle$letter$\rangle \mid \langle$digit$\rangle \mid \langle$letter$\rangle\langle$remaining$\rangle \mid \langle$digit$\rangle\langle$remaining$\rangle$
3. $\langle$letter$\rangle$ $::= a \mid b \mid c \mid \cdots \mid z$
4. $\langle$digit$\rangle$ $::= 0 \mid 1 \mid 2 \mid 3 \mid 4 \mid 5 \mid 6 \mid 7 \mid 8 \mid 9$

Again we see that the productions "remaining $\mapsto$ letter remaining" and "remaining $\mapsto$ digit remaining", occurring in BNF statement 2, are recursive and normal.

Syntax Diagrams

A second alternative method for displaying the productions in some type 2 grammars is the **syntax diagram**. This is a pictorial display of the productions that allows the user to view the substitutions dynamically, that is, to view them as movement through the diagram. We will illustrate, in Fig. 2, the diagrams that result from translating typical sets of productions, usually all of the productions appearing on the right-hand side of some BNF statement.

A BNF statement that involves just a single production, such as $\langle w \rangle ::= \langle w_1 \rangle\langle w_2 \rangle\langle w_3 \rangle$, will result in the diagram shown in Fig. 2(a). The symbols (words) that make up the right-hand side of the production are drawn in sequence from left to right. The arrows indicate the direction in which to move to accomplish a substitution, while the label "w" indicates that we are substituting for the symbol w. Finally, the rectangles enclosing w_1, w_2, w_3 denote the fact that these are nonterminal symbols. If terminal symbols were present, they would instead be enclosed in circles or ellipses.

Figure 2(b) shows the situation when there are several productions with the same left-hand side. This figure is a syntax diagram translation of the following BNF specification:

$$\langle w \rangle ::= \langle w_1 \rangle\langle w_2 \rangle \mid \langle w_1 \rangle a \mid bc \langle w_2 \rangle$$

(where, by convention, a, b, and c must be terminal symbols). Here the diagram shows that when we substitute for w, by moving through the figure in the direction of the arrows, we may take any one of three paths. This corresponds to the three

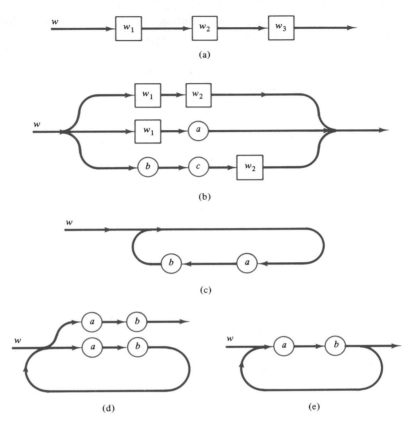

Figure 2

alternative substitutions for the symbol w. Now consider the following normal, recursive production, in BNF form:

$$\langle w \rangle ::= ab \langle w \rangle$$

The syntax diagram for this production is shown in Fig. 2(c). If we go through the loop once, we encounter a, then b, and we then return to the starting point designated by w. This represents the recursive substitution of abw for w. Several trips around the diagram represent several successive substitutions. Thus, if we traverse the diagram three times and return to the starting point, we see that w will be replaced by $abababw$ in three successive substitutions. This is typical of the way in which movement through a syntax diagram represents the substitution process.

The remarks above show how to construct a syntax diagram for a normal recursive production. Nonnormal recursive productions do not lead to the simple diagrams discussed above, but we may sometimes replace nonnormal, recursive productions by normal recursive productions, and obtain a grammar that produces the same language. Since recursive productions in regular grammars must be normal, syntax diagrams can always be used to represent regular grammars.

We also note that syntax diagrams for a language are by no means unique. They will not only change when different, equivalent productions are used, but they may be combined and simplified in a variety of ways. Consider the following BNF specification:

$$\langle w \rangle ::= ab \mid ab\langle w \rangle$$

If we construct the syntax diagram for w, using exactly the rules discussed above, we will obtain the diagram of Fig. 2(d). This shows that we can "escape" from w, that is, eliminate w entirely, only by passing through the upper path. On the other hand, we may first traverse the lower loop any number of times. Thus any movement through the diagram which eventually results in the complete elimination of w by successive substitutions will produce a string of terminal symbols of the form $(ab)^n$, $n \geq 1$.

It is easily seen that the simpler diagram of Fig. 2(e), produced by combining the paths of Fig. 2(d) in an obvious way, is an entirely equivalent syntax diagram. These types of simplifications are performed whenever possible.

Example 5   The syntax diagrams of Fig. 3(a) represent the BNF statements of Example 2, constructed with our original rules for drawing syntax diagrams. A slightly more aesthetic version is shown in Fig. 3(b).

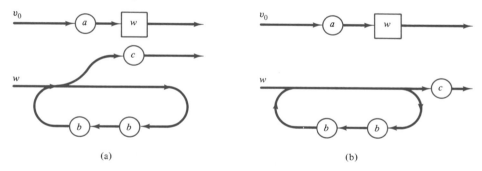

(a)   (b)

Figure 3

Example 6   Consider the BNF statements 1, 2, 3, and 4 of Example 4. The direct translation into syntax diagrams is shown in Fig. 4. In Fig. 5 we combine the first two diagrams of Fig. 4, and simplify the result. We thus eliminate the symbol "remaining", and we arrive at the customary syntax diagrams for identifiers.

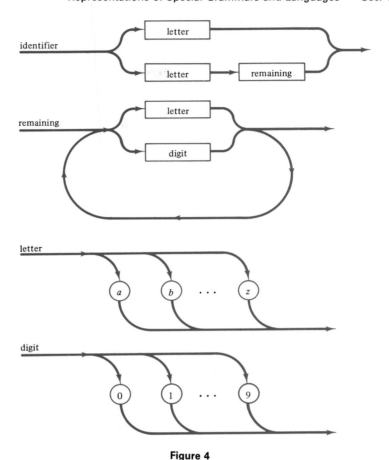

Figure 4

Example 7   The productions of Example 3, for well-formed decimal numbers, are shown in syntax diagram form in Fig. 6. We show in Fig. 7 the result of substituting the diagram for "unsigned-integer" into that for "decimal-number" and "decimal-fraction." In Fig. 8 the process of substitution is carried one step further. Although this is not usually done, it does illustrate the fact that one can be quite flexible in designing syntax diagrams.

　　　If we were to take the extreme case, and combine the diagrams of Fig. 8 into one huge diagram, that diagram would contain only terminal symbols. In that case a valid "decimal-number" would be any string that resulted from moving through the diagram, recording each symbol encountered, in the order in which it was encountered, and eventually exiting to the right.

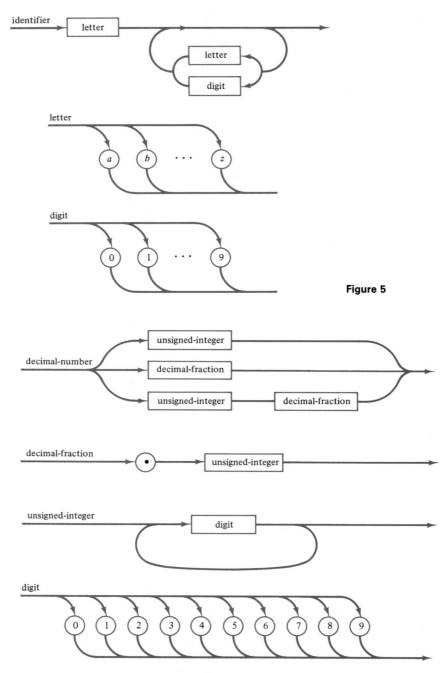

Figure 5

Figure 6

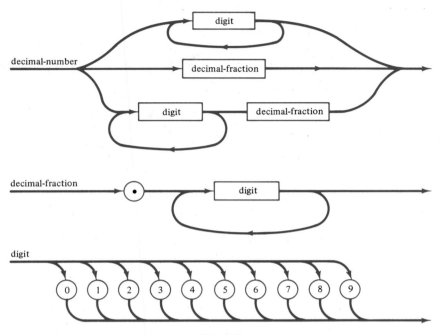

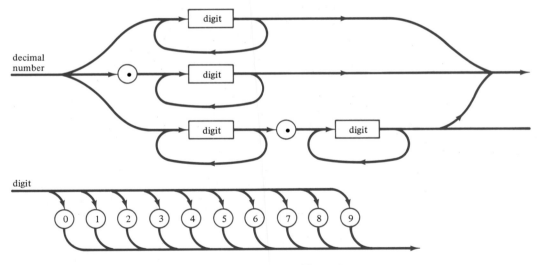

Figure 7

Figure 8

EXERCISE SET 10.2

In each of Exercises 1–5, we have referenced a grammar described in the exercises of a previous section. In each case give the BNF and corresponding syntax diagrams for the productions of the grammar.

1. Exercise 1 of Section 10.1.
2. Exercise 2 of Section 10.1.
3. Exercise 6 of Section 10.1.
4. Exercise 9 of Section 10.1.
5. Exercise 21 of Section 10.1.
6. Give the BNF for the productions of Exercise 3 of Section 10.1.
7. Give the BNF for the productions of Exercise 4 of Section 10.1.
8. Give the BNF for the productions of Exercise 5 of Section 10.1.
9. Give the BNF for the productions of Exercise 9 of Section 10.1.

In each of Exercises 12–16, we have referenced a grammar G, described in the exercises of a previous section. In each case find a regular expression that corresponds to the language $L(G)$.

12. Exercise 2 of Section 10.1
13. Exercise 3 of Section 10.1.
14. Exercise 5 of Section 10.1.
15. Exercise 6 of Section 10.1.
16. Exercise 9 of Section 10.1.
17. Find the regular expression that corresponds to the syntax diagram shown in Fig. 9.
18. Find the regular expression that corresponds to the syntax diagram shown in Fig. 10.

In Exercises 10 and 11, give a BNF representation for the syntax diagram shown. The symbols a, b, c, and d are supposed to be terminal symbols of some grammar. You may provide nonterminal symbols as needed (in addition to v_0), to use in the BNF productions. You may use several BNF statements if needed.

10. v_0

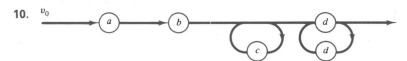

Figure 9

11. v_0

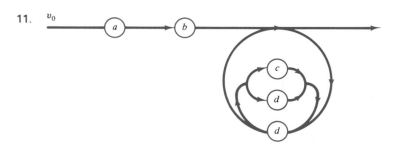

Figure 10

KEY IDEAS FOR REVIEW

☐ Phrase structure grammar: see page 257.

☐ Production: a statement $w \mapsto w'$, where $(w, w') \in \mapsto$.

☐ Direct derivability: see page 258.

☐ Terminal symbols: the elements of S.

☐ Nonterminal symbols: the elements of $V - S$.

☐ Derivation of a sentence: substitution process that produces a valid sentence.

☐ Language of a grammar G: set of all properly constructed sentences that can be produced from G.

☐ Types 0, 1, 2, 3 phrase structure grammars: see page 261.

☐ Context-free grammar: type 2 grammar.

☐ Regular grammar: type 3 grammar.

☐ Parsing: process of obtaining a derivation that will produce a given sentence.

☐ BNF notation: see page 263.

☐ Syntax diagram: see page 266.

REVIEW EXERCISES

1. Describe the language $L(G)$ produced by the grammar $G = (V, S, v_0, \mapsto)$ where $V = \{v_0, v_1, v_2, x, y, z\}$, $S = \{x, y, z\}$ and $\mapsto$ is given as follows:

$$v_0 \mapsto xv_0$$
$$v_1 \mapsto zv_2$$
$$v_0 \mapsto yv_1$$
$$v_1 \mapsto zzy_1$$
$$v_2 \mapsto y$$

2. Describe the language $L(G)$ produced by the grammar
$G = (V, S, v_0, \mapsto)$, where $V = \{v_0, v_1, a, b, c, d\}$, $S = \{a, b, c, d\}$ and $\mapsto$ is given as follows:

$$v_0 \mapsto abcv_0$$
$$v_0 \mapsto ddv_1$$
$$v_1 \mapsto cv_1$$
$$v_1 \mapsto ab$$
$$v_1 \mapsto d$$

3. Which of the following statements are true for the grammar of Exercise 2?

   (a) $cv_1 \Rightarrow cd$ (b) $cv_1 \Rightarrow cabd$

   (c) $v_0 \Rightarrow^\infty abcdv_1$ (d) $v_0 \Rightarrow^\infty abcddv_1$

   (e) $abv_0 \Rightarrow^\infty ababcv_0$ (f) $av_0 \Rightarrow^\infty addab$

4. Which of the following statements are true for the grammar of Exercise 1?

   (a) $v_0 \Rightarrow xxxv_0$ (b) $v_0 \Rightarrow^\infty xxxv_0$

   (c) $yv_1 \Rightarrow^\infty yyzv_2$ (d) $xyv_1 \Rightarrow xyzzv_1$

   (e) $xyv_1 \Rightarrow xyyzv_2$ (f) $zv_1x \Rightarrow^\infty zzyx$

5. Let $G = (V, S, v_0, \mapsto)$, where $V = \{v_0, v_1, s, t\}$, $S = \{s, t\}$, and $\mapsto$ is given by

$$v_0 \mapsto stv_0$$
$$v_0 \mapsto stv_1$$
$$v_1 \mapsto tv_1$$
$$v_1 \mapsto st$$
$$v_1 \mapsto t$$

State which of the following strings are in $L(G)$.

(a) *stststst*   (b) *st*

(c) *sttt*   (d) *tsts*

(e) *sts*   (f) *stttst*

6. Give syntax diagrams for the productions whose BNF form is as follows:

$$\langle v_0 \rangle ::= \langle v_0 \rangle a \mid \langle v_0 \rangle b \mid \langle v_1 \rangle \mid \langle v_1 \rangle c$$

$$\langle v_1 \rangle ::= \langle v_1 \rangle a \mid \langle v_1 \rangle cc \mid b \mid aa$$

7. Describe the productions given in Exercise 1 in BNF form.

8. Describe the productions given in Exercise 2 in BNF form.

9. Describe the productions given in Exercise 5 by syntax diagrams.

CHAPTER TEST

1. Describe the language $L(G)$ produced by the grammar $G = (V, S, v_0, \mapsto)$, where $V = \{v_0, v_1, m, n, p\}$, $S = \{m, n, p\}$ and $\mapsto$ is given as follows:

$$v_0 \mapsto mv_0$$
$$v_1 \mapsto mv_1$$
$$v_0 \mapsto nv_0$$
$$v_1 \mapsto nv_1$$
$$v_0 \mapsto pv_1$$
$$v_1 \mapsto p$$

2. Describe the productions of Problem 1 in BNF form.

3. Describe the productions of Problem 1 by syntax diagrams.

4. Let $G = \{V, S, v_0, \mapsto\}$, where $V = \{v_0, v_1, 0, 1\}$, $S = \{0, 1\}$, and $\mapsto$ is given by

$$v_0 \mapsto 00v_0$$
$$v_0 \mapsto 1v_1$$
$$v_1 \mapsto 00v_1$$
$$v_1 \mapsto 111v_1$$
$$v_1 \mapsto 0$$

Which of the following strings is in $L(G)$?

(a) 0010   (b) 0001

(c) 00001110   (d) 001001110

(e) 0011110   (f) 11111110

5. Describe the productions of Problem 4 in BNF form and with syntax diagrams.

6. Let $G = (V, S, v_0, \mapsto)$, where $V = \{v_0, v_1, v_2, 0, 1\}$, $S = \{0, 1\}$, and $\mapsto$ is given by

$$v_0 \mapsto 0v_1$$
$$v_0 \mapsto 1v_2$$
$$v_1 \mapsto 00v_1$$
$$v_1 \mapsto 11v_2$$
$$v_2 \mapsto 000v_2$$
$$v_2 \mapsto 1v_2$$
$$v_2 \mapsto 0$$
$$v_2 \mapsto 1$$

Answer each of the following questions true (T) or false (F). Each refers to the grammar above.

(a) G is a type 2 grammar.

(b) G is a type 3 grammar.

(c) 000111 is in $L(G)$.

(d) 11111 is in $L(G)$.

(e) $0v_1 \Rightarrow 00011v_2$

(f) $01v_1 \Rightarrow^\infty 0100111v_2$

11

Finite-State Machines

Prerequisites: Chapters 1, 4, 5, and 6.

We think of a machine as a system that can accept **input,** possibly produce **output,** and have some sort of internal memory that can keep track of certain information about previous inputs. The complete internal condition of the machine and all of its memory, at any particular time, is said to constitute the state of the machine at that time. The state in which a machine finds itself at any instant, summarizes its "memory" of past inputs, and determines how it will react to subsequent input. When more input arrives, the given state of the machine determines (with the input) the next state to be occupied, and any output that may be produced. If the number of states is finite, the machine is a finite-state machine.

A finite-state machine is a model that includes devices ranging from simple "flip-flops" all the way to entire computers. They can also be used to describe the effect of certain computer programs and other "nonhardware" systems. Finite-state machines are useful in the study of formal languages, and are often found in compilers and interpreters for various computer programming languages. There are several types and extensions of finite-state machines, and in this chapter we will introduce the most elementary versions.

11.1 Finite-State Machines

Suppose that we have a finite set $S = \{s_0, s_1, \ldots, s_n\}$, a finite set I, and for each $x \in I$, a function $f_x : S \to S$. Let $\mathcal{F} = \{f_x \mid x \in I\}$. The triple $(S, I, \mathcal{F})$ is called a **finite-state machine,** S is called the **state set** of the machine and the elements of S are called **states.** The set I is called the **input set** of the machine. For any input $x \in I$, the function f_x describes the effect that this input has on the states of the machine, and is called a **state transition function.** Thus if the machine is in state s_i and input x occurs, the next state of the machine will be $f_x(s_i)$.

Since the next state $f_x(s_i)$ is uniquely determined by the pair (x, s_i), there is a function $F : I \times S \to S$ given by

$$F(x, s_i) = f_x(s_i)$$

The individual functions f_x can all be recovered from a knowledge of F. Many authors will use a function $F : I \times S \to S$, instead of a set $\{f_x \mid x \in I\}$ to define a finite-state machine. The definitions are completely equivalent.

Example 1 Let $S = \{s_0, s_1\}$ and $I = \{0, 1\}$. Define f_0 and f_1 as follows:

$$f_0(s_0) = s_0, \qquad f_1(s_0) = s_1,$$
$$f_0(s_1) = s_1, \qquad f_1(s_1) = s_0$$

This finite-state machine has two states, s_0 and s_1, and accepts two possible inputs, 0 and 1. The input 0 leaves each state fixed, and the input 1 reverses states. One can think of this machine as a model for a circuit (or logical) device and visualize such a device as in Fig. 1. The output signals will, at any given time, consist of two voltages, one higher than the other. Either line 1 will be at the higher voltage and line 2 at the lower, or the reverse. The first set of output conditions will be denoted s_0, and the second will be denoted s_1. An input pulse, represented by the symbol 1, will reverse output voltages. The symbol 0 represents the absence of an input pulse, and so results in no change of output. This device is often called a **T flip-flop,** and is a concrete realization of the machine in this example.

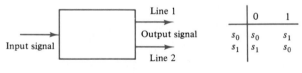

	0	1
s_0	s_0	s_1
s_1	s_1	s_0

Figure 1   **Figure 2**

We summarize this machine in Fig. 2. The table shown there lists the states down the side and inputs across the top. The column under each input gives the values of the function corresponding to that input, at each state shown on the left.

The arrangement illustrated in Fig. 2 for summarizing the effect of inputs on states, is called the **state transition table** of the finite-state machine. It can be used with any machine of reasonable size, and is a convenient method of specifying the machine.

Example 2 Consider the state transition table shown in Fig. 3. Here a and b are the possible inputs, and there are three states, s_0, s_1, and s_2. The table shows us that

$$f_a(s_0) = s_0, \qquad f_a(s_1) = s_2, \qquad f_a(s_2) = s_1$$

and

$$f_b(s_0) = s_1, \qquad f_b(s_1) = s_0, \qquad f_b(s_2) = s_2$$

If M is a finite-state machine with states S, inputs I, and state transition functions $\{f_x \mid x \in I\}$, we can determine a relation R_M on S in a natural way. If s_i, $s_j \in S$, we say that $s_i R_M s_j$ if there is an input x so that $f_x(s_i) = s_j$.

Thus $s_i R_M s_j$ means that if the machine is in state s_i there is some input $x \in I$ which, if received next, will put the machine in state s_j. The relation R_M permits us to describe the machine M as a labeled digraph (see Section 6.1) of the relation R_M on S, where each edge is labeled by the set of all inputs which cause the machine to change states as indicated by that edge.

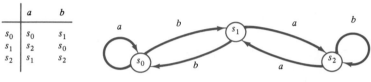

	a	b
s_0	s_0	s_1
s_1	s_2	s_0
s_2	s_1	s_2

Figure 3   **Figure 4**

Example 3 Consider the machine of Example 2. Figure 4 shows the digraph of the relation R_M, with each edge labeled appropriately. Notice that the entire structure of M can be recovered from this digraph, since edges and their labels indicate where each input sends each state.

Example 4 Consider the machine M whose state table is shown in Fig. 5(a). The digraph of R_M is then shown in Fig. 5(b), with edges labeled appropriately.

	a	b	c
s_0	s_0	s_0	s_0
s_1	s_2	s_3	s_2
s_2	s_1	s_0	s_3
s_3	s_3	s_2	s_3

(a)

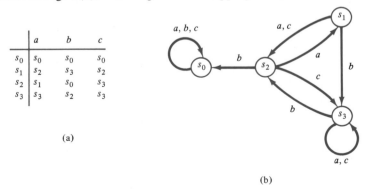

(b)

Figure 5

Note that an edge may be labeled by more than one input, since several inputs may cause the same change of state. The reader will observe that every input must be part of the label of exactly one edge out of each state. This is a general property which holds for the labeled digraphs of all finite-state machines. For brevity, we will refer to the labeled digraph of a machine M as the **digraph of M.**

It is possible to add a variety of extra features to a finite-state machine in order to increase the utility of the concept. A simple, yet very useful extension results in what is often called a **Moore machine,** or **recognition machine,** which is defined as a sequence $(S, I, \mathscr{F}, s_0, T)$, where $(S, I, \mathscr{F})$ constitutes a finite-state machine, $s_0 \in S$ and $T \subseteq S$. The state s_0 is called the **starting state** of M, and it will be used to represent the condition of the machine before it receives any input. The set T is called the set of **acceptance states** of M. These states will be used in Section 11.2 in connection with language recognition.

When the digraph of a Moore machine is drawn, the acceptance states are indicated with two concentric circles, instead of one. No special notation will be used on these digraphs for the starting state, but unless otherwise specified, this state will be named s_0.

Example 5 Let M be the Moore machine $(S, I, \mathscr{F}, s_0, T)$, where $(S, I, \mathscr{F})$ is the finite-state machine of Fig. 5, and $T = \{s_1, s_3\}$. Figure 6 shows the digraph of M.

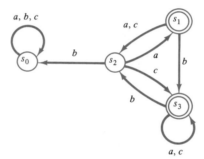

Figure 6

EXERCISE SET 11.1

In Exercises 1–4, draw the digraph of the machine whose state transition table is shown. Remember to label the edges with the appropriate inputs.

1.

	0	1
s_0	s_0	s_1
s_1	s_1	s_2
s_2	s_2	s_0

2.

	0	1	2
s_0	s_1	s_0	s_2
s_1	s_0	s_0	s_1
s_2	s_2	s_0	s_2

3.

	a	b
s_0	s_1	s_0
s_1	s_2	s_0
s_2	s_2	s_0

4.

	a	b
s_0	s_1	s_0
s_1	s_2	s_1
s_2	s_3	s_2
s_3	s_3	s_3

In Exercises 5–7, construct the state transition table of the finite-state machine whose digraph is shown.

5.

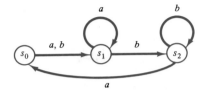

Figure 7

6.

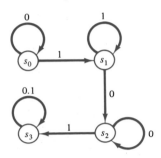

Figure 8

7.

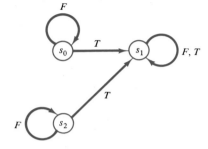

Figure 9

THEORETICAL EXERCISES

T1. Let $M = (S, I, \mathcal{F})$ be a finite-state machine. Define a relation R on I as follows: $x_1 \ R \ x_2$ if and only if $f_{x_1}(s) = f_{x_2}(s)$ for every s in S. Show that R is an equivalence relation on I.

T2. Let $(S, *)$ be a finite semigroup. Then we may consider the machine $(S, S, \mathcal{F})$ where $\mathcal{F} = \{f_x \mid x \in S\}$, and $f_x(y) = x * y$ for all $x, y \in S$. Thus we have a finite-state machine in which the state set and the input set are the same. Define a relation R on S as follows: $x \ R \ y$ if and only if there

is some $z \in S$ such that $f_z(x) = y$. Show that R is transitive.

T3. Consider a finite group $(S, *)$ and let $(S, S, \mathcal{F})$ be the finite-state machine constructed in Exercise T2. Show that if R is the relation defined in Exercise T2 then R is an equivalence relation.

T4. Let $I = \{0, 1\}$, $S = \{a, b\}$. Construct all possible state transition tables of finite-state machines that have S as state set and I as input set.

11.2 Semigroups, Machines, and Languages

Let $M = (S, I, \mathcal{F})$ be a finite-state machine with state set $S = \{s_0, s_1, \ldots, s_n\}$, input set I, and state transition functions $\mathcal{F} = \{f_x \mid x \in I\}$.

We will associate with M two monoids, whose construction we recall from Section 6.3. First, there is the free monoid I^* on the input set I. This monoid consists of all finite sequences (or "strings" or "words") from I, with catenation as its binary operation. The identity is the empty string Λ. Second, we have the monoid S^S, which consists of all functions from S to S, and which has function composition as its binary operation. That is if $w_1 = x_1 x_2 \cdots x_n$ and $w_2 = y_1 y_2 \cdots y_k$, then $w_1 * w_2 = w_1 \cdot w_2 = x_1 x_2 \cdots x_n y_1 y_2 \cdots y_k$. The identity in S^S is the function 1_S defined by $1_S(s) = s$, for all s in S.

If $w = x_1 x_2 \cdots x_n \in I^*$, we let $f_w = f_{x_n} \circ f_{x_{n-1}} \circ \cdots \circ f_{x_2} \circ f_{x_1}$, the composition of the functions $f_{x_1}, f_{x_2}, \ldots, f_{x_n}$. Also define f_Λ to be 1_S. In this way we assign an element f_w of S^S to each element w of I^*. If we think of each f_x as the "effect" of the input x on the states of the machine M, then f_w represents the combined effect of all the input letters in the word w, received in the sequence specified by w. We call f_w the **state transition function corresponding to w.**

Example 1   Let $M = (S, I, \mathscr{F})$, where $S = \{s_0, s_1, s_2\}$, $I = \{0, 1\}$, and $\mathscr{F}$ is given by the following state transition table.

	0	1
s_0	s_0	s_1
s_1	s_2	s_2
s_2	s_1	s_0

Let $w = 011 \in I^*$. Then

$$f_w(s_0) = (f_1 \circ f_1 \circ f_0)(s_0) = f_1(f_1(f_0(s_0)))$$
$$= f_1(f_1(s_0)) = f_1(s_1) = s_2$$

Similarly,

$$f_w(s_1) = f_1(f_1(f_0(s_1))) = f_1(f_1(s_2)) = f_1(s_0) = s_1$$

and

$$f_w(s_2) = f_1(f_1(f_0(s_2))) = f_1(f_1(s_1)) = f_1(s_2) = s_0$$

Example 2   Let us consider the same machine M as in Example 1, and examine the problem of computing f_w a little differently. In Example 1 we used the definition directly, and for a large machine we would program an algorithm to compute the values of f_w in just that way. However, if the machine is of moderate size, human beings may find another procedure to be preferable.

We begin by drawing the digraph of the machine M as shown in Fig. 1. We may use this digraph to compute word transition functions by just following the edges corresponding to successive input letter transitions. Thus to compute $f_w(s_0)$, we start at state s_0 and see that input 0 takes us to state s_0. The input 1 which follows takes us on to state s_1, and the final input of 1 takes us to s_2. Thus $f_w(s_0) = s_2$, as before.

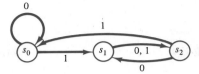

Figure 1

Let us compute $f_{w'}$ where $w' = 01011$. The successive transitions of s_0 are

$$s_0 \xrightarrow{\ 0\ } s_0 \xrightarrow{\ 1\ } s_1 \xrightarrow{\ 0\ } s_2 \xrightarrow{\ 1\ } s_0 \xrightarrow{\ 1\ } s_1$$

so $f_{w'}(s_0) = s_1$. Similar displays show that $f_{w'}(s_1) = s_2$ and $f_{w'}(s_2) = s_0$.

This method of interpreting word transition functions such as f_w and $f_{w'}$ is useful in designing machines that have word transitions possessing certain desired properties.

Theorem 1   If w_1 and w_2 are any two strings in I^*, then $f_{w_1 \cdot w_2} = f_{w_2} \circ f_{w_1}$.

Proof. Let $w_1 = x_1 x_2 \cdots x_k$ and $w_2 = y_1 y_2 \cdots y_m$ be strings in I^*. Then
$f_{w_1 \cdot w_2} = f_{x_1 x_2 \cdots x_k y_1 y_2 \cdots y_m} = (f_{y_m} \circ \cdots \circ f_{y_2} \circ f_{y_1}) \circ (f_{x_k} \circ \cdots \circ f_{x_2} \circ f_{x_1}) = f_{w_2} \circ f_{w_1}$

Example 3   Let $S = \{s_0, s_1, s_2\}$, $I = \{a, b, d\}$, and consider the finite-state machine $M = (S, I, \mathcal{F})$ defined by the digraph shown in Fig. 2. Compute the functions $f_{bad}, f_{add},$ and f_{badadd}, and verify that

$$f_{add} \circ f_{bad} = f_{badadd}$$

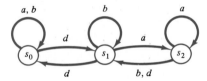

Figure 2

Solution. f_{bad} is computed by the following sequence of transitions:

$$s_0 \xrightarrow{\ b\ } s_0 \xrightarrow{\ a\ } s_0 \xrightarrow{\ d\ } s_1$$
$$s_1 \xrightarrow{\ b\ } s_1 \xrightarrow{\ a\ } s_2 \xrightarrow{\ d\ } s_1$$
$$s_2 \xrightarrow{\ b\ } s_1 \xrightarrow{\ a\ } s_2 \xrightarrow{\ d\ } s_1$$

Thus $f_{bad}(s_0) = s_1$, $f_{bad}(s_1) = s_1$, and $f_{bad}(s_2) = s_1$.
Similarly, for f_{add},

$$s_0 \xrightarrow{\ a\ } s_0 \xrightarrow{\ d\ } s_1 \xrightarrow{\ d\ } s_0$$

$$s_1 \xrightarrow{\ a\ } s_2 \xrightarrow{\ d\ } s_1 \xrightarrow{\ d\ } s_0.$$
$$s_2 \xrightarrow{\ a\ } s_2 \xrightarrow{\ d\ } s_1 \xrightarrow{\ d\ } s_0.$$

so $f_{add}(s_i) = s_0$ for $i = 0, 1, 2$. A similar computation shows that

$$f_{badadd}(s_0) = s_0, \qquad f_{badadd}(s_1) = s_0, \qquad f_{badadd}(s_2) = s_0$$

and the same formulas hold for $f_{add} \circ f_{bad}$. In fact,

$$(f_{add} \circ f_{bad})(s_0) = f_{add}(f_{bad}(s_0)) = f_{add}(s_1) = s_0$$
$$(f_{add} \circ f_{bad})(s_1) = f_{add}(f_{bad}(s_1)) = f_{add}(s_1) = s_0 \ .$$
$$(f_{add} \circ f_{bad})(s_2) = f_{add}(f_{bad}(s_2)) = f_{add}(s_1) = s_0$$

Example 4   Consider the machine whose digraph is shown in Fig. 3. Show that $f_w(s_0) = s_0$ if and only if w has $3n$ 1's for some $n \geq 0$.

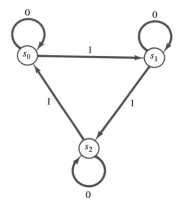

Figure 3

Solution. From Fig. 3 we see that $f_0 = 1_s$, so that the 0's in a string $w \in I^*$ have no effect on f_w. Thus if $\overline{w}$ is w with all 0's removed, then $f_w = f_{\overline{w}}$. Let $l(w)$ denote the **length** of w, that is, the number of digits in w. Then $l(\overline{w})$ is the number of 1's in w, for all $w \in I^*$. For each $n \geq 0$ consider the statement

$$P(n): \text{Let } w \in I^* \text{ and let } l(\overline{w}) = m.$$

(a) If $m = 3n$, then $f_w(s_0) = s_0$.

(b) If $m = 3n + 1$, then $f_w(s_0) = s_1$.

(c) If $m = 3n + 2$, then $f_w(s_0) = s_2$.

We prove by mathematical induction that $P(n)$ is true for all $n \geq 0$.

Basis Step. Suppose that $n = 0$. In case (a) $m = 0$, therefore w has no 1's and $f_w(s_0) = 1_s(s_0) = s_0$. In case (b) $m = 1$, so $\overline{w} = 1$ and $f_w(s_0) = f_{\overline{w}}(s_0) = f_1(s_0) = s_1$. Finally, in case (c) $m = 2$, so $\overline{w} = 11$, and $f_w(s_0) = f_{\overline{w}}(s_0) = f_{11}(s_0) = f_1(s_1) = s_2$.

Induction Step. Suppose that $P(k)$ is true for some $k \geq 0$. We must show that $P(k + 1)$ is true. Let $w \in I^*$, and denote $l(\overline{w})$ by m. In case (a) $m = 3(k + 1) = 3k + 3$; therefore, $\overline{w} = 111 \cdot w'$, where $l(w') = 3k$. Then $f_{w'}(s_0) = s_0$ by induction hypothesis, and $f_{111}(s_0) = s_0$ by direct computation, so $f_{\overline{w}}(s_0) = f_{w'}(f_{111}(s_0)) = f_{w'}(s_0) = s_0$. Cases (b) and (c) are handled in the same way, therefore $P(k + 1)$ is true.

By mathematical induction, $P(n)$ is true for all $n \geq 0$, so $f_w(s_0) = s_0$ if and only if the number of 1's in w is a multiple of 3.

Suppose now that $(S, I, \mathscr{F}, s_0, T)$ is a Moore machine. As in Chapter 10, we may think of certain subsets of I^* as "languages" with "words" from I. Using M, we can define such a subset, which we will denote by $L(M)$, and call the **language of the machine M**. Define $L(M)$ to be the set of all $w \in I^*$ such that $f_w(s_0) \in T$. In other words, $L(M)$ consists of all strings which, when used as input to the machine, cause the starting state s_0 to move to an "acceptance state" in T. Thus, in this sense, M "accepts" the string. It is for this reason that the states in T were named acceptance states in Section 11.1.

Example 5   Let $M = (S, I, \mathscr{F}, s_0, T)$ be the Moore machine in which $(S, I, \mathscr{F})$ is the finite-state machine whose digraph is shown in Fig. 3, and $T = \{s_1\}$. The discussion of Example 4 shows that $f_w(s_0) = s_1$ if and only if the number of 1's in w is of the form $3n + 1$ for some $n \geq 0$. Thus $L(M)$ is exactly the set of all strings with $3n + 1$ 1's for some $n \geq 0$.

Example 6   Consider the Moore machine M whose digraph is shown in Fig. 4. Here state s_0 is the starting state, and $T = \{s_2\}$. What is $L(M)$? Clearly, the input set is $I = \{a, b\}$. Observe that, in order for a string w to cause a transition from s_0 to s_2, w must contain at least two b's. After reaching s_2, any additional letters have no effect. Thus $L(M)$ is the set of all strings having two or more b's. We see, for example, that $f_{aabaa}(s_0) = s_1$, so $aabaa$ is rejected. On the other hand, $f_{abaab}(s_0) = s_2$, so $abaab$ is accepted.

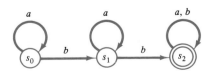

Figure 4

EXERCISE SET 11.2

In Exercises 1–5, we refer to the finite state machine whose state transition table is given below.

	0	1
s_0	s_0	s_1
s_1	s_1	s_2
s_2	s_2	s_3
s_3	s_3	s_0

1. List the values of the transition function f_w for $w = 01001$.

2. List the values of the transition function f_w for $w = 11100$.

3. Describe the set of binary words (sequences of 0's and 1's) w having the property that $f_w(s_o) = s_0$.

4. Describe the set of binary words w having the property that $f_w = f_{010}$.

5. Describe the set of binary words w having the property that $f_w(s_0) = s_2$.

In Exercises 6–10, we refer to the finite-state machine whose digraph is shown in Fig. 5.

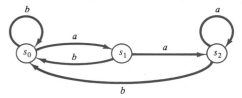

Figure 5

6. List the values of the transition function f_w for $w = abba$.

7. List the values of the transition function f_w for $w = babab$.

8. Describe the set of words w having the property that $f_w(s_0) = s_2$.

9. Describe the set of words w having the property that $f_w(s_0) = s_0$.

10. Describe the set of words w having the property that $f_w = f_{aba}$.

11. Describe the language recognized by the Moore machine whose digraph is shown in Fig. 6.

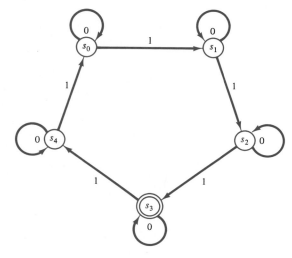

Figure 6

12. Describe the language recognized by the Moore machine whose state table is shown below, if $T = \{s_2\}$ and s_0 is the starting state.

	0	1
s_0	s_1	s_2
s_1	s_1	s_2
s_2	s_2	s_1

11.3 Machines and Grammars (Optional)

Let $M = (S, I, \mathscr{F}, s_0, T)$ be a Moore machine. In Section 11.2 we defined the language, $L(M)$, of the machine M. It is natural to ask if there is a connection between such a language and the languages of phrase structure grammars, discussed in Section 10.1. The following theorem, due to S. Kleene, describes the connection.

Theorem 1   Let I be a set and let $L \subseteq I^*$. Then L is a type 3 language, that is, $L = L(G)$, where G is a type 3 grammar having I as its set of terminal symbols, if and only if $L = L(M)$ for some Moore machine $M = (S, I, \mathscr{F}, s_0, T)$.

We will not give a complete and detailed proof of Theorem 1. However, it is easy to give a construction which produces a type 3 grammar from a given Moore machine. This is done in such a way that the grammar and the machine have the same language. Let $M = (S, I, \mathscr{F}, s_0, T)$ be a given Moore machine. We construct a type 3 grammar $G = (V, I, s_0, \mapsto)$ as follows. Let $V = I \cup S$; that is, I will be the set of terminal symbols for G, while S will be the set of nonterminal symbols. Let s_i and s_j be in S, and $x \in I$. Productions are then created according to the following rules.

Productions arising from a Moore machine $M = (S, I, \mathscr{F}, s_0, T)$

1. Write $s_i \mapsto x s_j$ if $f_x(s_i) = s_j$, that is, if the input x takes state s_i to state s_j.
2. Write $s_i \mapsto x$ if $f_x(s_i) \in T$, that is, if the input x takes state s_i to some acceptance state.

Now let $\mapsto$ be the relation determined by the two conditions above, and take this relation as the production relation of G.

The grammar G constructed above has the same language as M. Suppose, for example, that $w = x_1 x_2 x_3 \in I^*$. The string w is in $L(M)$ if and only if $f_w(s_0) = f_{x_3}(f_{x_2}(f_{x_1}(s_0))) \in T$. Let $a = f_{x_1}(s_0)$, $b = f_{x_2}(a)$, and $c = f_{x_3}(b)$, where $c = f_w(s_0)$ is in T. Then the rules given above for constructing $\mapsto$ tell us that

1. $s_0 \mapsto x_1 a$
2. $a \mapsto x_2 b$
3. $b \mapsto x_3$

are all productions in G. The last one occurs because $c \in T$. If we begin with s_0, and substitute, using (1), (2), and (3) in succession, we see that $s_0 \Rightarrow^* x_1 x_2 x_3 = w$ (see Section 10.1), so $w \in L(G)$. A similar argument works for any string in $L(M)$, so $L(M) \subseteq L(G)$. If we reverse the argument given above, we can see that we also have $L(G) \subseteq L(M)$. Thus M and G have the same language.

Example 1   Consider the Moore machine M shown in Fig. 4 of Section 11.2. Construct a type 3 grammar G such that $L(G) = L(M)$.

Solution. Let $I = \{a, b\}$, $S = \{s_0, s_1, s_2\}$, and $V = I \cup S$. We construct the grammar $(V, I, s_0, \mapsto)$, where $\mapsto$ is described below.

$$\mapsto: \quad s_0 \mapsto a s_0 \qquad s_2 \mapsto b s_2$$
$$s_0 \mapsto b s_1 \qquad s_1 \mapsto b$$
$$s_1 \mapsto a s_1 \qquad s_2 \mapsto a$$
$$s_1 \mapsto b s_2 \qquad s_2 \mapsto b$$
$$s_2 \mapsto a s_2$$

The production relation $\mapsto$ is constructed as we indicated previously; therefore, $L(M) = L(G)$.

Example 2 Consider the Moore machine whose digraph is shown in Fig. 1. Describe, in English, the language $L(M)$. Then describe the productions of the corresponding grammar G in BNF form.

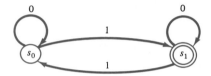

Figure 1

Solution. It is clear that 0's in the input string have no effect on the states. If an input string w has an odd number of 1's, then $f_w(s_0) = s_1$. If w has an even number of 1's, then $f_w(s_0) = s_0$. Since $T = \{s_1\}$, we see that $L(M)$ consists of all w in I^* which have an odd number of 1's as components.

Now, the type 3 grammar constructed from M is $G = (V, I, s_0, \mapsto)$ with $V = I \cup S$. The BNF of the relation $\mapsto$ is

$$\langle s_0 \rangle ::= 0\langle s_0 \rangle \mid 1\langle s_1 \rangle \mid 1$$
$$\langle s_1 \rangle ::= 0\langle s_1 \rangle \mid 1\langle s_0 \rangle \mid 0$$

Occasionally, we may need to determine the function performed by a given Moore machine, as we did in Examples 1 and 2. More commonly, however, it is necessary to construct a machine that will perform a given task. This task may be defined by giving an ordinary English description, or a type 3 grammar, perhaps in BNF or with a syntax diagram. There are systematic, almost mechanical ways to construct such a machine. Most of these use the concept of "nondeterministic machines," and employ a tedious translation process from such machines to the Moore machines that we have discussed. The reader may find discussions of these methods in more advanced texts.

If the task of the machine is not too complex, we may use simple reasoning to construct the machine in steps, usually in the form of its digraph.

Example 3 Construct a Moore machine M that will accept exactly the string 001 from input strings of 0's and 1's. In other words, $I = \{0, 1\}$ and $L(M) = \{001\}$.

Solution. We must begin with a starting state s_0. If w is an input string of 0's and 1's, and if w begins with a 0, then w "may" be accepted (depending on the remainder of its components). Thus one step toward acceptance has been taken, and there needs to be a state s_1 that corresponds to this step. We therefore begin as in Fig. 2(a). If we next receive another 0, we have progressed one more step toward acceptance. We therefore construct another state s_2 and let 0 give a transition from s_1 to s_2. State s_1 represents the condition "first component of input is a 0," whereas state s_2 represents the condition "first two components of the input are respectively 00." This situation is shown in Fig. 2(b). Finally, if the third input component is a 1, we move to an acceptance state, as shown in Fig. 2(c). Any other beginning sequence of input digits, or any additional digits will move us to a "failure state" s_4 from which there is no escape. Thus Fig. 2(d) shows the completed machine.

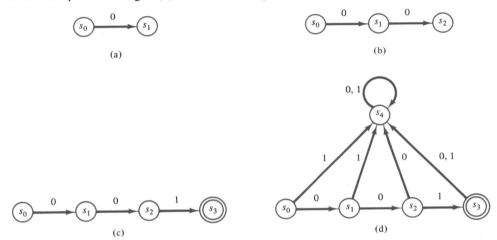

Figure 2

The process illustrated in Example 3 is difficult to precisely describe or generalize. We try to construct states representing each successive stage of input complexity, leading up to an acceptable string. There must also be states indicating the ways in which a promising input pattern may be destroyed when a certain component is received. If the machine is to recognize several, essentially different, types of input, then we will need to construct separate "branches" corresponding to each type of input. This process may result in some redundancy, but the machine can be simplified later.

Example 4 Let $I = \{0, 1\}$. Construct a Moore machine which accepts those input sequences w that contain the string 01 or the string 10 anywhere within them. In other words, we are to accept exactly those strings which do not consist entirely of 0's, or entirely of 1's.

Solution. This is a simple example in which, whatever input digit is received first, a string will be accepted if and only if the other digit is eventually received. There must be a starting state s_0, states s_1 and s_2 corresponding respectively to first receiving a 0 or 1, and (acceptance) states s_3 and s_4, which will be reached if and when the other digit is received. Having once reached an acceptance state, the machine stays in that state. Thus we construct the digraph of this machine as shown in Fig. 3.

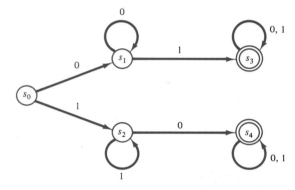

Figure 3

In Example 3, once an acceptance state is reached, any additional input will cause a permanent transition to a nonaccepting state. In Example 4, once an acceptance state is reached, any additional input will have no effect. Sometimes the situation is between these two extremes. As input is received, the machine may repeatedly enter and leave acceptance states. Consider the Moore machine M whose digraph is shown in Fig. 4. This machine is a slight modification of the finite-state machine given in

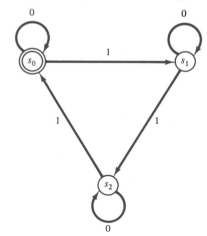

Figure 4

Example 4 of Section 11.2. We know from that example that $w \in L(M)$ if and only if the number of 1's in w is of the form $3k$, $k \geq 0$. As input components are received, M may enter and leave s_0 repeatedly. The conceptual states "one 1 has been received" and "four 1's have been received" may both be represented by s_1. When constructing machines, one should keep in mind the fact that a state, previously defined to represent one conceptual input condition, may be used for a new input condition, if these two conditions represent the same "degree of progress" of the input stream toward acceptance. The next example illustrates this fact.

Example 5   Construct a Moore machine that accepts exactly those input strings of x's and y's that end in yy.

Solution. Again we need a starting state s_0. If the input string begins with a y, we progress one step to a new state s_1 ("last input component received is a y"). On the other hand, if the input begins with an x, we have made no progress toward acceptance. Thus we may suppose that M is again in state s_0. This situation is shown in Fig. 5(a). If, while in state s_1, a y is received, we progress to an acceptance state s_2 ("last two components of input received were y's"). If instead the input received is an x, we must again receive two y's in order to be in an acceptance state. Thus we may again regard this as a return to state s_0. The situation at this point is shown in Fig. 5(b). Having reached state s_2, an additional input of y will have no effect, but an input of x will necessitate two more y's for acceptance. Thus we can again regard M as being in state s_0. The final Moore machine is shown in Fig. 5(c).

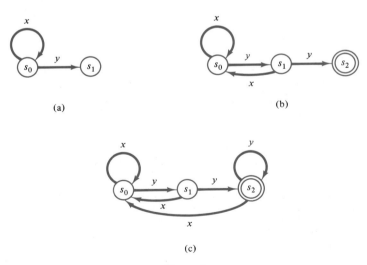

(a)

(b)

(c)

Figure 5

EXERCISE SET 11.3

In Exercises 1–4, describe (in words) the language recognized by the Moore machines whose digraphs are given

1.

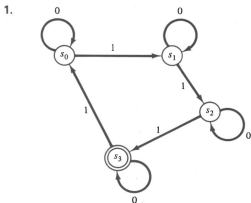

Figure 6

2.

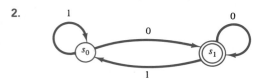

Figure 7

3.

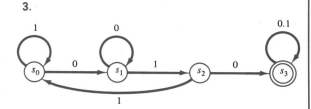

Figure 8

4.

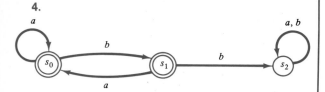

Figure 9

In Exercises 5 and 6, describe (in words) the language recognized by the Moore machines whose state tables are given. The starting state is s_0, and the set T, of acceptance states, is shown.

5.

	0	1
s_0	s_1	s_0
s_1	s_1	s_2
s_2	s_1	s_0

$T = \{s_2\}$

6.

	0	1
s_0	s_0	s_1
s_1	s_0	s_1

$T = \{s_1\}$

7. Let M be the Moore machine of Exercise 1. Construct a type 3 grammar $G = (V, I, s_0, \mapsto)$, such that $L(M) = L(G)$.

8. Let M be the Moore machine of Exercise 3. Construct a type 3 grammar $G = (V, I, s_0, \mapsto)$, such that $L(M) = L(G)$. Describe $\mapsto$ in BNF.

9. Let M be the Moore machine of Exercise 4. Construct a type 3 grammar $G = (V, I, s_0, \mapsto)$, such that $L(M) = L(G)$. Describe $\mapsto$ in BNF.

10. Let M be the Moore machine of Exercise 5. Construct a type 3 grammar $G = (V, I, s_0, \mapsto)$, such that $L(M) = L(G)$. Describe $\mapsto$ by a syntax diagram.

In Exercises 11–24, construct the digraph of a Moore machine that accepts the input strings described, and no others.

11. Inputs a, b: strings where the number of b's is divisible by 3.

12. Inputs a, b: strings where the number of a's is even and the number of b's is a multiple of 3.

13. Inputs x, y: strings that have an even number of y's.

14. Inputs 0, 1: strings that contain 0011.

15. Inputs 0, 1: strings that end with 0011.

16. Inputs $\square$, $\triangle$: strings that contain $\square\triangle$ or $\triangle\square$.

17. Inputs $+$, $\times$: strings that contain $+\times\times$ or $\times++$.

18. Inputs w, z: strings that contain wz or zzw.

19. Inputs a, b: strings that contain ab and end in bbb.

20. Inputs $+$, $\times$: strings that end in $+\times\times$.

21. Inputs w, z: strings that end in wz or zzw.

22. Inputs 0, 1, 2: string 0120 is the only string recognized.

23. Inputs a, b, c: strings aab or abc are to be recognized.

24. Inputs x, y, z: strings xzx or yx or zyx are to be recognized.

In Exercises 25–28, construct the state table of a Moore machine that recognizes the given input strings, and no others.

25. Inputs 0, 1: strings ending in 0101.

26. Inputs a, b: strings where the number of b's is divisible by 4.

27. Inputs x, y: strings having exactly two x's.

28. Inputs a, b: strings that do not have two successive b's.

THEORETICAL EXERCISE

T1 Let $M = (S, I, \mathcal{F}, s_0, T)$ be a Moore machine. Define a relation R on S as follows: $s_i \, R \, s_j$ if and only if $f_w(s_i)$ and $f_w(s_j)$ either both belong to T or neither does, for every $w \in I^*$. Show that R is an equivalence relation on S.

KEY IDEAS FOR REVIEW

☐ Finite-state machine: $(S, I, \mathcal{F})$, where S is a finite set of states, I is a set of inputs, and $\mathcal{F} = \{f_x \mid x \in I\}$.

☐ State transition table: see page 276.

☐ $R_M : s_i \, R_M \, s_j$ if there is an input x so that $f_x(s_i) = s_j$.

☐ State transition function f_w, $w = x_1 x_2 \cdots x_n : f_w = f_{x_n} \circ \cdots \circ f_{x_2} \circ f_{x_1}$, $f_\Lambda = 1_s$.

☐ Theorem: Let $M = (S, I, \mathcal{F})$ be a finite state machine. Then if w_1 and w_2 are in I^*, $f_{w_1 \cdot w_2} = f_{w_2} \circ f_{w_1}$.

☐ Recognition of Moore machine: $M = (S, I, \mathcal{F}, s_0, T)$, where $s_0 \in S$ is the starting state and $T \subseteq S$ is the set of acceptance states.

☐ Language accepted by M: $L(M) = \{w \in I^* \mid f_w(s_0) \in T\}$.

☐ Theorem: Let I be a set and $L \subseteq I^*$. Then L is a type 3 language, that is, $L = L(G)$ where G is a type 3 grammar having I as its set of terminal symbols, if and only if $L = L(M)$ for some Moore machine $M = (S, I, \mathcal{F}, s_0, T)$.

REVIEW EXERCISES

1. Draw the digraph of the machine whose state transition table is given to the right. Remember to label the edges with appropriate inputs.

	x	y
s_0	s_1	s_2
s_1	s_2	s_0
s_2	s_0	s_1

2. Construct the state transition table for the finite-state machine whose digraph is shown in the following figure.

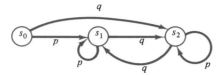

3. For the machine of Exercise 1, list the values of the transition function f_w for $w = xxyyx$.

4. For the machine of Exercise 2, list the values of the transition function f_w for $w = pqpqpq$.

5. For the machine of Exercise 2, describe the set of words w, composed of p's and q's, having the property that

$$f_w(s_0) = s_2$$

6. Describe the language recognized by the Moore machine whose digraph is shown.

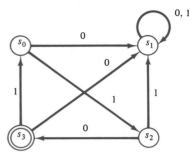

(Exercises 7, 8, and 9 are for the optional section)

7. For the machine shown in Exercise 6, construct a type 3 grammar $G = (V, I, s_0, \mapsto)$ so that $L(M) = L(G)$. Describe the productions in BNF form.

In Exercises 8 and 9, construct the digraph of a Moore machine that accepts the inputs described, and no others.

8. Inputs 0, 1: strings that contain 010.

9. Inputs a, b, c: strings a or bcc are to be recognized.

CHAPTER TEST

1. Construct the state transition table for the finite-state machine whose digraph is:

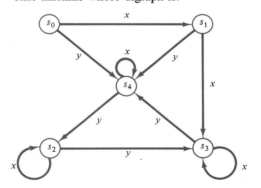

2. Draw the labeled digraph of the machine whose state transition table is

	0	1
s_0	s_3	s_3
s_1	s_0	s_0
s_2	s_1	s_2
s_3	s_2	s_3

3. For the machine of Problem 1, list the values of the transition function f_w for $w = xxxyx$.

4. For the machine of Problem 2, describe the binary words w having the property that $f_w(s_2) = s_0$.

5. Describe the language recognized by the following Moore machine.

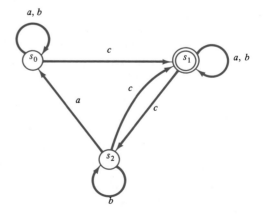

6. For each of the following questions, answer true (T) or false (F).

   (a) The Moore machine of Problem 5 accepts the string *aaccc*.

   (b) The Moore machine of Problem 5 accepts the string *bcabba*.

   (c) For the machine of Problem 5, if $w = ccaab$, then $f_w(s_1) = s_2$.

   (d) A Moore machine can have more than one acceptance state.

   (e) For a Moore machine, the starting state cannot be an acceptance state.

12

Elements of Probability

Prerequisites: Chapters 1 and 2

In this section we present a brief introduction to probability theory. Most of the results are applications of the material on counting the number of elements in a set or a sequence; that is, they are applications of the material on permutations and combinations (see Section 2.2)

12.1 Sample Spaces and Events

Many experiments do not yield exactly the same results when performed repeatedly. For example, if we toss a coin we are not sure whether we will get heads or tails, and if we toss a die we have no way of knowing which of the six possible numbers will turn up. Experiments of this type are called **probabilistic**, in contrast to **deterministic** experiments, whose outcome is always the same.

Sample Spaces

The set A consisting of all the outcomes of an experiment is called the **sample space** of the experiment. With a given experiment, we can often associate more than one sample space depending on what the observer chooses to record as an outcome.

Example 1   Suppose that a nickel and a quarter are tossed in the air. We describe three possible sample spaces that can be associated with this experiment:

1. If the observer decides to record as an outcome the number of heads observed, then the sample space is

$$A = \{0, 1, 2\}$$

2. If the observer decides to record the sequence of heads (H) and tails (T) observed, listing the condition of the nickel first and then that of the quarter, then the sample space is

$$A = \{HH, HT, TH, TT\}$$

3. If the observer decides to record the fact that the coins match (M) or do not match (D), then the sample space is

$$A = \{M, D\}$$

We thus see that in addition to describing the experiment, we must indicate exactly what the observer wishes to record. Then the set of all outcomes of this type becomes the sample space A for the experiment.

A sample space may contain a finite number or an infinite number of outcomes.

Example 2 Determine the sample space for an experiment consisting of tossing a six-sided die twice and recording the sequence of numbers showing on the top face of the die at each toss.

Solution. An outcome of the experiment in this case will consist of an ordered pair of numbers (n, m), where n and m can be 1, 2, 3, 4, 5, or 6. Thus the sample space A consists of the product set

$$\{1, 2, 3, 4, 5, 6\} \times \{1, 2, 3, 4, 5, 6\}$$

which has 36 elements.

Example 3 An experiment consists of drawing three coins in succession from a box containing four pennies and five dimes, and recording the sequence of results. Determine the sample space of this experiment.

Solution. An outcome can be recorded as a sequence of length three constructed from the letters P (penny) and D (dime). Thus the sample space A is

$$\{(P, P, P), (P, P, D), (P, D, P), (P, D, D),$$
$$(D, P, P), (D, P, D), (D, D, P), (D, D, D)\}$$

Events

A statement about the outcome of an experiment, which for each particular outcome will be either true or false, is said to describe an **event**. Thus for Example 2, the statements "Each of the numbers recorded is less than 3" and "The sum of the two numbers recorded is 4" would describe events. The event described by a statement is taken to be the set of all outcomes for which the statement is true; that is, an event can be regarded as a subspace of the sample space. Thus the event E described by the first statement given above is

$$E = \{(1, 1), (1, 2)\ (2, 1), (2, 2)\}$$

Similarly, the event F described by the second statement is

$$F = \{(1, 3), (2, 2), (3, 1)\}$$

Example 4   Consider the experiment described in Example 2. Determine the events described by each statement.
 (a) The sum of the numbers showing on the top faces is 8.
 (b) The sum of the numbers showing on the top faces is at least 10.

Solution. (a) The event consists of all ordered pairs from the set $\{1, 2, 3, 4, 5, 6\} \times \{1, 2, 3, 4, 5, 6\}$ whose sum is 8. Thus the event is

$$\{(2, 6), (3, 5), (4, 4), (5, 3), (6, 2)\}$$

 (b) The event consists of all ordered pairs whose sum is 10, 11, or 12. Thus the event is

$$\{(4, 6), (5, 5), (5, 6), (6, 4), (6, 5), (6, 6)\}$$

If A is a sample space of an experiment, then A itself is an event called the **certain event** and the empty subset of A is called the **impossible event**.
 Since events are sets, we can combine them by applying the operations of union, intersection, and complementation to form new events. The sample space A is the universal set for these events. Thus if E and F are events, we can form the new events $E \cup F$, $E \cap F$, and $\overline{E}$. What do these new events mean? An outcome of the experiment belongs to $E \cup F$ when it belongs to E or F (or both). In other words, the event $E \cup F$ occurs exactly when E or F occurs. Similarly, the event $E \cap F$ occurs if and only if both E and F occur. Finally, $\overline{E}$ occurs if and only if E does not occur.

Example 5   Consider the experiment of tossing a die and recording the number on the top face. Let E be the event that the number is even and let F be the event that the number is prime. Then

$$E = \{2, 4, 6\} \quad \text{and} \quad F = \{2, 3, 5\}$$

The event that the number showing is either even or a prime is

$$E \cup F = \{2, 3, 4, 5, 6\}$$

The event that the number showing is an even prime is

$$E \cap F = \{2\}$$

Finally, the event that the number showing is not even is

$$\overline{E} = \{1, 3, 5\}$$

and the event that the number showing is not prime is

$$\overline{F} = \{1, 4, 6\}$$

The events E and F are said to be **mutually exclusive** or **disjoint** if $E \cap F = \varnothing$. If E and F are mutually exclusive events, then E and F cannot both occur at the same time; if E occurs, then F does not occur, and if F occurs, then E does not. If $E_1, E_2, \ldots E_n$ are all events then we will say that these sets are **mutually exclusive** or **disjoint** if any two of them are mutually exclusive. Again, this means that at most one of the events can occur on any given outcome of the experiment.

EXERCISE SET 12.1

In Exercises 1–6, describe the associated sample space.

1. In a class of 10 students, the instructor records the number of students present on a given day.

2. A coin is tossed three times and the sequence of heads (H) and tails (T) is recorded.

3. A marketing research firm conducts a survey of a group of people. People are classified according to the following characteristics.

   Sex: male (m), female (f)

   Income: low (l), average (a), upper (u)

   Smoker: yes (y), no (n)

A person is selected at random and classified accordingly.

4. Two letters are selected simultaneously at random from the letters a, b, c, d.

5. A silver urn and copper urn each contain blue, red, and green balls. An urn is chosen at random and then a ball is selected at random from this urn.

6. A box contains 12 items, four of which are defective. An item is chosen at random and not replaced. This is continued until all four defective items have been selected. The total number of items selected is recorded.

7. Suppose that the sample space of an experiment is $\{1, 2, 3\}$. Determine all possible events.

8. An experiment consists of tossing a die and recording the number on the top face. Determine each of the following events:

   (a) *E*: The number tossed is at least 4.

   (b) *F*: The number tossed is less than 3.

   (c) *G*: The number tossed is either divisible by 3 or is a prime. (Recall that the number 1 is not a prime.)

9. A card is selected at random from a standard deck. Let *E*, *F*, and *G* be the events

   E: The card is black.

   F: The card is in the suit of diamonds.

   G: The card is an ace.

   Describe the following events in words.

   (a) $E \cup G$ (b) $E \cap G$ (c) $\overline{E} \cap G$

   (d) $E \cup F \cup G$ (e) $E \cup \overline{F} \cup G$

10. A die is tossed twice and the numbers showing on the top faces are recorded in sequence. Determine the elements in each of the given events.

   (a) At least one of the numbers showing is a 5.

   (b) An 8 turns up. At least one of the numbers showing is an 8.

   (c) The sum of the numbers showing is less than 7.

   (d) The sum of the numbers showing is greater than 8.

11. A die is tossed and the number showing on the top face is recorded. Let *E*, *F*, and *G* be the events

   E: The number tossed is at least 3.

   F: The number tossed is at most 3.

   G: The number tossed is divisible by 2.

   (a) Are *E* and *F* mutually exclusive? Justify your answer.

   (b) Are *F* and *G* mutually exclusive? Justify your answer.

   (c) Is $E \cup F$ the certain event? Justify your answer.

   (d) Is $E \cap F$ the impossible event? Justify your answer.

THEORETICAL EXERCISE

T1. Let *E* be an event for an experiment with sample space *A*. Show that

   (a) $E \cup \overline{E}$ is the certain event.

   (b) $E \cap \overline{E}$ is the impossible event.

12.2 Assigning Probabilities to Events

In probability theory, we assume that each event *E* has been assigned a number $p(E)$ called the **probability of the event *E*.** We now look at probabilities. We will investigate ways in which they can be assigned, properties that they must satisfy, and the meaning that can be assigned to them.

The number $p(E)$ reflects our assessment of the likelihood that the event *E* will occur. More precisely, suppose that the underlying experiment is performed repeatedly, and that after *n* such performances, the event *E* has occurred n_E times. Then the fraction $f_E = n_E/n$, called the **frequency of occurrence of *E* in *n* trials,** is a measure of the likelihood that *E* will occur. When we assign the probability $p(E)$ to the event *E*, it means that in our judgment or experience, we believe that the fractions f_E will tend ever closer to a certain number as n becomes large, and that $p(E)$ is this number. Thus probabilities can be thought of as idealized frequencies of occurrence of events, to which actual frequencies of occurrence will tend as the experiment is performed repeatedly.

Example 1   Suppose that an experiment is performed 2000 times, that the frequency of occurrence f_E of an event E is recorded after 100, 500, 1000, and 2000 trials, and that Table 1 summarizes the results. Based on this table, it appears that the frequency f_E approaches $\frac{1}{2}$ as n becomes large. It could therefore be argued that $p(E)$ should be set equal to $\frac{1}{2}$. On the other hand, one might require more extensive evidence before assigning $\frac{1}{2}$ as the value of $p(E)$. In any case, this sort of evidence can never "prove" that $p(E) = \frac{1}{2}$. It serves only to make this a plausible assumption.

Table 1

Number of repetitions of the experiment	n_E	$f_E = n_E/n$
100	48	0.48
500	259	0.518
1000	496	0.496
2000	1002	0.501

If probabilities assigned to various events are to meaningfully represent frequencies of occurrence of the events, as explained above, then they cannot be assigned in a totally arbitrary way. They must satisfy certain conditions. In the first place, since every frequency f_E must satisfy the inequalities $0 \le f_E \le 1$, it is only reasonable to assume that

P1: $0 \le p(E) \le 1$ for every event E in A.

Also, since the event A must occur every time (every outcome belongs to A), and the event $\varnothing$ cannot occur, we assume that

P2: $p(A) = 1$ and $p(\varnothing) = 0$

Finally, if $E_1, E_2, \ldots, E_k$ are mutually exclusive events, then

$$n_{(E_1 \cup E_2 \cup \cdots \cup E_k)} = n_{E_1} + n_{E_2} + \cdots + n_{E_k}$$

since only one of these events can occur at a time. If we divide both sides of this equation by n, we see that the frequencies of occurrence must satisfy a similar equation. We therefore assume that

$$P3: \quad p(E_1 \cup E_2 \cup \cdots \cup E_k) = p(E_1) + p(E_2) + \cdots + p(E_k)$$

whenever the events are mutually exclusive.

If probabilities are assigned to all events in such a way that P1, P2, and P3 are always satisfied, then we have a **probability space.** We summarize in the following box.

Axioms for a Probability Space

P1: $0 \le p(E) \le 1$ for every event E in A.
P2: $p(A) = 1$ and $p(\varnothing) = 0$.
P3: $p(E_1 \cup E_2 \cup \cdots \cup E_k) = p(E_1) + p(E_2) + \cdots + p(E_k)$
whenever the events are mutually exclusive.

It is important to realize that mathematically, no demands are made on the probability space except those given by the probability axioms P1, P2, and P3. Probability theory begins with all probabilities assigned, and then investigates consequences of and relations between these probabilities. No mention is made of how the probabilities were assigned. However, the mathematical conclusions will be useful in an actual situation only if the probabilities assigned reflect what actually occurs in that situation.

Experimentation is not the only way to determine reasonable probabilities for events. The probability axioms and the nature of the experiment can sometimes provide logical arguments for choosing certain probabilities.

Example 2 Consider the experiment of tossing a coin and recording whether heads or tails results. Consider the events "heads turns up" (E) and "tails turns up" (F). The mechanics of the toss are not controllable in detail. Thus in the absence of any defect in the coin that might unbalance it, one may argue that E and F are equally likely to occur. There is a symmetry in the situation that makes it impossible to prefer one outcome over another. This argument lets us compute what the probabilities of E and F must be.

We have assumed that $p(E) = p(F)$, and it is clear that E and F are disjoint and $A = E \cup F$. Thus, using properties P2 and P3, we see that

$$1 = p(A) = p(E) + p(F) = 2p(E) \qquad [\text{since } p(F) = p(E)]$$

This shows that $p(E) = \frac{1}{2} = p(F)$. One may often assign appropriate probabilities to certain events by combining the symmetry of situations with the axioms of probability.

Finally, we will show that the problem of assigning probabilities to events can be reduced to the consideration of the simplest cases. Let A be a probability space.

Since we are assuming that A is finite, we may suppose that $A = \{x_1, x_2, \ldots, x_m\}$. Then each event $\{x_k\}$, consisting of just one outcome, is called an **elementary event.** For simplicity, let us write $p_k = p(\{x_k\})$. Then p_k is called the **elementary probability corresponding to the outcome x_k.** Since the elementary events are mutually disjoint and their union is A, the axioms of probability tell us that

EP1: $0 \le p_k \le 1$ for all k

EP2: $p_1 + p_2 + \cdots + p_n = 1$

If E is any event in A, say $E = \{x_{i_1}, x_{i_2}, \ldots, x_{i_m}\}$, then we can write

$$E = \{x_{i_1}\} \cup \{x_{i_2}\} \cup \cdots \cup \{x_{i_m}\}.$$

This means, by probability axiom P2, that

$$p(E) = p_{i_1} + p_{i_2} + \cdots + p_{i_m}.$$

Thus if we know the elementary probabilities, then we can compute the probability of any event E as follows.

To compute the probability of an event E

Step 1. List all outcomes belonging to E.

Step 2. Add the elementary probabilities corresponding to these outcomes.

Example 3   Suppose that an experiment has sample space $A = \{1, 2, 3, 4, 5, 6\}$ and that the elementary probabilities have been determined as follows.

$$p_1 = \tfrac{1}{12}, \qquad p_2 = \tfrac{1}{12}, \qquad p_3 = \tfrac{1}{3}, \qquad p_4 = \tfrac{1}{6}, \qquad p_5 = \tfrac{1}{4}, \qquad p_6 = \tfrac{1}{12}$$

Let E be the event "The outcome is an even number." Compute $p(E)$.

Solution. Since $E = \{2, 4, 6\}$, we see that

$$p(E) = p_2 + p_4 + p_6 = \tfrac{1}{12} + \tfrac{1}{6} + \tfrac{1}{12} = \tfrac{1}{3}.$$

In a similar way we can determine the probability of any event in A.

Thus we see that the problem of assigning probabilities to all events in a consistent way can be reduced to the problem of finding numbers $p_1, p_2, \ldots, p_m$ that satisfy EP1 and EP2. Again, mathematically speaking, there are no other restrictions on the p_k. However if the mathematical structure that results is to be useful in a particular situation, then the p_k's must reflect the actual behavior occurring in that situation.

EXERCISE SET 12.2

In Exercises 1–4, list the elementary events for the given experiment.

1. A number is picked at random from the box containing the numbers 1, 2, 3, and 4.

2. A pair of dice are tossed and the sum of the numbers showing on the top faces is recorded.

3. A vowel is selected at random from the set of all vowels {a, e, i, o, u}.

4. A card is selected at random and it is recorded whether the card is a club, diamond, spade, or heart.

5. When a certain defective die is thrown, the numbers from 1 to 6 will occur with the following probabilities:

$$p(\{1\}) = \tfrac{2}{18} \qquad p(\{2\}) = \tfrac{3}{18} \qquad p(\{3\}) = \tfrac{4}{18}$$
$$p(\{4\}) = \tfrac{3}{18} \qquad p(\{5\}) = \tfrac{4}{18} \qquad p(\{6\}) = \tfrac{2}{18}$$

Find the probability that
(a) an odd number is tossed.
(b) a prime number is tossed.
(c) a number less than 5 is tossed.
(d) a number greater than 3 is tossed.

6. Repeat Exercise 5 assuming that the die is not defective.

7. Suppose that E and F are mutually exclusive events such that $p(E) = 0.3$ and $p(F) = 0.4$.
(a) Find the probability that E does not occur.
(b) Find the probability that E and F occur.
(c) Find the probability that E or F occurs.
(d) Find the probability that E does not occur or that F does not occur.

8. Consider an experiment with sample space $A = \{x_1, x_2, x_3, x_4\}$ for which

$$p_1 = \tfrac{2}{7}, \qquad p_2 = \tfrac{3}{7}, \qquad p_3 = \tfrac{1}{7}, \qquad p_4 = \tfrac{1}{7}$$

Find the probability of the given event
(a) $E = \{x_1, x_2\}$
(b) $F = \{x_1, x_3, x_4\}$

9. There are four candidates for president: A, B, C, and D. Suppose that A is twice as likely to be elected as B, B is three times as likely as C, and C and D are equally likely to be elected. What is the probability that each candidate will be elected?

12.3 Equally Likely Outcomes

Let us assume that all outcomes in a sample space A are equally likely to occur. This is, of course, an assumption, and therefore cannot be proved. We would make such an assumption if experimental evidence or symmetry arguments indicated that it was appropriate in a particular situation (see Example 2 of Section 12.2). This situation arises frequently, and in this section we investigate the consequences. One additional piece of terminology is customary, and we introduce it now. Sometimes probability experiments involve choosing an object, in a nondeterministic way, from some collection. If the choice is made in such a way that all objects have an equal probability of being chosen, we say that we have made a **random selection** or **chosen an object**

at random from the collection. We will often use this terminology to specify examples of experiments with equally likely outcomes.

Suppose that $|A| = n$, that is, the experiment has n equally likely outcomes. We have assumed that the elementary probabilities are all equal, and since they must add to 1, this means that all elementary probabilities equal $1/n$. Now let E be an event that contains k elements, say $E = \{x_1, x_2, \ldots, x_k\}$. Since all elementary probabilities equal $1/n$, we must have

$$p(E) = \frac{1}{n} + \frac{1}{n} + \cdots + \frac{1}{n} \text{ (a total of } k \text{ summands)} = \frac{k}{n}$$

Since $k = |E|$, we have the following principle: If all outcomes are equally likely, then for every event E,

$$p(E) = \frac{|E|}{|A|} = \frac{\text{number of outcomes in } E}{\text{total number of outcomes}}$$

In this case, the computation of probabilities reduces to counting numbers of elements in sets. For this reason, methods of counting developed previously are quite useful. Let us review some formulas for counting that we have seen before, and add a few extensions.

Sometimes the occurrence of an event can be specified in stages or steps. In that case we can often determine the number of outcomes in the event by considering the stages and applying the following general result.

Formula 1   **The Extended Multiplication Principle**
Suppose that tasks $T_1, T_2, \ldots, T_k$ are to be performed in sequence. If T_1 can be performed in n_1 ways, and for each of these ways T_2 can be performed in n_2 ways, and for each of these $n_1 n_2$ ways of performing $T_1 T_2$ in sequence, T_3 can be performed in n_3 ways, and so on, then the sequence $T_1 T_2 \cdots T_k$ can be performed in exactly $n_1 n_2 \cdots n_k$ ways (See Theorem 2 of Section 2.2).

Another interesting event is one that is the union of two, possibly simpler events. Then the following result from set theory can be applied.

Formula 2   **The Addition Principle**
If E and F are finite sets, then

$$|E \cup F| = |E| + |F| - |E \cap F|$$

(see Theorem 1 of Section 2.1).

In the special case when E and F are mutually exclusive events, Formula 2 states that the number of outcomes in the union of two events is the sum of the number of outcomes in each event. This special case generalizes easily, as follows, and allows us to decompose some events of interest into the union of disjoint events that may be easier to analyze.

Formula 3 **The Decomposition Principle**

If $E_1, E_2, \ldots, E_k$ are mutually exclusive events in a finite sample space, then

$$\left| E_1 \cup E_2 \cup \cdots \cup E_k \right| = \left| E_1 \right| + \left| E_2 \right| + \cdots + \left| E_k \right|$$

Finally, we recall the formulas for permutations and combinations given in Section 2.2. The student should review the definitions in that section.

Formulas 4   (a) The number of sequences of length k that can be constructed from a set with n elements, allowing repetition of elements in a sequence, is n^k.

(b) Let P_k^n be the number of permutations of n objects taken k at a time, that is, the number of sequences of length k constructed from a set of n elements, if no repetition of elements is allowed in a sequence. Then

$$P_k^n = \frac{n!}{(n-k)!}$$

(c) Let $\binom{n}{k}$ (also written C_k^n) be the number of combinations of n objects taken k at a time, that is, the number of k-element subsets of a set with n elements. Then

$$\binom{n}{k} = \frac{n!}{k!(n-k)!}$$

Computing the probability of events, in the equally likely outcome case, almost always involves the use of one or more of the counting formulas above. These computations are often difficult for students to master. One of the reasons is that a particular problem may require the use of several of the above formulas in combination. Problems such as these are most easily handled by decomposing them into natural parts and trying to identify the correct formulas for solving each part, and combining the results.

We now give some examples of such probability computations, and in each case we identify the counting formulas used in the solution. The student is advised to observe closely the situations in which the various formulas are appropriate.

Example 1   Choose four cards at random from a standard 52-card deck. What is the probability that four kings will be chosen?

Solution. The outcomes of this experiment are four-card subsets of the deck. By definition of random choice, we have assumed that any subset of four cards has an equal chance of being chosen. Let E be the event that all four cards are kings. Then for the probability of E we have

$$p(E) = \frac{\text{number of outcomes in } E}{\text{total number of outcomes}}$$

By Formula 4(c) we see that the total number of outcomes, that is, the total number of four-element subsets of a deck with 52 cards, is

$$\binom{52}{4} = \frac{52!}{4! \, 48!} = \frac{52 \cdot 51 \cdot 50 \cdot 49}{1 \cdot 2 \cdot 3 \cdot 4} = 270{,}725$$

On the other hand, E contains only one outcome. Thus

$$p(E) = \frac{1}{270{,}725} = 0.000003694$$

Example 2   A box contains six red balls and four green balls. Four balls are selected at random from the box. What is the probability that two of the selected balls will be red and two will be green?

Solution. We first determine the total number of outcomes of the experiment. This is simply the number of ways of selecting four objects out of 10. Since the order does not matter, Formula 4(c) tells us that this number is

$$\binom{10}{4} = \frac{10 \cdot 9 \cdot 8 \cdot 7}{1 \cdot 2 \cdot 3 \cdot 4} = 210$$

Now the event E, that two of the selected balls are red and two are green, can be thought of as the result of two consecutive tasks.

Task 1. Choose two red balls from the six red balls in the box.

Task 2. Choose two green balls from the four green balls in the box.

By Formula 4(c), task 1 can be performed in $\binom{6}{2} = 15$ ways and task 2 can be performed in $\binom{4}{2} = 6$ ways. Thus, by Formula 1, event E can occur in $15 \cdot 6 = 90$ ways, and therefore

$$p(E) = \frac{90}{210} = \frac{3}{7}$$

Note that we needed a combination of Formulas 1 and 4(c) for this solution.

Example 3 A fair, six-sided die is tossed three times in succession, and the resulting sequence of numbers is recorded. What is the probability of the event E, that either all three numbers are equal or none of them is a 4?

Solution. Since the die is assumed fair, all outcomes are equally likely. First, we compute the total number of outcomes of the experiment. This is the number of sequences of length 3, allowing repetititions, that can be constructed from the set $\{1, 2, 3, 4, 5, 6\}$. By Formula 4(a), this number is $6^3 = 216$.

Now event E cannot, as was the case in Example 2, be described as the result of performing two successive tasks. We can, however, write it as the union of two simpler events. Let F be the event that all three numbers recorded are equal, and let G be the event that none of the numbers recorded is a 4. Then $E = F \cup G$. By the addition principle, $|F \cup G| = |F| + |G| - |F \cap G|$.

Now there are only six outcomes in which all numbers are equal, so $|F| = 6$. The event G consists of all sequences of length 3 that can be formed from the set $\{1, 2, 3, 5, 6\}$. By Formula 4(a), this number is 5^3. Thus $|G| = 5^3 = 125$. Finally, the event $F \cap G$ consists of all sequences for which all three numbers are equal and none is a 4. Clearly, there are only five ways for this to happen, so $|F \cap G| = 5$. By the addition principle,

$$|E| = |F \cup G| = 6 + 125 - 5 = 126$$

Thus, by combining Formulas 2 and 4(a), we have shown that

$$p(E) = \frac{126}{216} = \frac{7}{12}$$

Example 4 Consider again the experiment of Example 3, in which four balls are selected at random from a box containing six red balls and four green balls.

(a) If E is the event that no more than two of the balls are red, compute the probability of E.

(b) If F is the event that no more than three of the balls are red, compute the probability of F.

Solution. (a) Here E can be decomposed as the union of mutually exclusive events. Let E_0 be the event that none of the chosen balls is red, let E_1 be the event that exactly one of the chosen balls is red, and let E_2 be the event that exactly two of the chosen balls are red. Then E_0, E_1, and E_2 are mutually disjoint and $E = E_0 \cup E_1 \cup E_2$. By Formula 3, $|E| = |E_0| + |E_1| + |E_2|$. If none of the balls is red, then all four must be green. Since there are only four green balls in the box, there is only one way for event E_0 to occur. Thus $|E_0| = 1$. If one ball is red, the other three must be green. To make such a choice, we must choose one red ball from a set of six, and then three green balls from a set of four. The first choice can be made in $\binom{6}{1}$ ways and the second in $\binom{4}{3}$ ways by Formula 4(c). Thus by Formula 1, the number of outcomes in E_1 is

$$\binom{6}{1}\binom{4}{3} = 6 \cdot 4 = 24.$$

In exactly the same way, by combining Formulas 4(c) and 1, we can show that the number of outcomes in E_2 is (verify) $\binom{6}{2}\binom{4}{2} = 15 \cdot 6 = 90$. Thus, by Formula 3, $|E| = 1 + 24 + 90 = 115$. On the other hand, the total number of ways of choosing four balls is $\binom{10}{4} = 210$, so

$$p(E) = 115/210 = 23/42.$$

(b) We could compute $|F|$ in the same way that we computed $|E|$ in part (a), by decomposing F into four mutually exclusive events. The analysis would, however, be even lengthier than that of part (a). We choose instead to illustrate another approach that is frequently useful. Let $\overline{F}$ be the complementary event to F; that is, $\overline{F}$ occurs if and only if F does not occur. Since F and $\overline{F}$ are mutually exclusive and their union is A, we must have $p(F) + p(\overline{F}) = 1$. In other words,

$$p(F) = 1 - p(\overline{F})$$

This formula holds for any event F, and it can be used when the complementary event is easier to analyze. This is the case here, since $\overline{F}$ is the event that all four balls chosen are red. These four red balls can be chosen from the six in the box in a total of $\binom{6}{4} = 15$ ways, so $p(\overline{F}) = 15/210 = 1/12$. This means that

$$p(F) = 1 - 1/12 = 11/12.$$

In any problem in which the computation of the probability of some event is very complex, the student should consider whether it would be easier to compute the probability of the complementary event, and then use the approach described above.

EXERCISE SET 12.3

1. Suppose that a fair die is tossed and the number showing on the top face is recorded. Let E, F, and G be the following events:

$$E = \{1, 2, 3, 5\}$$

$$F = \{2, 4\}$$

$$G = \{1, 4, 6\}$$

Compute the probability of the indicated event.

(a) $E \cup F$ (b) $E \cap F$ (c) $\overline{E} \cap F$
(d) $E \cup G$ (e) $\overline{E} \cup \overline{G}$ (f) $\overline{E} \cap \overline{F}$

2. Suppose that three balls are selected at random from an urn containing seven red balls and five black balls.
   (a) Compute the probability that all three balls are red.
   (b) Compute the probability that at least two balls are black.

(c) Compute the probability that at most two balls are black.

(d) Compute the probability that at least one ball is red.

3. Suppose that four balls are randomly selected from an urn containing five red balls and six white balls. What is the probability that two balls will be red and two balls will be white?

4. Suppose that a five-card hand is selected at random from a standard 52-card deck. What is the probability that the hand has

(a) exactly three spades?

(b) two diamonds and three clubs?

(c) at least one heart?

(d) at most two diamonds?

5. Consider the experiment of tossing three fair coins and recording the sequence of heads (H) and tails (T).

(a) What is the probability that two tails will be tossed?

(b) What is the probability that at least one head will be tossed?

6. Suppose that two fair dice are tossed and the numbers on the top faces are recorded. What is the probability that

(a) a 4 was tossed?

(b) a prime number was tossed?

(c) the sum of the numbers tossed is less than 5?

(d) the sum of the numbers tossed is at least 7?

7. A family-planning counselor records the sex of the children in a three-child family. Assume that any birth is equally likely to result in a male or a female child.

(a) Letting m denote male and f female, list the elements in the sample space.

(b) What is the probability of having three boys?

(c) What is the probability of having two girls followed by a boy?

(d) What is the probability of having two girls and a boy in any order?

8. A woman's wardrobe contains five pairs of gloves. If the woman selects two gloves at random, what is the probability that the gloves will be a matching pair?

KEY IDEAS FOR REVIEW

☐ Sample space: the set of all outcomes of an experiment.

☐ Event: a subset of the sample space.

☐ Certain event: the event certain to occur, the entire sample space.

☐ Impossible event: the empty subset of the sample space.

☐ Mutually exclusive events: any two events E and F with $E \cap F = \varnothing$.

☐ f_E: the frequency of occurence of the event E in n trials.

☐ $p(E)$: the probability of the event E.

☐ Probability space: page 300

☐ Elementary event: event consisting of just one outcome.

☐ Random selection: page 302

☐ Extended multiplication principle: page 303

☐ Addition principle: If E and F are finite sets, then $|E \cup F| = |E| + |F| - |E \cap F|$

☐ The decomposition principle: page 304

REVIEW EXERCISES

In Exercises 1 and 2, describe the associated sample space.

1. A programming firm has an IBM computer and an Apple computer. For each of these computers they have the PASCAL, FORTRAN, COBOL, FORTH, and APL languages. For a particular application, one of the computers is chosen at random and then one of the languages is chosen at random.

2. A letter is selected at random from eight letters in the word AARDVARK.

3. Let S be a sample space containing n elements. How many events are there for this experiment?

4. A medical team classifies people according to their weight, income, and smoking habits. For a person selected at random, let E, F, and G be the following events:
   E: A person is overweight: yes (y); no (n).
   F: A person's income is: low (l), average (a), high (h).
   G: A person smokes: yes (y); no (n)
   List the elements in each of the following events.

   (a) $E \cup F$
   (b) $\bar{E} \cap F$
   (c) $(E \cup G) \cap F$

5. A coin is tossed three times and the sequence of heads and tails is recorded. Determine the elements in the following events.
   (a) The number of tails is at least one.
   (b) The number of tails is exactly two.

6. The outcome of a particular game of chance is an integer from 1 to 5. Integers 1, 2, and 3 are equally likely to occur, and integers 4 and 5 are equally likely to occur. The probability that the outcome is greater than 2 is $\frac{1}{2}$. Find the probability of each possible outcome.

7. Consider a loaded die in which the outcome is three times as likely to be odd as it is to be even, the even numbers are equally likely to occur, and the odd numbers are equally likely to occur.
   (a) What is the probability that the number 3 will occur?
   (b) What is the probability that the number 2 will show up?

8. A fair coin is tossed five times. What is the probability of obtaining three heads and two tails?

9. Suppose that a number from 1 to 12 is chosen at random. What is the probability that the number is
   (a) greater than 5?
   (b) greater than 3 or less than 8?
   (c) greater than 3 and less than 8?
   (d) divisible by 3?
   (e) greater than 4 and divisible by 2?

10. A fair die is rolled three times in succession. Find the probability that the three resulting numbers
    (a) contain exactly two 3's.
    (b) form a strictly increasing sequence.
    (c) contain at least one 3.
    (d) contain at most one 3.
    (e) contain no 3's.

11. Suppose that two cards are selected at random from a standard 52-card deck. What is the probability that both cards are less than 10 and neither one is red?

CHAPTER TEST

In Problems 1 and 2, describe the associated sample space.

1. A nickel and a dime are tossed and the sequence of heads and tails is recorded.

2. A die is tossed and we observe the number showing on the top face. We record y if this number is divisible by 3 and n if it is not divisible by 3.

3. The alumni office of a certain university classifies its graduates according to the following characteristics:

Sex: male (*m*), female (*f*)

Income: low (*l*), average (*a*), high (*h*)

Education: undergraduate (*u*), graduate (*g*)

For one of the alumni selected at random, determine the elements in the following events.

(a) The event E that the person was a graduate student.

(b) The event F that the person is either a male or has average income.

(c) The event G that the person is a female who was an undergraduate student.

4. Let $S = \{1, 2, 3, 4, 5, 6\}$ be the sample space of an experiment and let

$$E = \{1, 3, 4, 5\}, \quad F = \{2, 3\}, \quad G = \{4\}$$

be events.

(a) Compute the events $E \cup F$, $E \cap F$, and $\overline{F}$.

(b) Compute the events $\overline{E} \cup F$ and $\overline{F} \cap G$.

5. A basket contains three apples, five bananas, 4 oranges, and six pears. A piece of fruit is chosen at random from the basket. Compute the probability that

(a) An apple or a pear is chosen.

(b) The fruit chosen is not an orange.

6. Two cards are chosen at random from a standard 52-card deck. What is the probability that both are red or both are kings?

7. For each of the following parts, answer true (T) or false (F).

(a) If E and F are two events in a probability space, and if $p(E) = p(F) = \frac{1}{3}$, then $p(E \cup F) = \frac{2}{3}$.

(b) If an event in a finite probability space has probability 1, then the event is certain to occur.

(c) If two events in a probability space are disjoint, and one has probability zero, then the other has probability 1.

(d) The probabilities of all elementary events must add to 1.

(e) If a fair coin is tossed twice in succession, the probability that the outcomes are different is $\frac{1}{2}$.

Appendix: Logic

Logic is the discipline that deals with the methods of reasoning. On an elementary level, logic provides rules and techniques for determining whether a given argument is valid. Logical reasoning is used in mathematics to prove theorems, in computer science to verify the correctness of programs and to prove theorems, in the natural and physical sciences to draw conclusions from experiments, in the social sciences, and in our everyday lives to solve a multitude of problems. Indeed, we are constantly using logical reasoning. In this appendix we review a few of the basic ideas.

A **statement** or **proposition** is a declarative sentence that is either true or false, but not both.

Example 1 Which of the following are statements?

(a) $2 + 3 = 5$
(b) The earth is round.
(c) $3 - x = 5$
(d) Do you speak English?
(e) Take two aspirins.
(f) The temperature on the surface of the planet Venus is 800°F.
(g) The sun will come out tomorrow.

Solution. (a) and (b) are statements that happen to be true.

(c) is a declarative sentence, but it is not a statement, since it is true or false depending on the value of x that is used.

(d) is a question, so it is not a statement.

(e) is not a statement, it is a command.

(f) is a declarative sentence whose truth or falsity we do not know at this time; however, we can in principle determine if it is true or false, so it is a statement.

(g) is a statement since it is either true or false, but not both, although we would have to wait until tomorrow to find out if it is true or false.

Logical Connectives and Compound Statements

In mathematics, the letters x, y, z, . . . often denote variables that can be replaced by real numbers, and these variables can be combined by the familiar operations $+$, $\times$, $-$, and $\div$. In logic, the letters p, q, r, . . . denote **propositional variables**, that is, variables that can be replaced by statements. Thus we can write

p: The sun is shining today.
q: It is cold.

Statements or propositional variables can be combined by logical connectives to obtain **compound statements**. For example, we may combine the statements above by the connective *and* to form the compound statement

The sun is shining today *and* it is cold

or

p and q

The truth value of a compound statement depends only on the truth values of the statements being combined and on the types of connectives being used. We shall now look at the most important connectives.

If p is a statement, the **negation** of p is the statement *not p*, denoted by $\sim p$. Thus $\sim p$ is the statement "it is not the case that p." From this definition, it follows that if p is true, then $\sim p$ is false, and if p is false, then $\sim p$ is true. The truth value of $\sim p$, relative to p, is given in Table 1. Such a table, giving the truth values of a compound statement in terms of the truth values of its component parts, is called a **truth table**. Strictly speaking, *not* is not a connective, since it does not join two statements, and $\sim p$ is not really a compound statement.

Table 1

p	$\sim p$
T	F
F	T

Example 2   Give the negations of the following statements.

(a) p: $2 + 3 > 1$
(b) q: It is cold.

Solution. (a) $\sim p$: $2 + 3$ is not greater than 1. That is,

$$\sim p: \quad 2 + 3 \leq 1$$

Since p is true in this case, $\sim p$ is false.
 (b) $\sim q$:   It is not the case that it is cold, or more simply

$$\sim q: \quad \text{It is not cold.}$$

If p and q are statements, the **conjunction** of p and q is the compound statement "p and q", denoted by $p \wedge q$. The connective *and* is denoted by the symbol $\wedge$. The compound statement $p \wedge q$ is true when both p and q are true; otherwise, it is false. The truth values of $p \wedge q$ in terms of the truth values of p and of q are given in the truth table shown in Table 2.

Observe that in giving the truth table of $p \wedge q$ we need to look at four possible cases. This follows from the fact that p and q can each be true or false.

Table 2

p	q	$p \wedge q$
T	T	T
T	F	F
F	T	F
F	F	F

Example 3   Form the conjunction of p and q:

(a) p: It is snowing; q: I am cold.
(b) p: $2 < 3$; q: $-5 > -8$
(c) p: It is snowing; q: $3 < 5$

Solution. (a) $p \wedge q$:   It is snowing and I am cold.
(b) $p \wedge q$: $2 < 3$ and $-5 > -8$
(c) $p \wedge q$:   It is snowing and $3 < 5$.

Example 3(c) shows that in logic, unlike in every day English, we may join two totally unrelated statements by the connective *and*.

If p and q are statements, the **disjunction** of p and q is the compound statement "p or q," denoted by $p \vee q$. The connective *or* is denoted by the symbol $\vee$. The compound statement $p \vee q$ is true if at least one of p or q is true; it is false when both p and q are false. The truth values of $p \vee q$ are given in the truth table shown in Table 3.

Table 3

p	q	$p \vee q$
T	T	T
T	F	T
F	T	T
F	F	F

Example 4   Form the disjunction of p and q.

(a) p: 2 is a positive integer q: $\sqrt{2}$ is a rational number.
(b) p: $2 + 3 \neq 5$ q: London is the capital of France.

Solution. (a) $p \vee q$:   2 is a positive integer or $\sqrt{2}$ is a rational number. Since p is true, the disjunction $p \vee q$ is true, even though q is false.

(b) $p \vee q$: $2 + 3 \neq 5$ or London is the capital of France. Since p and q are both false, $p \vee q$ is false.

Example 4(b) shows that in logic, unlike in ordinary English, we may join two totally unrelated statements by the connective *or*.

The connective *or* is more complicated than the connective *and* because it is used in two different ways in English. Suppose that we say "I drove to work or I took the train to work." In this compound statement we have the disjunction of the statements p: "I drove to work" and q: "I took the train to work." Of course, exactly one of the two possibilities occurred. Both could not have occurred, so the connective or is being used in an *exclusive* sense. On the other hand, consider the disjunction "I passed mathematics or I failed French." In this case, at least one of the two possibilities occurred. However, both could have occurred, so the connective *or* is being used in an *inclusive* sense. In mathematics and computer science we agree to always use the connective *or* in the inclusive manner.

If p and q are statements, the compound statement if p then q, denoted by $p \rightarrow q$, is called a **conditional statement** or **implication**. Statement p is called the **antecedent**, and statement q is called the **consequent**. The connective *if* $\cdots$ *then* is denoted by the symbol $\rightarrow$.

Example 5   Write the implication $p \rightarrow q$.

(a) p: I am hungry; q: I will eat.
(b) p: It is snowing; q: $3 + 5 = 8$.

Solution. (a) If I am hungry, then I will eat.
(b) If it is snowing, then $3 + 5 = 8$.

Example 5(b) shows that in logic we use conditional statements in a more general sense than is customary. Thus in English, when we say "if p then q," we are tacitly assuming that there is a cause-and-effect relationship between p and q. That is, we would never use the statement in Example 5(b) in ordinary English, since there is no way that statement p can have any effect on statement q.

In logic, implication is used in a much weaker sense. To say that the compound statement $p \rightarrow q$ is true simply asserts that if p is true, then q will also be found to be true. In other words, $p \rightarrow q$ says only that we will not have p true and q false at the same time. It does not say that p "causes" q in the usual sense. Table 4 describes the truth values of $p \rightarrow q$ in terms of the truth values of p and q. Notice that $p \rightarrow q$ is considered false only if p is true and q is false. In particular, if p is false, then $p \rightarrow q$ is true for any q. This fact is sometimes described by the statement: " A false hypothesis implies any conclusion," but this statement is misleading, as it seems to imply that if the hypothesis is false, then the conclusion must be true (a clearly silly statement). Similarly, if q is true, then $p \rightarrow q$ will be true for any statement p. The implication

"If $2 + 2 = 5$, then I am king of England"

is true, simply because p ("$2 + 2 = 5$") is false, so it is not the case that p is true and q is false simultaneously.

Table 4

p	q	$p \rightarrow q$
T	T	T
T	F	F
F	T	T
F	F	T

In the English language, and in mathematics, each of the following expressions is an equivalent form of the conditional statement $p \rightarrow q$:

p implies q

q, if p

p only if q

p is a sufficient condition for q

q is a necessary condition for p

If $p \rightarrow q$ is an implication, then the **converse** of $p \rightarrow q$ is the implication $q \rightarrow p$, and the **contrapositive** of $p \rightarrow q$ is the implication $\sim q \rightarrow \sim p$.

Example 6 Give the converse and contrapositive of the implication "If it is raining, then I get wet."

Solution. We have p: it is raining and q: I get wet. Then the converse is $q \rightarrow p$: If I get wet, then it is raining. The contrapositive is $\sim q \rightarrow \sim p$: If I do not get wet, then it is not raining.

If p and q are statements, the compound statement p if and only if q, denoted by $p \leftrightarrow q$, is called an **equivalence**. The connective *if and only if* is denoted by the symbol $\leftrightarrow$. The truth values of $p \leftrightarrow q$ are given in Table 5. Observe that $p \leftrightarrow q$ is true only when both p and q are true or when both are false. The equivalence $p \leftrightarrow q$ is also stated as p is a necessary and sufficient condition for q.

Table 5

p	q	$p \leftrightarrow q$
T	T	T
T	F	F
F	T	F
F	F	T

Example 7 Is the following equivalence true?

$$3 > 2 \quad \text{if and only if} \quad 0 < 3 - 2$$

Solution. Let p be the statement $3 > 2$ and let q be the statement $0 < 3 - 2$. Since p and q are both true, we conclude that $p \leftrightarrow q$ is true.

In general, a compound statement may have many component parts, each of which is itself a statement, represented by some propositional variable. The statement

$$s: \quad p \rightarrow (q \wedge (p \rightarrow r))$$

involves three propositions, p, q, and r, each of which may independently be true or false. There are altogether $2^3 = 8$ possible combinations of truth values for p, q, and r, and the truth table for s must give the truth or falsity of s in all these cases. If a compound statement s contains n component statements, there will be 2^n entries needed in the truth table for s. Such a truth table may be systematically constructed in the following way.

Steps for Constructing General Truth Tables

Step 1. The first n columns of the table are labeled by the component propositional variables. Further columns are constructed for all intermediate combinations of statements, culminating in the given statement.

Step 2. Under each of the first n headings, we list the 2^n possible n-tuples of truth values of the component statement s. Each n-tuple is listed on a separate row.

Step 3. For each row we compute, in sequence, all remaining truth values.

Example 8   Compute the truth table of the statement $(p \rightarrow q) \leftrightarrow (\sim q \rightarrow \sim p)$

Solution. The following table is constructed using steps 1, 2, and 3 above.

p	q	$p \rightarrow q$	$\sim q$	$\sim p$	$\sim q \rightarrow \sim p$	$(p \rightarrow q) \leftrightarrow (\sim q \rightarrow \sim p)$
T	T	T	F	F	T	T
T	F	F	T	F	F	T
F	T	T	F	T	T	T
F	F	T	T	T.	T	T

A statement that is true for all possible values of its propositional variables is called a **tautology**. A statement that is always false is called a **contradiction** or an **absurdity**, and a statement that can be either true or false, depending on the truth values of its propositional variables, is called a **contingency**.

Example 9
(a) The statement in Example 8 is a tautology.
(b) The statement $p \wedge \sim p$ is an absurdity (verify).
(c) The statement $(p \rightarrow q) \wedge (p \vee q)$ is a contingency.

In Table 6 we have listed a number of important tautologies involving equivalences, all of which can be proved by the use of a truth table. When an equivalence is shown to be a tautology, it means that its two component parts are always either both true or both false, for any values of the propositional variables. Thus the two sides are simply different ways of making the same statement and we say that they are logically **equivalent**.

Tautologies 6, 10, and 11 of Table 6 give equivalent statements for the negation of the compound statements $p \wedge q, p \vee q, p \to q$, and $p \leftrightarrow q$. Tautology 8 of Table 6 shows us that any implication is logically equivalent with its contrapositive. We will use this property in proofs. When attempting to prove an implication, it is equivalent, and sometimes simpler, to prove the contrapositive.

Table 6 Tautologies

1. $p \wedge p \leftrightarrow p$	$p \vee p \leftrightarrow p$	(Idempotent property)
2. $p \wedge q \leftrightarrow q \wedge p$	$p \vee q \leftrightarrow q \vee p$	(Commutative property)
3. $(p \wedge q) \wedge r \leftrightarrow p \wedge (q \wedge r)$	$(p \vee q) \vee r \leftrightarrow p \vee (q \vee r)$	(Associative property)
4. $p \wedge (q \vee r) \leftrightarrow (p \wedge q) \vee (p \wedge r)$	$p \vee (q \wedge r) \leftrightarrow (p \vee q) \wedge (p \vee r)$	(Distributive property)
5. $\sim\sim p \leftrightarrow p$		
6. $\sim(p \wedge q) \leftrightarrow (\sim p) \vee (\sim q)$	$\sim(p \vee q) \leftrightarrow (\sim p) \wedge (\sim q)$	(De Morgan's laws)
7. $(p \to q) \leftrightarrow ((\sim p) \vee q)$		
8. $(p \to q) \leftrightarrow (\sim q \to \sim p)$		
9. $(p \leftrightarrow q) \leftrightarrow [(p \to q) \wedge (q \to p)]$		
10. $\sim(p \to q) \leftrightarrow p \wedge \sim q$		
11. $\sim(p \leftrightarrow q) \leftrightarrow [(p \wedge \sim q) \vee (q \wedge \sim p)]$		

Table 7 lists several important tautologies that are implications. These are used extensively in proving results in mathematics and computer science and we will illustrate them below.

Table 7

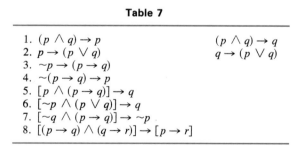

1. $(p \wedge q) \to p$ $(p \wedge q) \to q$
2. $p \to (p \vee q)$ $q \to (p \vee q)$
3. $\sim p \to (p \to q)$
4. $\sim(p \to q) \to p$
5. $[p \wedge (p \to q)] \to q$
6. $[\sim p \wedge (p \vee q)] \to q$
7. $[\sim q \wedge (p \to q)] \to \sim p$
8. $[(p \to q) \wedge (q \to r)] \to [p \to r]$

Methods of Proof

If an implication $p \to q$ is a tautology where p and q may be compound statements involving any number of logical variables, we say that q **logically follows** from p. Suppose that an implication of the form $(p_1 \wedge p_2 \wedge \cdots \wedge p_n) \to q$ is a tautology. Then this implication is true regardless of the truth values of any of its components.

In this case, we say that q **logically follows** from $p_1, p_2, \ldots, p_n$. When q logically follows from $p_1, p_2, \ldots, p_n$, we write

$$p_1$$

$$p_2$$

$$\cdot$$
$$\cdot$$
$$\cdot$$

$$\underline{p_n}$$

$$\therefore q$$

(where the symbol $\therefore$ means "therefore"). This means that if we know that p_1 is true, p_2 is true, $\ldots$, and p_n is true, then we know q is true.

Virtually all mathematical theorems are composed of implications of the type

$$(p_1 \wedge p_2 \wedge \cdots \wedge p_n) \rightarrow q.$$

The p_i's are called the **hypotheses**, and q is called the **conclusion**. To "prove the theorem" means to show that the *implication* is a tautology. Note that we are not trying to show that q (the conclusion) is true, but only that q will be true if the p_i are all true. For this reason, mathematical proofs often begin with the statement, "suppose that $p_1, p_2, \ldots,$ and p_n are true" and conclude with the statement "therefore, q is true." The proof does not show that q is true but simply shows that q has to be true if the p_i are all true.

Arguments based on tautologies represent universally correct methods of reasoning. Their validity depends only on the form of the statements involved and not on the truth values of the variables they contain. Such arguments are often called **rules of inference**. The various steps in the mathematical proof of a theorem must follow from the use of various rules of inference, and a mathematical proof of a theorem must begin with the hypotheses, proceed through various steps, each justified by some rule of inference, and arrive at the conclusion.

Example 10   According to item 8 of Table 7, $[(p \rightarrow q) \wedge (q \rightarrow r)] \rightarrow (p \rightarrow r)$ is a tautology. Thus the argument

$$p \rightarrow q$$

$$\underline{q \rightarrow r}$$

$$\therefore p \rightarrow r$$

is universally valid, and so is a rule of inference.

Example 11 Is the following argument valid?

> If you invest in the stock market, then you will get rich.
> If you get rich, then you will be happy.
> ――――――――――――――――――――――――――――――――――――
> ∴ If you invest in the stock market, then you will be happy.

Solution. The argument is of the form

$$p \rightarrow q$$
$$q \rightarrow r$$
$$\overline{}$$
$$\therefore p \rightarrow r$$

Hence the argument is valid, although the conclusion may be false.

Example 12 The tautology $(p \leftrightarrow q) \leftrightarrow [(p \rightarrow q) \wedge (q \rightarrow p)]$ is given in Table 6, number 9. Thus both of the following arguments are valid:

$$p \leftrightarrow q \qquad\qquad\qquad p \rightarrow q$$
$$\therefore (p \rightarrow q) \wedge (q \rightarrow p) \qquad q \rightarrow p$$
$$\therefore p \leftrightarrow q$$

Some mathematical theorems are equivalences, that is, they are of the form $p \leftrightarrow q$. They are usually stated "p if and only if q." By Example 12, the proof of such a theorem is logically equivalent with proving both $p \rightarrow q$ and $q \rightarrow p$, and this is almost always the way in which equivalences are proved. We first asume that p is true, and show that q must then be true; next we assume that q is true and show that p must then be true.

A very important rule of inference is

$$p$$
$$p \rightarrow q$$
$$\overline{}$$
$$\therefore q$$

That is, "p is true, and $p \rightarrow q$ is true, so q is true." This follows from tautology number 5 in Table 7.

Some rules of inference were given Latin names by classical scholars. Tautology 5 is referred to as **modus ponens**, or loosely, the method of asserting.

Example 13 Is the following argument valid?

> Smoking is healthy.
> If smoking is healthy, then cigarettes are prescribed by physicians.
> ∴ Cigarettes are prescribed by physicians.

Solution. The argument is valid since it is of the form modus ponens, discussed above. However, the conclusion is false. Observe that the first premise p: "smoking is healthy" is false. The second premise $p \rightarrow q$ is then true and $[p \wedge (p \rightarrow q)]$, the conjunction of the two premises is false.

Example 14 Is the following argument valid?

> If taxes were lowered, then income rises.
> Income rises.
> ∴ Taxes were lowered.

Solution. Let p and q be the statements

$$p: \text{ taxes were lowered} \qquad q: \text{ income rises}$$

Then the argument is of the form

$$p \rightarrow q$$
$$\underline{q}$$
$$\therefore p$$

Assume that $p \rightarrow q$ and q are both true. Now $p \rightarrow q$ may be true with p being false. Then the conclusion p is false. Hence the argument is not valid. Another approach is to ascertain whether the statement $[(p \rightarrow q) \wedge q]$ logically implies the statement p. A truth table shows that this is not the case.

An important technique of proof called the **indirect method** of proof follows from the tautology

$$(p \rightarrow q) \leftrightarrow [(\sim q) \rightarrow (\sim p)]$$

This states, as we have mentioned previously, that an implication is equivalent with its contrapositive. Thus to prove $p \rightarrow q$ indirectly, we assume that q is false (the statement $\sim q$) and show that p is then false (the statement $\sim p$).

Example 15   Let n be an integer. Prove that if n^2 is odd, then n is odd.

Solution. Let p and q be the statements "n^2 is odd" and "n is odd," respectively. We have to prove that $p \rightarrow q$ is true. Instead, we prove the contrapositive $\sim q \rightarrow \sim p$. Thus suppose that n is not odd, so that n is even. Then $n = 2k$, where k is an integer. We have $n^2 = (2k)^2 = 4k^2 = 2(2k^2)$, so n^2 is even. We thus show that if n is even, then n^2 is even, which is the contrapositive of the given statement. Hence the given statement has been proved.

Another important technique of proof is **proof by contradiction**. This method is based on the tautology

$$[(p \rightarrow q) \wedge (\sim q)] \rightarrow (\sim p)$$

Thus the rule of inference

$$p \rightarrow q$$
$$\underline{\sim q}$$
$$\therefore \sim p$$

is valid. Informally, this states that if a statement p implies a false statement q, then p must be false. This is often applied to the case where q is an absurdity or contradiction, that is, a statement which is always false, no matter what truth values its propositional variables have. An example is given by taking q as the contradiction $r \wedge (\sim r)$. Thus any statement that implies a contradiction must be false. In order to use proof by contradiction, suppose we wish to show that a statement q logically follows from statements $p_1, p_2, \ldots, p_n$. Assume that $\sim q$ is true (that is, q is false) as an extra hypothesis, and that $p_1, p_2, \ldots, p_n$ are also true. If this enlarged hypothesis $p_1 \wedge p_2 \wedge \cdots \wedge p_n \wedge (\sim q)$ implies a contradiction, then at least one of the statements $p_1, p_2, \ldots, p_n, (\sim q)$ must be false. This means that if all p_i's are true, then $\sim q$ must be false, so q must be true. Thus q follows from $p_1, p_2, \ldots, p_n$. This is proof by contradiction.

Example 16   Prove that there is no rational number p/q whose square is 2. (In other words, $\sqrt{2}$ is irrational.)

Proof. Assume that $(p/q)^2 = 2$ for some integers p and q, with no common factors. (If the latter condition is not true, we can make it true by canceling common factors). Then $p^2 = 2q^2$, so p^2 is even. This implies that p is even, since the square of an odd number is odd. Thus $p = 2n$ for some integer n. We see that $2q^2 = p^2 = (2n)^2 = 4n^2$, and so $q^2 = 2n^2$. Thus, q^2 is even, and so q is even. We see that p and q are both even, and therefore have 2 as a common factor. This is a contradiction to the assumptions. Thus the assumptions must be false.

We have presented several rules of inference and logical equivalences which correspond to valid techniques of proof. In order to prove a theorem of the (typical) form $(p_1 \wedge p_2 \wedge \cdots \wedge p_n) \to q$, we begin with the hypotheses $p_1, p_2, \ldots, p_n$ and show that some result r_1, logically follows. Then, using $p_1, p_2, \ldots, p_n, r_1$, we show that some other statement r_2 logically follows. We continue this process, producing intermediate statements $r_1, r_2, \ldots, r_k$, called **steps in the proof**, until we can finally show that the conclusion q logically follows from $p_1, p_2, \ldots, p_n$, $r_1, \ldots, r_k$. Each logical step must be justified by some valid technique of proof, based on the rules of inference we have developed, or on other rules which come from tautological implications we have not discussed. At any stage, we may replace a statement that needs to be derived by its contrapositive statement, or by any other equivalent form.

In practice, the construction of proofs is an art, and must be learned in part from observation and experience. The choice of intermediate steps and methods of deriving them is a creative activity which cannot be precisely described.

Example 17   Let n and m be integers. Prove that $n^2 = m^2$ if and only if $n = m$ or $n = -m$.

Let us analyze this proof as we present it. Suppose that p is the statement "$n^2 = m^2$,"q is the statement "$n = m$," and r is the statement "$n = -m$." Then we are asked to prove the theorem

$$p \leftrightarrow (q \vee r)$$

We know from previous discussion that we may instead prove that the two statements $s: p \to (q \vee r)$ and $t: (q \vee r) \to p$ are true. Let us begin by showing t to be true. Thus we assume that either $q: n = m$ or $r: n = -m$ is true. If q is true, then $n^2 = m^2$, and if r is true, then $n^2 = (-m)^2 = m^2$, so in either case p is true. We have therefore shown that the implication $t: q \vee r \to p$ is true.

Now we must prove that $s: p \to q \vee r$ is true; that is, we assume p and try to prove q or r. If p is true, then $n^2 = m^2$, so $n^2 - m^2 = (n - m)(n + m) = 0$. If r_1 is the intermediate statement $(n - m)(n + m) = 0$, we have shown that $p \to r_1$ is true. We now show that $r_1 \to (q \vee r)$ is true, by showing that the contrapositive statement $\sim(q \vee r) \to (\sim r_1)$ is true. Now $\sim(q \vee r)$ is equivalent with $(\sim q) \wedge (\sim r)$, so we show that $(\sim q) \wedge (\sim r) \to (\sim r_1)$. Thus, if $(\sim q): n \neq m$ and $(\sim r): n \neq -m$ are true, then $(n - m) \neq 0$ and $(n + m) \neq 0$, so $(n - m)(n + m) \neq 0$ and r_1 is false. We have therefore shown that $r_1 \to (q \vee r)$ is true. Finally, from the truth of $p \to r_1$ and $r_1 \to (q \vee r)$, we can conclude that $p \to (q \vee r)$ is true, and we are done.

We do not usually analyze proofs in this detailed a manner. We have done so only to illustrate that proofs are devised by piecing together equivalences and valid steps resulting from rules of inference.

As a final remark, we caution the reader that many mathematical theorems actually state that some statement is true for all objects of a certain type. Sometimes this is not evident. Thus the theorem given in Example 17 really states that for all integers n and m, $n^2 = m^2$ if and only if $n = m$ or $n = -m$.

Similarly, the statement "If x and y are real numbers, and $x \neq y$, then $x < y$ or $y < x$" is a statement about all real numbers x and y. To prove such a theorem, we must make sure that the steps in the proof are each valid for every number. We could not assume, for example, that x was 2, or that y was π, or $\sqrt{3}$.

On the other hand, in order to prove that such a theorem is false, we need only find a single example where the property fails.

Example 18   Prove or disprove the statement that if x and y are real numbers,

$$(x^2 = y^2) \rightarrow (x = y)$$

To prove this result, we would need to provide steps, each of which would be true for all x and y. To disprove the result, we only need to find one example for which the implication is false. This is because the theorem claims to be true for all x and y.

Since $(-3)^2 = 3^2$, but $-3 \neq 3$, the theorem is false. This is called a **counterexample**, and any other counterexample would do just as well.

In summary, if a theorem claims that a property holds for all objects of a certain type, then to prove it, we must use steps that are valid for all objects of that type, and that do not make reference to any particular object. To disprove the theorem, we need only show one counterexample, that is, one particular object or set of objects for which the claim fails.

EXERCISE SET

1. Which of the following are statements?
   - (a) Is 2 a positive number?
   - (b) $x^2 + x + 1 = 0$
   - (c) Study logic.
   - (d) There will be snow in January.
   - (e) If stocks fall, then I will lose money.

2. Give the negations of the following statements.
   - (a) $2 + 7 \leq 11$
   - (b) 2 is an even integer and 8 is an odd integer.
   - (c) It will rain tomorrow or it will not snow tomorrow.
   - (d) If you drive, then I will walk.

3. In each of the following, form the conjunction and the disjunction of p and q.
   - (a) p: $3 + 1 < 5$; q: $7 = 3 \times 6$
   - (b) p: I am rich; q: I am happy.
   - (c) p: I will drive my car; q: I will be late.

4. Determine the truth of falsity of each of the following statements.
   - (a) $2 < 3$ and 3 is a positive integer.
   - (b) $2 \geq 3$ and 3 is a positive integer.
   - (c) $2 < 3$ and 3 is not a positive integer.
   - (d) $2 \geq 3$ and 3 is not a positive integer.

5. Determine the truth or falsity of each of the following statements.
  (a) $2 < 3$ or 3 is a positive integer.
  (b) $2 \geq 3$ or 3 is a positive integer.
  (c) $2 < 3$ or 3 is not a positive integer.
  (d) $2 \geq 3$ or 3 is not a positive integer.

6. Which of the following statements is the negation of the statements "2 is even and -3 is negative"?
  (a) 2 is even and -3 is not negative.
  (b) 2 is odd and -3 is not negative.
  (c) 2 is even or -3 is not negative.
  (d) 2 is odd or -3 is not negative.

7. Which of the following statements is the negation of the statement "2 is even or -3 is negative"?
  (a) 2 is even or -3 is not negative.
  (b) 2 is odd or -3 is not negative.
  (c) 2 is even and -3 is not negative.
  (d) 2 is odd and -3 is not negative.

8. State the converse of each of the following implications.
  (a) If $2 + 2 = 4$, then I am not King of England.
  (b) If I am not President of the United States, then I will walk to work.
  (c) If I am late, then I did not take the train to work.
  (d) If I have time and I am not too tired, then I will go to the store.
  (e) If I have enough money, then I will buy a car and I will buy a house.

9. State the contrapositive of each implication in Exercise 8.

10. Determine the truth or falsity of each of the following statements.
  (a) If 2 is even, then New York has a large population.
  (b) If 2 is even, then New York has a small population.
  (c) If 2 is odd, then New York has a large population.
  (d) If 2 is odd, then New York has a small population.

In Exercises 11 and 12, let p, q, and r be the following statements;
  p: I will study discrete structures.
  q: I will go to a movie.
  r: I am in a good mood.

11. Write the following statements in terms of p, q, and r and logical connectives.
  (a) If I am not in a good mood, then I will go to a movie.
  (b) I will not go to a movie, and I will study discrete structures.
  (c) I will go to a movie only if I will not study discrete structures.
  (d) If I will not study discrete structures, then I am not in a good mood.

12. Write the English sentences corresponding to the following statements.
  (a) $(\sim p) \wedge q$
  (b) $r \rightarrow (p \vee q)$
  (c) $(\sim r) \rightarrow ((\sim q) \vee p)$
  (d) $(q \wedge (\sim p)) \leftrightarrow r$

13. By examining truth tables, determine whether each of the following is a tautology, a contingency, or an absurdity.
  (a) $p \wedge \sim p$
  (b) $p \rightarrow (q \rightarrow p)$
  (c) $q \rightarrow (q \rightarrow p)$
  (d) $q \vee (\sim q \wedge p)$
  (e) $(q \wedge p) \vee (q \wedge \sim p)$
  (f) $(p \wedge q) \rightarrow p$
  (g) $p \rightarrow (q \wedge p)$

In Exercises 14–20, state whether the arguments given are valid or are not valid. If they are valid, identify the tautology or tautologies used.

14. If I drive to work, then I will arrive tired.
I am not tired when I arrive at work.

∴ I do not drive to work.

15. If I drive to work, then I arrive at work tired.
I arrive at work tired.

∴ I drive to work.

16. If I drive to work, then I will arrive tired.
    I do not drive to work.

    ∴ I will not arrive tired.

17. If I drive to work, then I will arrive tired.
    I drive to work.

    ∴ I will arrive tired.

18. I will become famous or I will not become a writer.
    I will become a writer.

    ∴ I will become famous.

19. I will become famous or I will be a writer.
    I will not be a writer.

    ∴ I will become famous.

20. If I try hard and I have talent, then I will become
    a musician.
    If I become a musician, then I will be happy.

    ∴ If I will not be happy, then I did not try hard, or
    I do not have talent.

KEY IDEAS FOR REVIEW

☐ Statement: declarative sentence that is either true or false, but not both.

☐ Propositional variable: letter denoting a statement.

☐ Compound statement: statement obtained by combining two or more statements by a logical connective.

☐ Logical connectives: not ($\sim$), and ($\wedge$), or ($\vee$), if then ($\rightarrow$), if and only if ($\leftrightarrow$).

☐ Conjunction: $p \wedge q$ (p and q).

☐ Disjunction: $p \vee q$ (p or q).

☐ Conditional statement or implication: $p \rightarrow q$ (if p then q); p is the antecedent or hypothesis, q is the consequent or conclusion.

☐ Converse of $p \rightarrow q$: $q \rightarrow p$.

☐ Contrapositive of $p \rightarrow q$: $\sim q \rightarrow \sim p$.

☐ Equivalence: $p \leftrightarrow q$.

☐ Tautology: statement that is true for all possible values of its propositional variables.

☐ Absurdity: statement that is false for all possible values of its propositional variables.

☐ Contingency: statement that may be true or false, depending on the truth values of its propositional variables.

☐ Logically equivalent statements p and q: $p \leftrightarrow q$.

☐ Methods of proof:
   q logically follows from p: see page 318.
   Rules of inference: see page 319.
   Modus ponens: see page 320.
   Indirect method: see page 321.
   Proof by contradiction: see page 322.

☐ Counterexample: single instance that disproves a theorem or proposition.

Answers
to Odd-Numbered
Exercises

CHAPTER 1

Exercise Set 1.1, page 6

1. (a) T (b) F (c) T (d) F (e) T (f) F
3. (a) {A, R, D, V, K} (b) {B, O, K} (c) {M, I, S, P}
5. (a) $\{x \mid x$ is a positive even integer and $x \leq 10\}$
   (b) $\{x \mid x$ is a vowel of the English alphabet$\}$
   (c) $\{x \mid x$ is a square of a positive integer $y \leq 6\}$
   (d) $\{x \mid x \in Z$ and $|x| \leq 2\}$.
7. (b) (c) and (e).
9. $\varnothing$ {ALGOL}, {BASIC}, {FORTRAN}, {ALGOL, BASIC}, {ALGOL, FORTRAN}, {BASIC, FORTRAN}, {ALGOL, BASIC, FORTRAN}
11. (a) T (b) F (c) T (d) F (e) T (f) T (g) T (h) T
13. (a) $\subseteq$ (b) $\subseteq$ (c) $\not\subseteq$ (d) $\subseteq$ (e) $\not\subseteq$ (f) $\subseteq$
15. (a) F (b) T (c) F (d) T (e) T (f) F

Exercise Set 1.2, page 11

1. {2, 1}   3. $\{x \mid x \in Z^+$ and x is odd$\}$ 5. $\{a, b, c, d\}$ 7. 2, 4, 8, 16
9. $-1, 10, 27, 50$ 11. $x, y, z; x, x, y, y, z, z; x, y, z, x, y, z, x, \ldots$
13. $2n - 1$ 15. $(-1)^{n+1}$ 17. $3n - 2$

Exercise Set 1.3, page 20

1. (a) $\{a, b, c, d, e, f, g\}$ (b) $\{a, c, d, e, f, g\}$ (c) $\{a, c\}$ (d) $\{f\}$
   (e) $\{a, b, c\}$ (f) $\{d, e, f, h, k\}$ (g) $\{a, b, c, d, e, f\}$ (h) $\{b, f, g\}$
3. (a) $\{a, b, c, d, e, f, g\}$ (b) $\varnothing$ (c) $\{a, c, g\}$
   (d) $\{a, c, f\}$ (e) $\{h, k\}$ (f) $\{a, b, c, d, e, f, h, k\}$

5. (a) $\{1, 2, 4, 5, 6, 8, 9\}$ (b) $\{1, 2, 3, 4, 6, 8\}$ (c) $\{1, 2, 4\}$ (d) $\varnothing$
    (e) $\{1, 6, 8\}$ (f) $\{5, 9\}$ (g) $\{3, 5, 7, 9\}$ (h) $\{1, 5, 6, 8, 9\}$
7. (a) $\{1, 2, 3, 4, 5, 6, 8, 9\}$ (b) $\{1, 2, 3, 4, 5, 7, 8, 9\}$ (c) $\{2, 4\}$
    (d) $\varnothing$ (e) $\{1, 2, 4, 6, 8\}$ (f) $\{\varnothing\}$
9. (a) $\{b, d, e, h\}$ (b) $\{b, c, d, f, g, h\}$ (c) $\{b, d, h\}$
    (d) $\{b, c, d, e, f, g, h\}$ (e) $\varnothing$ (f) $\{c, f, g\}$
11. (a) $\{x \mid x$ is a real number and $x \neq \pm 1\}$.   (b) $\{x \mid x$ is a real number and $x \neq -1, 4\}$.
    (c) $\{x \mid x$ is a real number and $x \neq \pm 1, 4\}$.   (d) $\{x \mid x$ is a real number and $x \neq -1\}$.
13. (a) T   (b) T   (c) F   (d) F
15. (a) 10101011   (b) 00001010   (c) 10101010   (d) 11011111   (e) 00110000

Review Exercises, page 22

1. (a) $\{1, 2, 3\}$ (b) $\{2, 3, 5, 7, 11, 13, 17, 19, 23, 29, 31, 37, 41, 43, 47\}$
    (c) $\{2\}$ (d) $\{1, 2, 3, 4, 6\}$
    (e) $\{\pm \frac{1}{3}, \pm \frac{1}{4}, \pm \frac{1}{5}, \pm \frac{1}{6}, \pm \frac{1}{7}, \pm \frac{1}{8}, \pm \frac{1}{9}, \pm \frac{2}{5}, \pm \frac{2}{7}, \pm \frac{2}{9}, \pm \frac{3}{7}, \pm \frac{3}{8}, \pm \frac{4}{9}\}$
    (f) $\{a, n, t, i, d, s, e, b, l, h, m, r\}$
3. (a) $\not\subseteq$ (b) $\subseteq$ (c) $\not\subseteq$ (d) $\in$
    (e) $\notin$ (f) $\notin$ (g) $\subseteq$ (h) $\not\subseteq$
5. $n^2 - 1$
7. (a) $\varnothing$ (b) $\{1, 3, 4\}$ (c) $\{3\}$ (d) $\{2, 4, 6\}$
9. $(A \cap B) \cup C$

Chapter Test, page 23

1. $x = 2, y = 3$ or $x = 3, y = 2$
2. (c) (d)
3. (a) $\{1, 2, 9\}$ (b) $\{5, 6, 9\}$ (c) $\{6\}$ (d) $\{9\}$
4. $\{0, 1, 7\}$
5. $a_n = 3n - 2$
6. (a) F   (b) T   (c) T   (d) F   (e) F   (f) F   (g) T

CHAPTER 2

Exercise Set 2.1, page 27

3. 13   7. $|B| = 6$ 9. $B \subseteq A$ 11. $|B| = 7$ 13. $|B| = 4$ 15. (a) 20   (b) 325

Exercise Set 2.2, page 35

1. 67,600   3. 16   5. 1296   7. (a) 24   (b) 120   (c) 720
9. (a) $12! = 479,001,600$ (b) $2 \cdot (6!)^2 = 1,036,800$ 11. 120   13. 120
15. (a) 1   (b) 35   (c) 6   (d) n (e) $\dfrac{n(n-1)}{2}$ (f) $\dfrac{n(n+1)}{2}$
17. 980   19. $C_7^{12} \cdot C_7^{15} = 5,096,520$ 21. $C_5^{26} \cdot C_3^{26} = 171,028,000$
23. (a) $C_5^8 = 56$ (b) $C_5^7 = 21$ (c) $C_2^8 \cdot C_3^7 = 980$ (d) $C_3^8 \cdot C_2^7 = 1176$

Review Exercises, page 40

1. $|A| = 5, |B| = 5, |C| = 5$
   $|A \cap B| = 1, |A \cap C| = 0, |B \cap C| = 2, |A \cap B \cap C| = 0, |A \cup B \cup C| = 12$

3. 23   5. (a) 120   (b) 24   (c) 720   7. $\binom{26}{4}\binom{26}{2} = 4{,}858{,}750$

9. $\binom{6}{2}\binom{9}{6} = 1260$

Chapter Test, page 41

1. 4   2. 6840   3. $\binom{26}{2}\binom{26}{3} = 2925$ 4. 650,000

5. $\dfrac{100}{6} = 16 + \text{a remainder of } 4$ 6. (a) T   (b) F   (c) T   (d) F

CHAPTER 3

Exercise Set 3.1, page 48

23. $\{x \mid x = 3k, \, k \in Z, \, k \geq 0\}$ 25. 0, 1, 1, 3, 5, 11

Exercise Set 3.2, page 57

1. $20 = 6 \cdot 3 + 2$ 3. $3 = 0 \cdot 22 + 3$ 5. $8 = (-2)(-3) + 2$ 7. (b), (c), (h)
9. $1; \, 1 = 11(32) - 13(27)$ 11. $8; \, 8 = 1(88) - 2(40)$ 13. $2; \, 2 = 12(34) - 7(58)$
15. 216   17. 1225

Exercise Set 3.3, page 64

1. (a) $a_{12} = -2, a_{22} = 1, a_{23} = 2$ (b) $b_{11} = 3, b_{31} = 4$
   (c) $c_{13} = 4, c_{21} = 5, c_{33} = 8$ (d) 2, 6, 8
3. $a = 3, b = 1, c = 8, d = -2$
5. (a) $\begin{bmatrix} 4 & 0 & 2 \\ 9 & 6 & 2 \\ 3 & 2 & 4 \end{bmatrix}$ (b) $\mathbf{AB} = \begin{bmatrix} 7 & 13 \\ -3 & 0 \end{bmatrix}$ $\mathbf{BA} = \begin{bmatrix} 4 & 1 & -2 \\ 10 & 3 & -1 \\ 16 & 5 & 0 \end{bmatrix}$

   (c) Not possible   (d) $\begin{bmatrix} 21 & 14 \\ -7 & 17 \end{bmatrix}$

7. (a) Not possible   (b) Not possible   (c) $\begin{bmatrix} 10 & 0 & -25 \\ 40 & 14 & 12 \end{bmatrix}$ (d) Not possible

9. (a) $\begin{bmatrix} 22 & 34 \\ 3 & 11 \\ -31 & 3 \end{bmatrix}$ (b) Not possible   (c) $\begin{bmatrix} 25 & 5 & 26 \\ 20 & -3 & 32 \end{bmatrix}$ (d) Not possible

11. $\begin{bmatrix} -8 & 9 \\ 1 & 22 \\ 10 & 19 \end{bmatrix}$ 13. (a) $\begin{bmatrix} 27 & 0 & 0 \\ 0 & -8 & 0 \\ 0 & 0 & 64 \end{bmatrix}$ (b) $\begin{bmatrix} 3^k & 0 & 0 \\ 0 & (-2)^k & 0 \\ 0 & 0 & 4^k \end{bmatrix}$

15.
$$\mathbf{A} = \begin{bmatrix} 1 & 0 \\ 0 & 0 \end{bmatrix} \quad \mathbf{B} = \begin{bmatrix} 0 & 0 \\ 0 & 1 \end{bmatrix}$$

Exercise Set 3.4, page 69

1. (a)
$$\mathbf{A} \vee \mathbf{B} = \begin{bmatrix} 1 & 1 \\ 0 & 1 \end{bmatrix}, \mathbf{A} \wedge \mathbf{B} = \begin{bmatrix} 1 & 0 \\ 0 & 1 \end{bmatrix}, \mathbf{A} \odot \mathbf{B} = \begin{bmatrix} 1 & 1 \\ 0 & 1 \end{bmatrix}$$

(b)
$$\mathbf{A} \vee \mathbf{B} = \begin{bmatrix} 1 & 1 & 1 \\ 0 & 1 & 1 \\ 1 & 0 & 1 \end{bmatrix}, \mathbf{A} \wedge \mathbf{B} = \begin{bmatrix} 1 & 0 & 0 \\ 0 & 0 & 1 \\ 1 & 0 & 0 \end{bmatrix}, \mathbf{A} \odot \mathbf{B} = \begin{bmatrix} 1 & 1 & 1 \\ 1 & 0 & 1 \\ 1 & 1 & 1 \end{bmatrix}$$

3. (a)
$$\mathbf{A} \vee \mathbf{B} = \begin{bmatrix} 1 & 1 \\ 1 & 1 \end{bmatrix}, \mathbf{A} \wedge \mathbf{B} = \begin{bmatrix} 0 & 0 \\ 1 & 0 \end{bmatrix}, \mathbf{A} \odot \mathbf{B} = \begin{bmatrix} 1 & 0 \\ 1 & 0 \end{bmatrix}$$

(b)
$$\mathbf{A} \vee \mathbf{B} = \begin{bmatrix} 1 & 1 & 1 \\ 1 & 1 & 1 \\ 1 & 0 & 1 \end{bmatrix}, \mathbf{A} \wedge \mathbf{B} = \begin{bmatrix} 1 & 0 & 0 \\ 0 & 0 & 1 \\ 1 & 0 & 0 \end{bmatrix}, \mathbf{A} \odot \mathbf{B} = \begin{bmatrix} 1 & 1 & 1 \\ 1 & 0 & 0 \\ 1 & 1 & 1 \end{bmatrix}$$

Review Exercises, page 70

3. $\mathbf{A}_2 = \begin{bmatrix} 2 & 0 \\ -1 & 2 \end{bmatrix}, \mathbf{A}_3 = \begin{bmatrix} 3 & 0 \\ -3 & 3 \end{bmatrix}, \mathbf{A}_4 = \begin{bmatrix} 7 & 0 \\ -7 & 7 \end{bmatrix}$

5. GCD = 30, LCM = 900

7. (a) Impossible   (b) $\begin{bmatrix} 3 & 10 \\ 30 & 14 \end{bmatrix}$ (c) Impossible   9. $\begin{bmatrix} 18 & 29 \\ -4 & 102 \end{bmatrix}$

11.
$$\mathbf{A} \vee \mathbf{B} = \begin{bmatrix} 1 & 1 & 1 \\ 1 & 1 & 1 \\ 1 & 1 & 1 \end{bmatrix}, \mathbf{A} \wedge \mathbf{B} = \begin{bmatrix} 1 & 0 & 0 \\ 0 & 0 & 1 \\ 0 & 0 & 0 \end{bmatrix}, \mathbf{A} \odot \mathbf{B} = \begin{bmatrix} 1 & 1 & 1 \\ 1 & 0 & 1 \\ 1 & 1 & 1 \end{bmatrix}$$

Chapter Test, page 71

2. GCD = 7, LCM = 7140   3. GCD = 7, 7 = 27(525) − 23(616)

4. $\begin{bmatrix} 0 & 2 \\ 12 & -2 \end{bmatrix}$ 5. $\begin{bmatrix} 3 & 11 \\ 0 & 8 \\ 10 & 5 \end{bmatrix}$

6.
$$\mathbf{A} \vee \mathbf{B} = \begin{bmatrix} 1 & 1 & 1 \\ 1 & 1 & 0 \\ 1 & 0 & 1 \end{bmatrix}, \mathbf{A} \wedge \mathbf{B} = \begin{bmatrix} 1 & 0 & 0 \\ 1 & 1 & 0 \\ 0 & 0 & 1 \end{bmatrix}, \mathbf{A} \odot \mathbf{B} = \begin{bmatrix} 1 & 1 & 1 \\ 1 & 1 & 0 \\ 1 & 1 & 1 \end{bmatrix}$$

7. (a) F   (b) T   (c) F   (d) T   (e) F

CHAPTER 4

Exercise Set 4.1, page 76

1. (a) $x = 4$ (b) $y = 3$ (c) $x = 2$ (d) $y = $ ALGOL

3. (a) {(a, 4), (a, 5), (a, 6), (b, 4), (b, 5), (b, 6)}

    (b) {(4, a), (5, a), (6, a), (4, b), (5, b), (6, b)}

    (c) $\{(a, a), (a, b), (b, a), (b, b)\}$

    (d) $\{(4, 4), (4, 5), (4, 6), (5, 4), (5, 5), (5, 6), (6, 4), (6, 5), (6, 6)\}$

5. (a) 6   (b) $\{(m, s), (m, m), (m, l), (f, s), (f, m), (f, l)\}$ **7.** 676

9. $\{(a, 1, r), (a, 1, s), (a, 2, r), (a, 2, s), (b, 1, r), (b, 1, s), (b, 2, r), (b, 2, s), (c, 1, r),$
    $(c, 1, s), (c, 2, r), (c, 2, s)\}$

11.

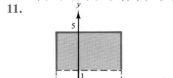

13. No

15. $\{\{a\}, \{b\}, \{c\}, \{d\}\}, \{\{a, b\}, \{c, d\}\}, \{\{a, c\}, \{b, d\}\}, \{\{a, d\}, \{b, c\}\}, \{\{a, b\}, \{c\}, \{d\}\},$
    $\{\{a, c\}, \{b\}, \{d\}\}, \{\{a, d\}, \{b\}, \{c\}\}, \{\{b, c\}, \{a\}, \{d\}\}, \{\{b, d\}, \{a\}, \{c\}\},$
    $\{\{c, d\}, \{a\}, \{b\}\}, \{\{a, b, c\}, \{d\}\} \{\{a, b, d\}, \{c\}\}, \{\{a, c, d\}, \{b\}\}, \{\{b, c, d\}, \{a\}\}, \{\{a, b, c, d\}\}$

Exercise Set 4.2, page 82

1. (b),   (d),   (e),   (f)

3. Dom $(R) = A$, Ran $R = \{1, 2\}$

$$\mathbf{M}_R = \begin{bmatrix} 1 & 1 & 0 \\ 1 & 0 & 0 \\ 0 & 1 & 0 \\ 1 & 0 & 0 \end{bmatrix}$$

5. Dom $(R) = \{$IBM, Univac, Commodore, Atari$\}$
    Ran $(R) = \{370, 1110, 4000, 800\}$

$$\mathbf{M}_R = \begin{bmatrix} 1 & 0 & 0 & 0 & 0 & 0 & 0 \\ 0 & 1 & 0 & 0 & 0 & 0 & 0 \\ 0 & 0 & 1 & 0 & 0 & 0 & 0 \\ 0 & 0 & 0 & 0 & 0 & 1 & 0 \\ 0 & 0 & 0 & 0 & 0 & 0 & 0 \end{bmatrix}$$

7. Dom $(R) = A = $ Ran (R)

$$\mathbf{M}_R = \begin{bmatrix} 1 & 0 & 0 & 0 & 0 \\ 0 & 1 & 0 & 0 & 0 \\ 0 & 0 & 1 & 0 & 0 \\ 0 & 0 & 0 & 1 & 0 \\ 0 & 0 & 0 & 0 & 1 \end{bmatrix}$$

9. Dom $(R) = A = $ Ran (R)

$$\mathbf{M}_R = \begin{bmatrix} 1 & 0 & 0 & 0 & 0 \\ 1 & 1 & 0 & 0 & 0 \\ 1 & 0 & 1 & 0 & 0 \\ 1 & 1 & 0 & 1 & 0 \\ 1 & 1 & 1 & 0 & 1 \end{bmatrix}$$

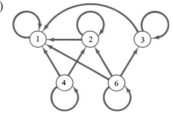

11. Dom $(R) = \{3, 5, 7, 9\}$, Ran $(R) = \{2, 4, 6, 8\}$

$$\mathbf{M}_R = \begin{bmatrix} 0 & 0 & 0 & 0 \\ 1 & 0 & 0 & 0 \\ 1 & 1 & 0 & 0 \\ 1 & 1 & 1 & 0 \\ 1 & 1 & 1 & 1 \end{bmatrix}$$

13. Dom $(R) = \{1, 2, 3, 4\}$, Ran $(R) = \{2, 3, 4, 5\}$

$$\mathbf{M}_R = \begin{bmatrix} 0 & 1 & 0 & 0 & 0 & 0 \\ 0 & 0 & 1 & 0 & 0 & 0 \\ 0 & 0 & 0 & 1 & 0 & 0 \\ 0 & 0 & 0 & 0 & 1 & 0 \\ 0 & 0 & 0 & 0 & 0 & 0 \\ 0 & 0 & 0 & 0 & 0 & 0 \end{bmatrix}$$

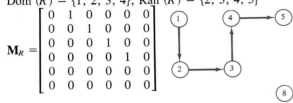

15. Dom $(R) = [-5, 5] =$ Ran (R)

17. (a) $\left\{ \dfrac{3\sqrt{3}}{2}, -\dfrac{3\sqrt{3}}{2} \right\}$ (b) $\varnothing$ (c) $\varnothing$

19. (a) $\{r, u\}$ (b) $\{r, u\}$ (c) $\{r, u, v\}$

21. $R = \{(a_1, b_1), (a_1, b_2), (a_1, b_5), (a_2, b_1), (a_2, b_3), (a_3, b_1), (a_3, b_3), (a_4, b_5), (a_5, b_1),$
    $(a_5, b_5), (a_5, b_6)\}$

Exercise Set 4.3, page 87

1.
$$\mathbf{M}_R = \begin{bmatrix} 1 & 0 & 0 & 0 & 1 \\ 1 & 0 & 1 & 1 & 0 \\ 0 & 1 & 0 & 0 & 0 \\ 0 & 0 & 1 & 0 & 0 \\ 0 & 0 & 0 & 1 & 0 \end{bmatrix}$$

3.

5. $R = \{(1, 1), (1, 3), (2, 3), (3, 1), (4, 1), (4, 2), (4, 4)\}$

7. $R = \{(1, 2), (1, 3), (1, 4), (2, 2), (2, 3), (4, 1), (4, 4), (4, 5)\}$

9. 2^{nm}

11.

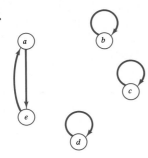

13.

$$\mathbf{M}_R = \begin{bmatrix} 0 & 1 & 0 & 0 & 0 \\ 0 & 1 & 1 & 0 & 0 \\ 0 & 0 & 0 & 1 & 0 \\ 0 & 0 & 0 & 1 & 0 \\ 1 & 0 & 0 & 1 & 0 \end{bmatrix}$$

15.

	Vertex 1	Vertex 2	Vertex 3	Vertex 4	Vertex 5
In-degree	1	2	1	3	0
Out-degree	1	2	1	1	2

Exercise Set 4.4, page 93

1. (1, 2), (1, 6), (2, 3), (3, 3), (3, 4), (4, 1), (4, 3), (4, 5), (6, 4)
3. (3, 3, 4, 1), (3, 3, 3, 3), (3, 3, 4, 5), (3, 4, 3, 4), (3, 4, 1, 2), (3, 3, 3, 4), (3, 4, 1, 6), (3, 4, 3, 3), (3, 3, 4, 3)
5. (6, 4, 1, 6) or (6, 4, 3, 4, 1, 6)
7. (1, 2, 3, 3), (1, 2, 3, 4), (1, 6, 4, 1), (1, 6, 4, 3), (1, 6, 4, 5), (2, 3, 3, 3), (2, 3, 3, 4), (2, 3, 4, 3), (2, 3, 4, 1), (2, 3, 4, 5), (3, 3, 3, 3), (3, 3, 3, 4), (3, 3, 4, 3), (3, 3, 4, 1), (3, 3, 4, 5), (3, 4, 1, 2), (3, 4, 1, 6), (3, 4, 3, 4), (3, 4, 3, 3), (4, 3, 3, 3), (4, 3, 4, 3), (4, 3, 4, 1), (4, 1, 6, 4), (4, 3, 4, 5), (4, 1, 2, 3), (6, 4, 3, 4), (6, 4, 3, 3), (6, 4, 1, 6), (6, 4, 1, 2)

9. $\begin{bmatrix} 0 & 0 & 1 & 1 & 0 & 0 \\ 0 & 0 & 1 & 1 & 0 & 0 \\ 1 & 0 & 1 & 1 & 1 & 0 \\ 0 & 1 & 1 & 1 & 0 & 1 \\ 0 & 0 & 0 & 0 & 0 & 0 \\ 1 & 0 & 1 & 0 & 1 & 0 \end{bmatrix}$

11. $\begin{bmatrix} 1 & 1 & 1 & 1 & 1 & 1 \\ 1 & 1 & 1 & 1 & 1 & 1 \\ 1 & 1 & 1 & 1 & 1 & 1 \\ 1 & 1 & 1 & 1 & 1 & 1 \\ 0 & 0 & 0 & 0 & 0 & 0 \\ 1 & 1 & 1 & 1 & 1 & 1 \end{bmatrix}$

13. (c, d, c), (c, e, f), (c, d, b) 15. (c, d, c) or (c, d, b, f, d, c)
17. (a, b, b), (a, b, f), (a, c, d), (a, c, e), (b, b, b), (b, b, f), (b, f, d), (c, d, c), (c, d, b), (c, e, f), (d, c, d), (d, c, e), (d, b, b), (d, b, f), (e, f, d), (f, d, b), (f, d, c)

19.

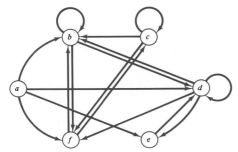

21. $\{(a, b), (a, c), (a, d), (a, e), (a, f), (b, b), (b, c), (b, d), (b, e), (b, f), (c, b), (c, c), (c, d), (c, e), (c, f),$
    $(d, b), (d, c), (d, e), (d, d), (d, f), (e, b), (e, c), (e, d), (e, e), (e, f), (f, b), (f, c), (f, d), (f, e), (f, f)\}$

25. 1, 2, 4, 3, 5, 6, 4

Review Exercises, page 95

1. (a) $\{(2, 0), (3, 1), (2, 4), (5, 0), (5, 1), (5, 4)\}$
   (b) $\{(0, 0), (0, 1), (0, 4), (1, 0), (1, 1), (1, 4), (4, 0), (4, 1), (4, 4)\}$
   (c) $\{(2, 2), (2, 5), (5, 2), (5, 5)\}$
   (d) $\{(0, 2), (0, 5), (1, 2), (1, 5), (4, 2), (4, 5)\}$

3. $\{(1, e, \alpha), (1, e, \beta), (1, f, \alpha), (1, f, \beta), (3, e, \alpha), (3, e, \beta), (3, f, \alpha), (3, f, \beta), (5, e, \alpha), (5, e, \beta),$
   $(5, f, \alpha), (5, f, \beta)\}$

5. $[-5, -2] \times [-3, 2]$

7. Possible answers:
   $\{\{a, b\}, \{c, d\}, \{e, f, g\}\}; \{\{a\}, \{b, c, d\}, \{e, f, g\}\}; \{\{a, b, c, d, e\}, \{f, g\}\}; \{\{a, b\}, \{c\}, \{d, e\}, \{f, g\}\}$

9. Dom $(R) = \{a, b, c\}$, Ran $(R) = \{2, 3, 5, 7\}$,
$$\mathbf{M}_R = \begin{bmatrix} 1 & 1 & 1 & 0 \\ 1 & 0 & 0 & 0 \\ 0 & 0 & 1 & 1 \end{bmatrix}$$

11. $R(1) = \{-\frac{3}{2}\}$
    $R(A_1) = \{-6, 0, \frac{3}{2}, \frac{9}{2}\}$

13. (a) A_1 (b) The set of all odd integers (c) A_1

15. $\{(a, a), (a, d), (b, a), (b, c), (c, c), (c, d), (d, a), (d, c), (d, d)\}$

17.
$$\mathbf{M}_R = \begin{bmatrix} 0 & 1 & 0 & 1 & 0 \\ 1 & 0 & 1 & 0 & 0 \\ 0 & 0 & 0 & 1 & 0 \\ 0 & 0 & 1 & 0 & 0 \\ 1 & 1 & 1 & 1 & 0 \end{bmatrix}$$

19. (a) 4, 4; 7, 7 (b) None (c) 3, 4, 5, 6; 3, 5, 6, 7; 3, 4, 4, 5
    (d) Possible answers: 5, 6, 4, 5; 5, 6, 7, 1, 5
    (e) Possible answers: 4, 4; 4, 5, 6, 4

Chapter Test, page 96

1. $\{(a, 1), (a, 2), (b, 1), (b, 2), (c, 1), (c, 2)\}$

2. Dom $(R) = \{1, 2, 3, 4\}$, Ran $(R) = \{a, b, c\}$,
$$\mathbf{M}_R = \begin{bmatrix} 1 & 1 & 1 \\ 1 & 0 & 0 \\ 0 & 1 & 0 \\ 1 & 0 & 1 \end{bmatrix}$$

3. (a) $\{4\}$ (b) $\{3\}$ (c) $\{4, 5, 6\}$

4. $\{(a, a), (a, d), (b, b), (b, c), (b, d), (c, a), (c, c), (c, d), (d, c), (d, d), (d, e), (e, a), (e, c), (e, e)\}$

5. (a) Possible answers: 1, 2, 6; 1, 2, 3; 1, 4, 4; 1, 4, 3
   (b) Possible answers: 4, 3, 4; 4, 4

6. (a) F (b) F (c) T (d) T (e) F

CHAPTER 5

Exercise Set 5.1, page 105

	Reflexive	Irreflexive	Symmetric	Asymmetric	Antisymmetric	Transitive
1.	Y	N	Y	N	N	Y
3.	N	N	N	N	N	N
5.	N	Y	Y	Y	Y	Y
7.	N	N	N	N	N	Y
9.	N	N	N	N	Y	Y
11.	N	Y	Y	N	N	N
13.	Y	N	N	N	N	N
15.	Y	N	N	N	Y	Y
17.	N	Y	Y	N	N	N
19.	N	N	Y	N	N	N

21.

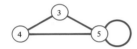

23. $\{(1, 2), (2, 1), (1, 5), (5, 1), (1, 6), (6, 1), (2, 3), (3, 2), (2, 7), (7, 2), (6, 5), (5, 6)\}$

Exercise Set 5.2, page 111

1. Yes **3.** Yes **5.** No **7.** No **9.** Yes **11.** Yes
13. (b) $\{\{(1, 1), (2, 2), (3, 3), (4, 4), (5, 5)\}, \{(1, 2), (2, 4)\}, \{(2, 1), (4, 2)\}, \{(1, 3)\}, \{(3, 1)\}, \{(1, 4)\},$
$\{(4, 1)\}, \{(1, 5)\}, \{(5, 1)\}, \{(2, 3)\}, \{(3, 2)\}, \{(2, 5)\}, \{(5, 2)\}, \{(3, 4)\}, \{(4, 3)\}, \{(3, 5)\}, \{(5, 3)\},$
$\{(4, 5)\}, \{(5, 4)\}\}$
15. $\{(1, 1), (1, 3), (1, 5), (3, 1), (3, 3), (3, 5), (5, 1), (5, 3), (5, 5), (2, 2), (2, 4), (4, 2), (4, 4)\}$

Exercise Set 5.3, page 122

1. (a) $\{(1, 3), (2, 2), (3, 1), (3, 3), (4, 1), (4, 2)\}$
  (b) $\{(1, 1), (1, 2), (1, 3), (2, 1), (2, 3), (3, 2), (3, 3), (4, 3)\}$
  (c) $\{(2, 3), (3, 2)\}$ (d) $\{(3, 1), (3, 2), (2, 3), (3, 3)\}$
3. (a) $\{(1, 3), (2, 1), (2, 2), (3, 2), (3, 3)\}$
  (b) $\{(1, 1), (1, 2), (2, 1), (2, 3), (3, 1), (3, 2), (3, 3)\}$
  (c) $\{(3, 1)\}$ (d) $\{(1, 2), (1, 3), (2, 3), (3, 3)\}$
5. (a) $\{(1, 4), (2, 1), (3, 1), (3, 2), (3, 3), (4, 2), (4, 3), (4, 4)\}$
  (b) $\{(1, 1), (1, 2), (2, 2), (2, 3), (2, 4), (4, 1)\}$
  (c) $\{(1, 1), (1, 2), (1, 3), (1, 4), (2, 2), (2, 3), (2, 4), (3, 1), (3, 2), (3, 4), (4, 1), (4, 3), (4, 4)\}$
  (d) $\{(1, 1), (2, 1), (4, 1), (2, 2), (3, 2), (4, 2), (2, 3), (1, 3), (4, 4), (1, 4), (3, 4)\}$.

7. (a) $\{(1, 1), (1, 4), (2, 2), (2, 3), (3, 3), (3, 4)\}$ (b) $\{(1, 2), (2, 4), (3, 1), (3, 2)\}$.

    (c) $\{(1, 1), (1, 2), (1, 3), (1, 4), (2, 1), (2, 4), (3, 1), (3, 2), (3, 3)\}$

    (d) $\{(1, 1), (1, 3), (2, 1), (2, 3), (3, 3), (4, 1), (4, 2)\}$

9. (a) $\{(3, 1), (1, 2), (3, 2), (1, 3), (2, 3), (4, 3), (2, 4)\}$

    (b)
                                         (c)
$$\mathbf{M}_R = \begin{bmatrix} 1 & 1 & 0 & 1 \\ 0 & 1 & 0 & 1 \\ 0 & 0 & 1 & 0 \\ 1 & 0 & 1 & 1 \end{bmatrix}$$

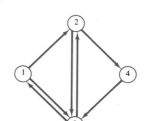

11. (a) $\begin{bmatrix} 0 & 0 & 0 \\ 0 & 0 & 1 \\ 0 & 0 & 0 \\ 0 & 1 & 0 \end{bmatrix}$ (b) $\begin{bmatrix} 1 & 1 & 1 \\ 1 & 1 & 1 \\ 0 & 1 & 1 \\ 1 & 1 & 1 \end{bmatrix}$ (c) $\begin{bmatrix} 0 & 0 & 0 & 1 \\ 1 & 1 & 0 & 1 \\ 0 & 1 & 1 & 1 \end{bmatrix}$ (d) $\begin{bmatrix} 0 & 1 & 0 \\ 0 & 1 & 0 \\ 1 & 0 & 1 \\ 1 & 0 & 1 \end{bmatrix}$

17. Partition corresponding to $R \cap S$ is $\{\{1, 2\}, \{3\}, \{4\}, \{5\}, \{6\}\}$.

19. (a) Yes   (b) No   (c) $\{(a, \square), (a, \diamondsuit), (a, \triangle), (c, \square), (b, \triangle), (c, \diamondsuit), (d, \triangle)\}$.

21. All pairs of people in A who are either brothers or sisters.

23. All pairs of people in A who have the same last name and receive the same benefits.

25. $R = \{(2, 6), (2, 12), (6, 2), (12, 2), (2, 2), (6, 6), (3, 3), (12, 12), (6, 12), (12, 6)\}$

    $S = \{(3, 6), (6, 3), (3, 12), (12, 3), (3, 3), (6, 6), (12, 12), (12, 6), (6, 12), (2, 2)\}$

    (a) $\{(2, 3), (3, 2), (3, 6), (3, 12), (12, 3), (6, 3)\}$

    (b) $\{(3, 3), (6, 6), (12, 12), (6, 12), (12, 6), (2, 2)\}$

    (c) $\{(2, 2), (2, 6), (2, 12), (3, 3), (3, 6), (3, 12), (6, 2), (6, 3), (6, 6), (6, 12), (12, 2), (12, 3), (12, 6),$
       $(12, 12)\}$.

    (d) S

27. (a) Yes   (b) Yes   (c) Yes

    (d) $\{(1, 1), (1, 2), (1, 3), (1, 4), (2, 4), (2, 1), (2, 2), (3, 1), (3, 2), (4, 1), (4, 2), (4, 3), (4, 4)\}$

    (e) $\{(1, 1), (1, 4), (1, 3), (2, 1), (2, 4), (3, 4), (4, 1), (4, 3), (4, 4)\}$

    (f) $\{(3, 1), (3, 2), (4, 1), (4, 2), (2, 4), (2, 1), (2, 2), (1, 1), (1, 2)\}$

    (g) $\{(3, 1), (3, 4), (4, 4), (2, 1), (2, 4), (1, 1), (1, 4)\}$

29. (a) $\begin{bmatrix} 1 & 1 & 0 & 1 & 1 \\ 0 & 1 & 0 & 1 & 1 \\ 1 & 1 & 0 & 1 & 1 \\ 1 & 1 & 0 & 1 & 1 \\ 1 & 1 & 0 & 0 & 1 \end{bmatrix}$ (b) $\begin{bmatrix} 1 & 0 & 1 & 1 & 1 \\ 1 & 1 & 0 & 1 & 1 \\ 1 & 0 & 1 & 1 & 1 \\ 1 & 1 & 1 & 1 & 1 \\ 0 & 0 & 0 & 1 & 1 \end{bmatrix}$

    (c) $\begin{bmatrix} 1 & 1 & 0 & 1 & 1 \\ 1 & 1 & 0 & 0 & 1 \\ 0 & 1 & 0 & 1 & 1 \\ 1 & 1 & 0 & 1 & 1 \\ 1 & 1 & 0 & 0 & 1 \end{bmatrix}$ (d) $\begin{bmatrix} 1 & 1 & 1 & 1 & 1 \\ 1 & 0 & 1 & 1 & 1 \\ 1 & 1 & 0 & 1 & 1 \\ 1 & 1 & 1 & 1 & 1 \\ 0 & 1 & 0 & 1 & 1 \end{bmatrix}$

Exercise Set 5.4, page 130

1. $\begin{bmatrix} 1 & 1 & 1 \\ 1 & 1 & 1 \\ 1 & 1 & 1 \end{bmatrix}$ 3. $\begin{bmatrix} 1 & 1 & 1 \\ 1 & 1 & 1 \\ 1 & 1 & 1 \end{bmatrix}$

5. $\begin{bmatrix} 1 & 0 & 0 & 1 \\ 1 & 1 & 0 & 1 \\ 0 & 0 & 1 & 0 \\ 0 & 0 & 0 & 1 \end{bmatrix}$ 7. $\begin{bmatrix} 1 & 0 & 0 & 1 \\ 0 & 1 & 1 & 0 \\ 0 & 1 & 1 & 0 \\ 1 & 0 & 0 & 1 \end{bmatrix}$

Review Exercises, page 131

1. Not reflexive, not irreflexive, not symmetric, not asymmetric, antisymmetric, transitive

3.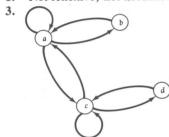

5. Yes   7. (b) $\{\{a, c, d\}, \{b\}\}$

9. (a) $\overline{R} = \{(b, c), (c, c), (d, b)\}$
    (b) $\overline{S} = \{(a, a), (a, b), (b, a), (b, b), (c, a), (c, b), (c, c), (d, b), (d, c)\}$
    (c) $R \cup S = \{(a, a), (a, b), (a, c), (a, d), (b, a), (b, b), (b, c), (c, a), (c, b), (c, d), (d, a), (d, c)\}$
    (d) $R \cap S = \{(a, c), (d, a)\}$
    (e) $R^{-1} = \{(a, a), (b, a), (c, a), (a, b), (b, b), (a, c), (b, c), (a, d), (c, d)\}$
    (f) $S^{-1} = \{(c, a), (c, b), (d, a), (d, c), (a, d)\}$

11.
$$\mathbf{M}_{R^{\infty}} = \begin{bmatrix} 1 & 1 & 0 & 0 \\ 1 & 1 & 0 & 0 \\ 1 & 1 & 1 & 1 \\ 1 & 1 & 1 & 1 \end{bmatrix}$$

Chapter Test, page 132

1. Not reflexive, not irreflexive, not symmetric, not asymmetric, not antisymmetric, not transitive
2. No
3. (b) $\{\{1, 2, 3\}, \{4\}\}$
4. $R = \{(a, a), (b, b), (b, c), (c, b), (c, c), (d, d), (d, e), (e, d), (e, e)\}$
5. (a) $\overline{S} = \{(a, a), (a, b), (a, x), (a, z), (b, a), (b, x), (b, y), (x, a), (x, x), (x, y), (x, z), (y, a), (y, b), (y, x),$
        $(y, y), (y, z), (z, b), (z, x), (z, y), (z, z)\}$
    (b) $R^{-1} = \{(a, a), (a, z), (y, a), (b, b), (z, b)\}$
    (c) $S \circ R = \{(a, b), (a, y), (x, b), (x, z), (y, a)\}$
    (d) $S \circ S^{-1} = \{(a, a), (b, b), (b, x), (x, b), (x, x), (z, z)\}$

6.

$$\mathbf{M}_{R^\infty} = \begin{bmatrix} 1 & 1 & 1 & 1 \\ 1 & 1 & 1 & 1 \\ 1 & 1 & 1 & 1 \\ 1 & 1 & 1 & 1 \end{bmatrix}$$

7. (a) F (b) T (c) F (d) T (e) T

CHAPTER 6

Exercise Set 6.1, page 142

1. Function, range $= \{1, 2\}$ **3.** Not a function **5.** Function

11. (a) 3 (b) 1 (c) $(x - 1)^2$ (d) $x^2 - 1$ (e) $x - 2$ (f) x^4

15. One-to-one and onto **17.** Not one-to-one, onto **19.** Not one-to-one, not onto

21. Not one-to-one, onto **23.** Yes

Exercise Set 6.2, page 151

1. (a), (c)

3. (a) $\begin{pmatrix} 1 & 2 & 3 & 4 & 5 & 6 \\ 3 & 4 & 1 & 2 & 6 & 5 \end{pmatrix}$ (b) $\begin{pmatrix} 1 & 2 & 3 & 4 & 5 & 6 \\ 2 & 5 & 6 & 3 & 1 & 4 \end{pmatrix}$

(c) $\begin{pmatrix} 1 & 2 & 3 & 4 & 5 & 6 \\ 5 & 2 & 1 & 6 & 3 & 4 \end{pmatrix}$ (d) $\begin{pmatrix} 1 & 2 & 3 & 4 & 5 & 6 \\ 4 & 5 & 1 & 2 & 6 & 3 \end{pmatrix}$

5. $\begin{pmatrix} 1 & 2 & 3 & 4 & 5 & 6 & 7 & 8 \\ 5 & 1 & 2 & 4 & 7 & 6 & 8 & 3 \end{pmatrix}$ **7.** $\begin{pmatrix} 1 & 2 & 3 & 4 & 5 & 6 & 7 & 8 \\ 5 & 1 & 3 & 2 & 6 & 7 & 4 & 8 \end{pmatrix}$

9. $\begin{pmatrix} a & b & c & d & e & f & g \\ f & c & d & e & b & g & a \end{pmatrix}$ **11.** $(6, 8) \circ (2, 3) \circ (1, 4, 5)$ **13.** $(1, 6, 3, 7, 2, 5, 4, 8)$

15. (a, g, e, c, b, d) **17.** $(2, 6) \circ (2, 8) \circ (2, 5) \circ (2, 4) \circ (2, 1)$ **19.** Even **21.** Even

Exercise Set 6.3, page 163

1. Yes **3.** No **5.** No **7.** No **9.** Commutative, associative

11. Not commutative, associative **13.** Commutative, associative

17. Monoid, identity $= 1$, commutative **19.** Semigroup, commutative

21. Monoid, identity $= 1$, commutative **23.** Commutative monoid; identity $= 0$ **25.** No **27.** No

29. Yes, abelian, identity $= -2$, $a^{-1} = -4 - a$

31. Yes, abelian, identity is the $m \times n$ zero matrix, $A^{-1} = -A$

Review Exercises, page 165

1. (a) Function; Dom $(R) = \{1, 2, 8, 9\}$, Ran $(R) = \{s, u, v\}$

(b) Function; Dom $(R) = \{2, 8\}$, Ran $(R) = \{t, v\}$

(c) Not a function

3. (a) $(f \circ f)(x) = 81x^9$ (b) $(f^{-1} \circ g)(x) = \dfrac{3x^3 - 1}{6}$ (c) $((g \circ g) \circ f)(x) = 108x^3 + 6$

5. (a) $\begin{pmatrix} 1 & 2 & 3 & 4 & 5 & 6 & 7 & 8 & 9 \\ 2 & 3 & 5 & 4 & 7 & 6 & 1 & 8 & 9 \end{pmatrix}$ (b) $\begin{pmatrix} 1 & 2 & 3 & 4 & 5 & 6 & 7 & 8 & 9 \\ 7 & 4 & 3 & 5 & 6 & 8 & 2 & 1 & 9 \end{pmatrix}$

(c) $\begin{pmatrix} 1 & 2 & 3 & 4 & 5 & 6 & 7 & 8 & 9 \\ 9 & 4 & 3 & 5 & 2 & 6 & 7 & 1 & 8 \end{pmatrix}$ (d) $\begin{pmatrix} 1 & 2 & 3 & 4 & 5 & 6 & 7 & 8 & 9 \\ 2 & 4 & 8 & 3 & 5 & 6 & 9 & 5 & 7 \end{pmatrix}$

7. (a) Even   (b) Even

9. (a) Not commutative, not associative
    (b) Commutative and associative
    (c) Not commutative, not associative

11. (a) Not a group   (b) Not a group   (c) Group

Chapter Test, page 167

1. (a) Function; not one to one, not onto   (b) Function; one to one, onto
    (c) Function; not one to one, not onto

2. (a) $\{(c, 1), (a, 2), (b, 3)\}$ (b) $\{(1, a), (2, a), (3, b)\} = f$ (c) $\{(1, a), (2, a), (3, b)\} = f$

3. (a) $\begin{pmatrix} 1 & 2 & 3 & 4 & 5 & 6 & 7 & 8 & 9 & 10 \\ 5 & 2 & 9 & 3 & 10 & 6 & 7 & 8 & 4 & 1 \end{pmatrix}$ (b) $\begin{pmatrix} 1 & 2 & 3 & 4 & 5 & 6 & 7 & 8 & 9 & 10 \\ 9 & 2 & 6 & 4 & 5 & 7 & 1 & 8 & 10 & 3 \end{pmatrix}$

    (c) $\begin{pmatrix} 1 & 2 & 3 & 4 & 5 & 6 & 7 & 8 & 9 & 10 \\ 9 & 3 & 4 & 1 & 5 & 8 & 6 & 7 & 10 & 2 \end{pmatrix}$

4. (a) $(2, 3, 5) \circ (4, 6)$ (b) $(1, 2, 3, 8) \circ (4, 7, 6) \circ (5, 9)$ (c) $(1, 2, 3) \circ (4, 5, 6, 7, 8)$

5. (a) Valid   (b) Invalid   (c) Invalid

6. (a) Monoid   (b) Semigroup   (c) Group

7. (a) F   (b) T   (c) T   (d) T   (e) T

CHAPTER 7

Exercise Set 7.1, page 179

1. No   3. Yes   5. Yes   7. No

9. $\{(a, a), (b, b), (c, c), (a, b)\}$
    $\{(a, a), (b, b), (c, c), (a, b), (a, c)\}$
    $\{(a, a), (b, b), (c, c), (a, b), (c, b)\}$
    $\{(a, a), (b, b), (c, c), (a, b), (b, c), (a, c)\}$
    $\{(a, a), (b, b), (c, c), (a, b), (c, b), (c, a)\}$

11.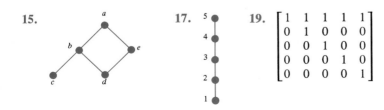

13. $\{(1, 1), (2, 2), (3, 3), (4, 4), (1, 3), (2, 3), (3, 4), (1, 4), (2, 4)\}$

15.

17.

19. $\begin{bmatrix} 1 & 1 & 1 & 1 & 1 \\ 0 & 1 & 0 & 0 & 0 \\ 0 & 0 & 1 & 0 & 0 \\ 0 & 0 & 0 & 1 & 0 \\ 0 & 0 & 0 & 0 & 1 \end{bmatrix}$

21. (a) T (b) T (c) F (d) F
23. Linearly ordered **25.**

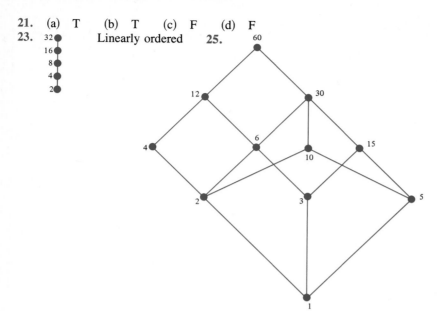

Exercise Set 7.2, page 185

1. Maximal elements: 3, 5; minimal elements: 1, 6 **3.** Maximal elements: e, f; minimal elements: a
5. No maximal or minimal elements **7.** Maximal elements: 1; minimal elements: none
9. Greatest element: f; least element: a **11.** No greatest element; no least element
13. No greatest element; no least element **15.** Greatest element: 72; least element: 2
17. (a) Upper bounds: f, g, h (b) Lower bounds: c, a, b (c) LUB : f (d) GLB : c
19. (a) Upper bounds: d, e, f (b) Lower bounds: b, a (c) LUB : d (d) GLB : b
21. (a) Upper bounds: none (b) Lower bounds: b (c) LUB : none (d) GLB : b
23. (a) Upper bounds: none (b) Lower bounds: 1, 2, 3 (c) LUB : none (d) GLB : 3
25. (a) Upper bounds: $\{x \mid x \geq 2\}$ (b) Lower bounds: $\{x \mid x \leq 1\}$ (c) LUB : 2 (d) GLB : 1
27. (a) Upper bounds: 12, 24, 48 (b) Lower bounds: 2 (c) LUB : 12 (d) GLB : 2

Exercise Set 7.3, page 194

1. Yes **3.** No **5.** Yes **7.** No
9. One element: ● two elements: ● ; three elements: ● ; four elements: ● , ● ;

(a)   (b)   (c)   (d)

five elements: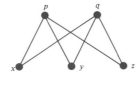

(e)   (f)   (g)   (h)   (j)

11. $1' = 105$; $3' = 35$; $5' = 21$; $7' = 15$; $15' = 7$; $35' = 3$; $21' = 5$; $105' = 1$

13. Nondistributive, noncomplemented   **15.** Distributive, noncomplemented

Exercise Set 7.4, page 203

1. No   **3.** No   **5.** Yes   **7.** Yes   **9.** Yes

Review Exercises, page 206

1. Not a partial order

3. $\{(a, a), (a, c), (c, c), (a, b), (b, b), (b, e), (e, e), (a, e), (b, d), (d, d), (a, d)\}$

5. Maximal elements: $\{y, z\}$

Minimal elements: $\{u, v, x\}$

7. Yes

9.

Chapter Test, page 206

1.

2.

3. $$\begin{bmatrix} 0 & 0 & 1 & 1 & 1 & 1 \\ 0 & 0 & 0 & 1 & 1 & 1 \\ 0 & 0 & 0 & 1 & 1 & 1 \\ 0 & 0 & 0 & 0 & 1 & 1 \\ 0 & 0 & 0 & 0 & 0 & 0 \\ 0 & 0 & 0 & 0 & 0 & 0 \end{bmatrix}$$

4. (a) {(1, 1, 1, 1, 0, 1, 0, 1), (1, 1, 1, 1, 1, 1, 0, 1), (1, 1, 1, 1, 0, 1, 1, 1), (1, 1, 1, 1, 1, 1, 1, 1)}
    (b) LUB: (1, 1, 1, 1, 0, 1, 0, 1); GLB: (0, 0, 0, 0, 0, 0, 0, 0)
    (c) (0, 0, 1, 1, 1, 1, 0, 0)
5. Yes; B_2
6. (a) T   (b) F   (c) T   (d) F   (e) T

CHAPTER 8

Exercise Set 8.1, page 212

1.

x	y	z	$f(x, y, z) = x \wedge (y \vee z')$
0	0	0	0
0	0	1	0
0	1	0	0
0	1	1	0
1	0	0	1
1	0	1	0
1	1	0	1
1	1	1	1

3.

x	y	z	$f(x, y, z) = (x \wedge y') \vee (y \wedge (x' \vee y))$
0	0	0	0
0	0	1	0
0	1	0	1
0	1	1	1
1	0	0	1
1	0	1	1
1	1	0	1
1	1	1	1

5.

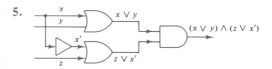

Exercise Set 8.2, page 222

1.

	y'	y
x'	1	0
x	0	1

3.

	y'	y'	y	y
x'	1	1	0	0
x	1	0	0	1

(brackets: z over middle columns, z' over outer columns)

5.

0	0	0	1
0	0	0	1
0	0	1	1
0	0	1	0

7. $(x' \wedge y') \vee (x \wedge y)$ **9.** $(x' \wedge z) \vee z'$

11. $(x' \wedge z') \vee (x' \wedge z \wedge w') \vee (x \wedge y' \wedge w')$

13. y' **15.** $(x' \wedge z) \vee (y \wedge z) \vee (x' \wedge y)$

17. $(x' \wedge y' \wedge w') \vee (x' \wedge y \wedge z' \wedge w) \vee (x' \wedge y \wedge z \wedge w') \vee (x \wedge y \wedge w')$

Review Exercises, page 224

1.

x	y	z	$f(x, y, z)$
0	0	0	0
0	0	1	1
0	1	0	0
0	1	1	0
1	0	0	0
1	0	1	1
1	1	0	1
1	1	1	1

3.

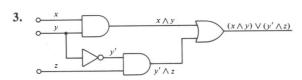

5.

	y'	y'	y	y
x'	1	1	0	1
x	0	1	1	0

(brackets: z over middle columns, z' over outer columns)

7. $z' \vee (y \wedge z)$

9. $(x' \wedge z') \vee (y' \wedge z) \vee (x \wedge z)$

Chapter Test, page 225

1.

x	y	z	$f(x, y, z)$
0	0	0	1
0	0	1	1
0	1	0	0
0	1	1	1
1	0	0	0
1	0	1	0
1	1	0	0
1	1	1	1

2.

3.

4. $(x \wedge z) \vee (y \wedge z')$

5. $(y' \wedge z') \vee (y' \wedge x)$

6.

x	y	z	$f(x, y, z)$
0	0	0	0
0	0	1	0
0	1	0	1
0	1	1	0
1	0	0	0
1	0	1	1
1	1	0	1
1	1	1	1

CHAPTER 9

Exercise Set 9.1, page 230

1. Tree; root is b. **3.** Tree; root is f. **5.** Not a tree. **7.** Tree; root is t.

9. 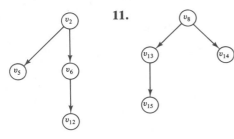 **11.**

Exercise Set 9.2, page 234

1. 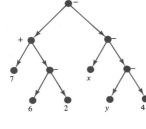 **3.**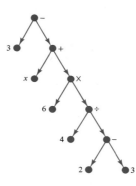

5. 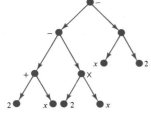 **7.**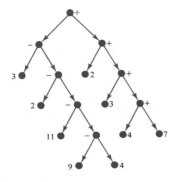

9.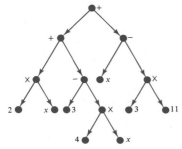

11.

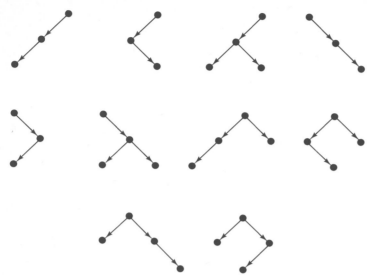

13. 721

Exercise Set 9.3, page 243

1. *xysztuv*   3. *abcghidkejf*   5. *ysxzvut*
7. *gchbiakdjef*   9. *syvutzx*   11. *ghcibkjfeda*
13. I NEVER SAW A PURPLE COW I HOPE I NEVER SEE ONE.
15.   17. 6 5 7 10 9 8 4 3 2 1   19.

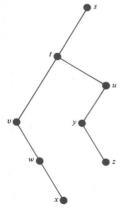

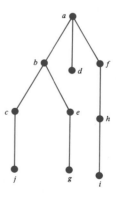

Review Exercises, page 252

1. R is a tree; root is c.
3.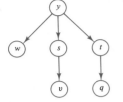

5.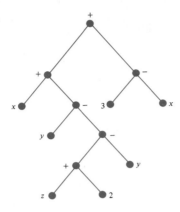

7. $x - yz \times x3$
9.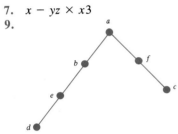

Chapter Test, page 253

1. Not a tree
2.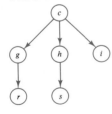

3.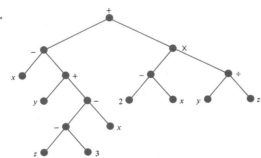

4. $iklj\,dbaecgfmhn$
5.

6. (a) F   (b) F   (c) T   (d) F   (e) F

CHAPTER 10

Exercise Set 10.1, page 262

1. $\{x^n y^m z \mid n \geq 0, m \geq 1\}$ **3.** $\{a^{2n}b \mid n \geq 0\} \cup \{a^{2n+1} \mid n \geq 0\}$

5. $\{((\cdots \underbrace{(a + a + \cdots + a)}_{n}) \cdots)}_{k} \mid k \geq 0, n \geq 3\}$
        $\underbrace{}_{k}$

7. $\{x^n y z^m \mid n \geq 1, m \geq 0\}$ **9.** (a), (c), (e), (h), (i)

11. (1) v_0 (2) v_0
               xv_0v_0 xv_0v_0
               xyv_0v_0 xv_0yv_0
               $xyzv_0$ xyv_0yv_0
               $xyzyv_0$ xyv_0yz
               $xyzyz$ $xyzyz$

13. (1) I   (2) I   **15.** $G = (V, S, v_0, \mapsto)$
               LW   LW $V = \{v_0, 0, 1\}, S = \{0, 1\}$
               aW   LDW $\mapsto : v_0 \mapsto 0v_0 1$
               aDW   LDDW $v_0 \mapsto 1v_0 0$
               a1W   LDDD $0v_0 \mapsto v_0 0$
               a1DW   L1DD $v_0 0 \mapsto 0v_0$
               a1DD   L10D $1v_0 \mapsto v_0 1$
               a10D   L100 $v_0 1 \mapsto 1v_0$
               a100   a100 $v_0 \mapsto \Lambda$

17. $G = (V, S, v_0, \mapsto)$ **19.** $G = (V, S, v_0, \mapsto)$ **21.** (a), (d)
        $V = \{v_0, a, b\}, S = \{a, b\}$ $V = \{v_0, x, y\}, S = \{x, y\}$
        $\mapsto : v_0 \mapsto av_0b$ $\mapsto : v_0 \mapsto v_0yy$
               $v_0 \mapsto ab$ $v_0 \mapsto xv_0$
                                                            $v_0 \mapsto xx$

Exercise Set 10.2, page 272

1. *BNF*
        $\langle v_0 \rangle ::= x\langle v_0 \rangle \mid y\langle v_1 \rangle$
        $\langle v_1 \rangle ::= y\langle v_1 \rangle \mid z$
        Syntax Diagram

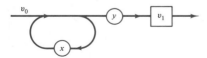 

3. *BNF*
        $\langle v_0 \rangle ::= a\langle v_1 \rangle$
        $\langle v_1 \rangle ::= b\langle v_0 \rangle \mid a$

Syntax Diagram

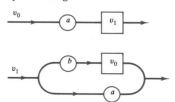

5. **BNF**

$\langle v_0 \rangle ::= aa\langle v_0 \rangle \mid b\langle v_1 \rangle$

$\langle v_1 \rangle ::= c\langle v_2 \rangle b \mid cb$

$\langle v_2 \rangle ::= bb\langle v_2 \rangle \mid bb$

Syntax Diagram

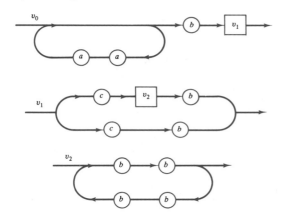

7. $\langle v_0 \rangle ::= x\langle v_0 \rangle\langle v_0 \rangle \mid y\langle v_0 \rangle \mid z$

9. $\langle I \rangle ::= \langle L \rangle\langle W \rangle \mid \langle L \rangle$

   $\langle W \rangle ::= \langle L \rangle\langle W \rangle \mid \langle D \rangle\langle W \rangle \mid \langle L \rangle \mid \langle D \rangle$

   $\langle L \rangle ::= a \mid b \mid c$

   $\langle D \rangle ::= 0 \mid 1 \mid 2 \mid 3 \mid 4 \mid 5 \mid 6 \mid 7 \mid 8 \mid 9$

11. $\langle v_0 \rangle ::= ab \mid ab\langle v_1 \rangle$

    $\langle v_1 \rangle ::= d \mid d\langle v_1 \rangle \mid d\langle v_2 \rangle\langle v_1 \rangle$

    $\langle v_2 \rangle ::= c \mid d$

Review Exercises, page 273

1. $L(G) = \{x^n y z^{2m+1} y \mid n \geq 0,\ m \geq 0\}$

3. (a), (d), (e), (f)

5. (a), (c), (f)

7. $\langle v_0 \rangle ::= x\langle v_0 \rangle \mid y\langle v_1 \rangle$

   $\langle v_1 \rangle ::= z\langle v_2 \rangle \mid zz\langle v_1 \rangle$

   $\langle v_2 \rangle ::= y$

9.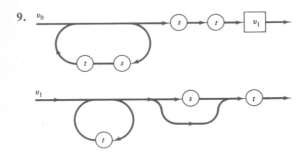

Chapter Test, page 274

1. $L(G) = \{m^k n^l p m^s n^t p \mid k \geq 0, l \geq 0, s \geq 0, t \geq 0\}$

2. $\langle v_0 \rangle \; :: \; = n\langle v_0 \rangle \mid m\langle v_0 \rangle \mid p\langle v_1 \rangle$
$\langle v_1 \rangle \; :: \; = m\langle v_1 \rangle \mid n\langle v_1 \rangle \mid p$

3.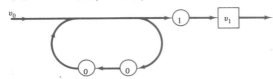

4. (a), (d), (e), (f)

5. $\langle v_0 \rangle \; :: \; = 00\langle v_0 \rangle \mid 1\langle v_1 \rangle$
$\langle v_1 \rangle \; :: \; = 00\langle v_1 \rangle \mid 111\langle v_1 \rangle \mid 0$

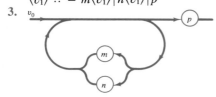

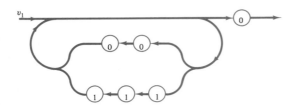

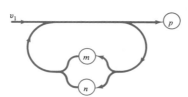

6. (a) T   (b) T   (c) T   (d) T   (e) F   (f) T

CHAPTER 11

Exercise Set 11.1, page 278

1. 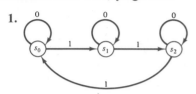   **3.**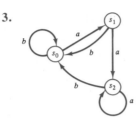

5.

	a	b
s_0	s_1	s_1
s_1	s_1	s_2
s_2	s_0	s_2

7.

	T	F
s_0	s_1	s_0
s_1	s_1	s_1
s_2	s_1	s_2

Exercise Set 11.2, page 284

1. $f_w(s_0) = s_2, f_w(s_1) = s_3, f_w(s_2) = s_0, f_w(s_3) = s_1$

3. The number of 1's is divisible by 4.   **5.** The number of 1's is of the form $2 + 4k$, $k \geq 0$.

7. $f_w(s_0) = s_0, f_w(s_1) = s_0, f_w(s_2) = s_0$ **9.** w must end in a "*b*."

11. Strings of 0's and 1's with $3 + 5n$ 1's, $n \geq 0$.

Exercise Set 11.3, page 290

1. All strings of 0's and 1's with the number of 1's equal to $3 + 4n$, $n \geq 0$.

3. All strings of 0's and 1's that contain the substring 010.

5. All strings of 0's and 1's that end with the string 01.

7. $G = (V, I, s_0, \mapsto), I = \{0, 1\}, V = \{s_0, s_1, s_2, s_3, 0, 1\}$

$\mapsto$:

$s_0 \mapsto 0s_0$
$s_0 \mapsto 1s_1$
$s_1 \mapsto 0s_1$
$s_1 \mapsto 1s_2$
$s_2 \mapsto 0s_2$
$s_2 \mapsto 1s_3$
$s_2 \mapsto 1$
$s_3 \mapsto 0s_3$
$s_3 \mapsto 0$
$s_3 \mapsto 1s_0$

9. $G = (V, I, s_0, \mapsto), I = \{a, b\}, V = \{s_0, s_1, s_2, a, b\}, \mapsto$ in BNF is

$\langle s_0 \rangle :: = a \langle s_0 \rangle \,|\, b \langle s_1 \rangle \,|\, a \,|\, b$
$\langle s_1 \rangle :: = a \langle s_0 \rangle \,|\, b \langle s_2 \rangle \,|\, a$
$\langle s_2 \rangle :: = a \langle s_2 \rangle \,|\, b \langle s_2 \rangle$

11.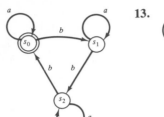

13.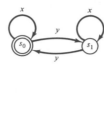

15.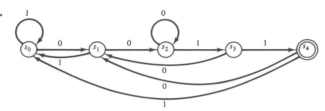

17.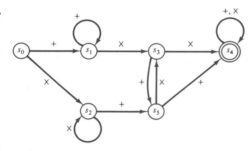

19.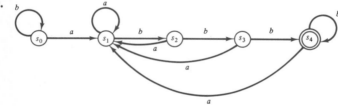

21.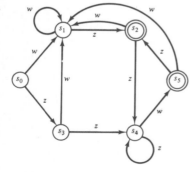

23.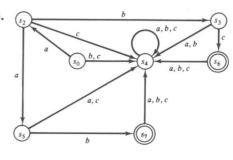

25.

	0	1
s_0	s_1	s_0
s_1	s_1	s_2
s_2	s_3	s_0
s_3	s_1	s_4
s_4	s_3	s_0

$T = \{s_4\}$

27.

	x	y
s_0	s_1	s_0
s_1	s_2	s_1
s_2	s_3	s_2
s_3	s_3	s_3

$T = \{s_2\}$

Review Exercises, page 291

1.

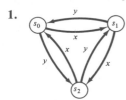

3. $f_w(s_0) = s_1, f_w(s_1) = s_2, f_w(s_2) = s_0$

5. w contains an odd number of q's.

7. $V = \{s_0, s_1, s_2, s_3, 0, 1\}$, $I = \{0, 1\}$
$$\langle s_0 \rangle ::= 0\langle s_1 \rangle \mid 1\langle s_2 \rangle$$
$$\langle s_1 \rangle ::= 0\langle s_1 \rangle \mid 1\langle s_1 \rangle$$
$$\langle s_2 \rangle ::= 1\langle s_1 \rangle \mid 0\langle s_3 \rangle \mid 0$$
$$\langle s_3 \rangle ::= 0\langle s_1 \rangle \mid 1\langle s_0 \rangle$$

9.

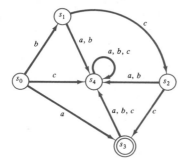

Chapter Test, page 292

1.

	x	y
s_0	s_1	s_4
s_1	s_3	s_4
s_2	s_2	s_3
s_3	s_3	s_4
s_4	s_4	s_2

2.

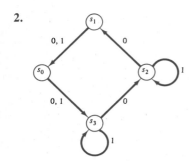

3. $f_w(s_0) = s_4, f_w(s_1) = s_4, f_w(s_2) = s_3, f_w(s_3) = s_4$
   $f_w(s_4) = s_2$
4. $w = 1^n\, 00$ or $w = 1^n\, 01$, where $n \geq 0$
5. $L(M)$ consists of all strings containing an odd number of c's.
6. (a) T (b) T (c) F (d) T (e) F

CHAPTER 12

Exercise Set 12.1, page 297

1. $\{0, 1, 2, 3, 4, 5, 6, 7, 8, 9, 10\}$
3. $(m, l, y), (m, l, n), (m, a, y), (m, a, n), (m, u, y), (m, u, n), (f, l, y), (f, l, n), (f, a, y), (f, a, n), (f, u, y),$
   $(f, u, n)\}$
5. $\{(s, b), (s, r), (s, g), (c, b), (c, r), (c, g)\}$
7. $\emptyset, \{1\}, \{2\}, \{3\}, \{1, 2\}, \{1, 3\}, \{2, 3\}, \{1, 2, 3\}$
9. (a) The card is black or an ace.
   (b) The card is a black ace.
   (c) The card is a red ace.
   (d) The card is black, or in the suit of diamonds, or an ace.
   (e) The card is an ace of clubs.
11. (a) No; $E \cap F = \{3\}$
    (b) No; $F \cap G = \{2\}$
    (c) Yes; $E \cup F = \{1, 2, 3, 4, 5, 6\}$
    (d) No; $E \cap F = \{3\}$

Exercise Set 12.2, page 302

1. $\{1\}, \{2\}, \{3\}, \{4\}$
3. $\{a\}, \{e\}, \{i\}, \{o\}, \{u\}$
5. (a) $\frac{5}{9}$ (b) $\frac{11}{18}$ (c) $\frac{2}{3}$ (d) $\frac{1}{2}$
7. (a) 0.7   (b) 0   (c) 0.7   (d) 1
9. $p(\{A\}) = \frac{6}{11}$ $p(\{B\}) = \frac{3}{11}$
   $p(\{C\}) = \frac{1}{11}$ $p(\{D\}) = \frac{1}{11}$

Exercise Set 12.3, page 307

1. (a) $\frac{5}{6}$ (b) $\frac{1}{6}$ (c) $\frac{1}{6}$ (d) 1   (e) $\frac{5}{6}$ (f) $\frac{1}{6}$

3. $\dfrac{\binom{3}{2}\binom{6}{2}}{\binom{11}{4}} = \dfrac{5}{11}$

5. (a) $\frac{3}{8}$ (b) $\frac{7}{8}$
7. (a) $\{(m,m,m), (m,m,f), (m,f,m), (m,f,f), (f,m,m), (f,m,f), (f,f,m), (f,f,f)\}$
   (b) $\frac{1}{8}$ (c) $\frac{1}{8}$ (d) $\frac{3}{8}$

Review Exercises, page 309

1. $\{$(IBM, PASCAL), (IBM, FORTRAN), (IBM, COBOL), (IBM, FORTH), (IBM, APL), (APPLE, PASCAL), (IBM, FORTRAN), (IBM, COBOL) (IBM, FORTH), (IBM, APL)$\}$
3. 2^n
5. (a) $\{$THH, HTH, HHT, THT, TTH, TTT, HTT$\}$ (b) $\{$TTH, THT, HTT$\}$
7. (a) $\frac{1}{4}$ (b) $\frac{1}{12}$
9. (a) $\frac{7}{12}$ (b) 1   (c) $\frac{1}{3}$ (d) $\frac{1}{3}$ (e) $\frac{1}{3}$

11. $\dfrac{\binom{18}{2}}{\binom{52}{2}} = \dfrac{3}{26}$

Chapter Test, page 309

1. $\{$HH, HT, TH, TT$\}$
2. $\{y, n\}$
3. (a) $E = \{(m,l,g), (m,a,g), (m,h,g), (f,l,g), (f,a,g), (f,h,g)\}$
   (b) $F = \{(m,l,u), (m,l,g), (m,a,u), (m,a,g), (m,h,u), (m,h,g), (f,a,u), (f,a,g)\}$
   (c) $G = \{(f,l,u), (f,a,u), (f,h,u)\}$
4. (a) $E \cup F = \{1,2,3,4,5\}, E \cap F = \{3\}, \bar{F} = \{1,4,5,6\}$
   (b) $\bar{E} \cup F = \{2,3,6\}, \bar{F} \cap G = \{4\}$
5. (a) $\frac{1}{2}$ (b) $\frac{7}{9}$

6. $\dfrac{\dbinom{26}{2} + \dbinom{4}{2} - 1}{\dbinom{52}{2}} = \dfrac{55}{221}$

7. (a) F   (b) T   (c) T   (d) T   (e) T

Appendix, page 324

1. (d) and (e)
3. (a) $3 + 1 < 5$ and $7 = 3 \times 6$; $3 + 1 < 5$ or $7 = 3 \times 6$
   (b) I am rich and I am happy; I am rich or I am happy.
   (c) I will drive my car and I will be late; I will drive my car or I will be late.
5. (a) T   (b) T   (c) T   (d) F   7. (d)
9. (a) If I am the King of England, then $2 + 2 \neq 4$.
   (b) If I will not walk to work, then I am President of the United States.
   (c) If I did take the train to work, then I am not late.
   (d) If I will not go to the store, then I do not have time or I am too tired.
   (e) If I will not buy a car or I will not buy a house, then I do not have enough money.
11. (a) $(\sim r) \to q$ (b) $(\sim q) \wedge p$ (c) $(\sim p) \to q$ (d) $(\sim p) \to (\sim r)$
13. (a) Absurdity   (b) Tautology   (c) Contingency   (d) Contingency
    (e) Contingency   (f) Tautology   (g) Contingency
15. Invalid   17. Valid, $((p \to q) \wedge p) \to q$ 19. Valid, $((p \vee q) \wedge \sim q) \to p$

Index

F

G

H

I